The Chemistry and Technology of Solid Rocket Propellants

(A Treatise on Solid Propellants)

The Chemistry and Technology of Solid Rocket Propellants

(A Treatise on Solid Propellants)

by

V.N. Krishnamurthy
T.L. Varghese

ALLIED PUBLISHERS PVT. LTD.

New Delhi • Mumbai • Kolkata • Lucknow • Chennai
Nagpur • Bangalore • Hyderabad • Ahmedabad

ALLIED PUBLISHERS PRIVATE LIMITED

Regd. Off.: 15 J.N. Heredia Marg, Ballard Estate, **Mumbai**–400001
Ph.: 022-22626476 • E-mail: mumbai.books@alliedpublishers.com

1/13-14 Asaf Ali Road, **New Delhi**–110002
Ph.: 011-23239001 • E-mail: delhi.books@alliedpublishers.com

17 Chittaranjan Avenue, **Kolkata**–700072
Ph.: 033-22129618 • E-mail: cal.books@alliedpublishers.com

751 Anna Salai, **Chennai**–600002
Ph.: 044-28523938 • E-mail: chennai.books@alliedpublishers.com

5th Main Road, Gandhinagar, **Bangalore**–560009
Ph.: 080-22262081 • E-mail: bngl.books@alliedpublishers.com

3-2-844/6 & 7 Kachiguda Station Road, **Hyderabad**–500027
Ph.: 040-24619079 • E-mail: hyd.books@alliedpublishers.com

60 Shiv Sunder Apartments (Ground Floor), Central Bazar Road,
Bajaj Nagar, **Nagpur**–440010

F-1 Sun House (First Floor), C.G. Road, Navrangpura,
Ellisbridge P.O., **Ahmedabad**–380006
Ph.: 079-26465916 • E-mail: ahmbd.books@alliedpublishers.com

Khasra No. 168, Plot No. 12-A, Opp. Wisdom Academy School,
Kamta, Surendra Nagar, **Lucknow**–227105
Ph.: 09335202549 • E-mail: lko.books@alliedpublishers.com

Website: www.alliedpublishers.com

ISBN: 978-93-85926-33-4

Published by Sunil Sachdev and printed by Ravi Sachdev at Allied Publishers Pvt. Ltd., (Printing Division), A-104 Mayapuri Phase II, New Delhi-110064

Dedicated

To

All those who strived to seek
and to develop
composite solid propellants
to propel
Indian Space Launch Vehicles
and sounding rockets

Foreword

On October 3rd 1957, the world's first successful artificial satellite, Sputnik-1, was launched by Soviet Union and began to orbit around the earth. Since then, breathtaking advances in space launchings are the evidence of the rapid development and growth of propellant technology. Both types of propellants—solid propellants with instant readiness, higher density and relative simplicity in rocket construction and liquid propellants with easier variation of thrust levels and higher specific impulse are used in today's launch vehicles.

Over the years from 1957 to 2015, the technology of solid propellants advanced rapidly. Solid Rocket Motors (SRM) have displaced liquid propellants in many tactical, strategic and space applications because of their lower cost and quicker reaction time. Most of the major space endeavors like space shuttle, Ariane, etc. are dependent to a large extent on solid propellants. Indian satellite launch vehicles like SLV-3, ASLV, PSLV and GSLV could complete their missions largely through the use of solid propellants. Space shuttle, Arianne V and GSLV Mk III have two solid boosters each weighing more than 500 t, 239 t and 209 t respectively. Today the utility of solid propellants have grown further. Solid propellants, find usage in missiles for defense purposes. They are also used in space programmes for critical tasks such as ignition, spin, separation assistance (ullage and retro rockets) and for escape in manned missions in case of malfunction of launch vehicles. The Hydroxyl Terminated Polybutadiene (HTPB) and ammonium perchlorate based propellant has become the work-horse propellant in solid rocket motors.

In independent India, experiments with solid propellants based rockets started with the flight of RH 75 sounding rockets using double base propellants in November 21st 1967 and with composite solid propellants in February 21st, 1969. The first solid propellant satellite launch vehicle SLV-3 was successfully flight tested on July 18th, 1980 by putting a 40 kg Rohini satellite into a near earth orbit. This was followed by ASLV, PSLV and GSLV flights with higher payload capabilities. PSLV has more than 30 successful launchings till date.

The concept of tiny rockets to space vehicles as self propelling phenomenon and ISRO's planning on solid propellant technology to maintain a lead over needs started together. In each of the tiny motors during early days, therefore, one saw a prelude of tomorrow's giant vehicles and the successful firing of each of these motors were burning away gradually, decades of nation's backwardness in this vital field. If propellant weight is considered as representative of the size of rocket growth of ISRO's rocket motors, which it is, the weights increased from a stagnant figure of 10 kg in 1967 to 220 kg in 1970, 700 kg in 1972 to 10 t in 1979, to 139 t in 1989 and to 209 t in 2010 as a result of ISRO's efforts. Dr. Krishnamurthy and Dr. Varghese along with their colleagues, under the leadership of Dr. Vasant Gowariker and Sri. M.R. Kurup, developed a series of propellants and scaled up to meet the requirements of the launch vehicle projects. The solid propellant development and characterization for various missions were completed well ahead of time. In fact, the solid booster motor of PSLV was the third largest booster in the world but first to fly with HTPB based propellant way back in 90s. Krishnamurthy and Varghese consolidated the efforts of Dr. Gowariker and Mr. Kurup and helped ISRO to realize the dreams of Dr. Sarabhai, the father of Indian Space Programme, by developing propellants to meet the goals of space programmes. I should particularly like to emphasise the extraordinary professionalism and insights that were brought to bear by Late Dr. Vasant Gowarikar, Late Dr. Kurup, Dr. Krishnamurthy and Dr. Varghese and their teams in the realisation of the different solid propellant modules. Achievement of world class performance levels including the specific impulse (-245-246 secs at sea level), and limited qualification tests leading to highest levels of reliability are two of the prominent features in this connection.

Krishnamurthy and Varghese have put down their long experience, expertise and knowledge in the form of a book entitled "The Chemistry and Technology of Solid Rocket Propellants" which will help those working on solid propellants and the younger generation to contribute further to the growth of this technology and through this to Indian Space Programme. As there is no structured course in our universities, this book will enable educational authorities to start a course so that the subject gets a focused attention and we get students with complete knowledge to work in Space and Defense Programmes. This work will serve as a rich source of information—not only on propellants and the ingredients that are used but also future directions in this restricted area including safety and handling and transport of these energetic materials. With their professional career of over thirty five years dedicated to solid propellants, they have brought out the growth of solid propellants in the world as well as in India in a simple and lucid manner in this book. I congratulate the authors for their outstanding effort of great value.

Dr. K. Kasturirangan
Former Secretary, Department of Space
& Chairman ISRO, Govt. of India

Preface

Space exploration has always fascinated the mankind. For some, space is a symbol of technological advancement and enrichment of human mind. For some, space becomes a precious resource to provide connectivity, cutting across the physical barriers and international boundaries as well as a vantage point to view the earth in totality. Even more spectacular is the benefits of technology in information services (such as telephone, television, weather forecasting, remote sensing, etc.), products that emerged from space programme (including pharmaceuticals, material processing etc.) and energy (solar energy conversion, nuclear waste disposal and microwave transmission). Accessibility to space will thus dictate our future progress and possibly even our survivability in the years to come.

The only means available for any object to be placed in the orbit is rocket propulsion which uses chemical propellants, may be solid, liquids or hybrids. Since October 4, 1957 when the first Russian launched first man-made earth satellite, Sputnik I, the possibility of space travel was no longer a speculation in the minds of common man. This is followed by spectacular achievements which include space walks, space docking, moon landing, mars orbiting and landing, other planets probe in the solar system, etc. All these have been possible because of developments of chemical rockets—liquids, solids and hybrids.

Solid propellants by their oblivious advantages like simplicity, ruggedness, instant readiness for flight, operational easiness, low cost, storage stability and high thrust capability have come to stay in launch vehicles and missiles. Present day commercial space launcher systems, generally use solid rocket propulsion during the launch and/or booster phases either as main engines in small launchers like Scout, SLV-3, ASLV and Orbital or as strap-on motors (space shuttle boosters, Arian V or GSLV Mk iii boosters). Solid propellants are also used in space programmes for inertial upper stage motors, critical tasks such as ignition, spin, separation assistance (ullage and retro rockets), de-boost motors as in the case of Chandrayaan mission and for escape in manned missions in case of malfunction of launch vehicle. Generally, solid propellant rocket motors were chosen over liquid propellant rockets for lower costs and quicker reaction time. The primary use of liquid propellant rockets appear to have shrunk to missions requiring multiple use, extremely high performance, many shutdown and restarts or throttling. The state of development of propellant, insulation, nozzle and thrust vector control system has reached a point that new progress can be undertaken with very high confidence regarding their outcome both from a schedule and performance point of view. Solid propellant propulsion system will continue to offer the designer of launch and development vehicles' option worthy of the most serious consideration.

Solid propellants are here to stay. Considerable development and evaluation work, however, remains to be done, including the meeting of new requirements not yet conceived. We had the privilege of working for Indian Space Prgramme by developing solid propellants to power Indian launch vehicles and putting India in the elite Space club. The country is self sufficient in this vital technology today, thanks to the developmental and indigenization efforts undertaken by our teams. The present book is the result of our efforts in the development programme. The book traces the history of the solid propellant development world over for understanding the subject from where we can build the Indian structure. The solid propellant development is an interdisciplinary programme which needs a good understanding of polymer science, oxidizer chemistry, materials science and chemical engineering practices. The coverage in this book is, therefore, aimed at giving the breadth which is needed by chemists and engineers working on modern propellant technology. It will also broaden the views of those already working in the field of propellant development, production and research. The real purpose of looking back, of course, is not only to obtain satisfaction from reflecting on past triumphs, rather it is to discover as many clues as possible to the likely developments of future.

Solid propellants utilize polymers with plasticizers and often providing a matrix with curing/cross-linking agents to contain other ingredients which participate in the combustion. The polymer plasticizer combine provides the mechanical properties required and allows processing of propellant into desired shape and sizes. The polymers and plasticizers serve as fuel as well as matrix to suspend oxidizer and metallic particles or may themselves be energetic. The book is an all in one solid propellant technology in nine chapters and the first chapter introduces the subject while tracing the history of solid propellants. The second chapter describes the ingredients used in solid propellants and the chemistry involved. The third chapter talks about the energetics of propellant formulations as way to finalize the ingredients for making a formulation. The fourth chapter briefly explains the processing aspects of composite propellants. The next chapter gives the importance of characterization and the methodology used in testing of uncured slurry and cured propellants. The sixth chapter gives the internal ballistics aspects of propellants. Advanced solid propellants for increasing the energy content of solid propellants is the subject of next chapter. Safety and quality aspects involved in propellant processing, storage and handling is covered in the eighth chapter. The last chapter on homogenous propellants gives the development and status of this class of propellants which is mainly used in guns, ammunition and missiles. The evolution of solid propellant technology in Indian Space Research Organization from beginning to end of 2014 which saw a sub orbital flight of the biggest booster for heavier payloads including manned missions.

We have tried to make the book easier to use by providing: a) a much more detailed table of contents, b) an expanded contents and glossary, c) appendix giving the evolution of solid propellant development in India, d) questions for each chapter and e) exhaustive references for further reading. The book has three major markets: it can be used as a course book for engineering graduates and post graduates in universities and Indian Institute of Technology centres where propellant technology becomes part of the curriculum; it can be given in one semester. The book can be used to indoctrinate scientists and engineers new to propellant area and to serve as a reference book to experienced persons working in propellant technology especially in ISRO centres, defence research centres like HEMRL, ARDE, etc., various ordnance factories and explosive factories in India, where one can look for some topic, data or equation. To researchers, it will serve as a source book from where they pursue their topics of research.

We wish to thank our colleagues of Propellant Engineering Division (PED), Polymers and Special Chemicals Division (PSC), Analytical and Spectroscopy Division (ASD), Rocket Propellant Plant (RPP), Propellant Fuel Complex (PFC), Solid Propellant Space Booster Plant (SPROB), Ammonium Perchlorate Experimental Plant (APEP), Solid Motor Project (SMP), and various launch vehicle projects SLV-3, ASLV, PSLV, GSLV for generating the figures and data relating to physical, thermal and mechanical characterization of propellants and ballistic evaluation of solid propellants during the development and production phase. The authors thank the colleagues in Propellant processing sections of PED, RPP and SPROB for processing of composite propellants formulations during development, scale-up and production. The authors gratefully acknowledge the valuable contributions of the following technologists and scientists towards solid propellant technology in ISRO and for some of the cited references in the text: Dr. K.V.C. Rao, Mr. K.S. Sastri, and Dr. K.N. Ninan (emeritus professor also) Deputy Directors of propellant entity, Vikram Sarabhai Space Centre (VSSC), Trivandrum; Mr. M.C. Uttam, late Mr. S.K. Athithan and V. Srinivasan, Deputy Directors of propellant entity, Satish Dhavan Space Centre (SDSC); M.C. Dattan, Director, SDSC, Dr. R.M. Muthiah, Mrs. Manjari Rajan, Mrs. Lalitha Ramachandran and late Dr. S.K. Nema, Group Directors of VSSC, Mr. S. Alwan, Mr. R. Sivaramakrishan, Mr. T.S. Ram and N. Muralidharan, General Managers of Rocket Propellant Plant, VSSC, Trivandrum and late Mr. N.V. Viswanathan, Head, Propellant Safety Group, VSSC, Trivandrum. We are thankful to the authors and editors of HEMCE conference proceedings for allowing to mention as references in this book. The authors thank Mr. P.N. Haridas of Liquid Propulsion Systems Centre and Dr. S.K. Manu of Vikram Sarabhai Space Centre of ISRO for the support for Electronic data processing. We also thank our former colleagues for the discussions and suggestions and Drs Raj Venkatesan, Jaishree Venkatesan (Community College, Texas, USA) and Anuj Varghese (Hyderabad University, Hyderabad) for going through the manuscript and valuable suggestions. Finally, we thank and appreciate the patience and perseverance shown by Mrs. Geetha Krishnamurthy, Mrs. Janaki Natarajan (VNK) and Mrs. Achamma Varghese (TLV) during the course of this mammoth job of preparing the draft for this book. We are also thankful to Dr. Vasant Ranchod Gowariker and Shri. Madhava R. Kurup, Former Directors of ISRO's centres, for introducing the field that made India join the Space club and all encouragement during the development and product phase of these propellants.

Finally, the authors thank Dr. K. Kasturirangan, former Chairman ISRO and Secretary, Department of Space, Government of India for writing the Foreword for the book and to CSR Murthy, Administrative Officer of Raman Research Institute, Bangalore for helping to get the Foreword from Dr. K. Kasturirangan.

V.N. Krishnamurthy
T.L. Varghese

Contents

Foreword .... *vii*

Preface .... *ix*

CHAPTER 1: Propellants—An Introduction 1–19

History of Rocketry, What are Propellants, Parts of a Rocket Motor, Newton's Third Law, Rocket Propulsion Principles, Vehicle Mass Ratio, Propellant Mass Fraction, Classification of Propellants—Liquid Bi-propellants, Liquid Monopropellants, Cryogenic Propellants, Hybrid Propellants, Fuel Rich Propellants, Gel Propellants, Homogeneous Propellants—Single Base Propellant, Double Base Propellant, Triple Base Propellants, Composite Propellants, Space Launch Vehicles, and their Classifications, Questions and References.

CHAPTER 2: Composite Propellants—Ingredients and their Functions ... 20–55

Oxidizer—Ammonium Perchlorate—Characteristics, Production, Other Perchlorates, Ammonium Nitrate, Fuel Binders, Salient Features of Binder, Thermoplastic Binders, Thermosetting Binders, Characteristics of Pre-polymers, Energetics of Polymeric Binders, Propellant Pre-polymers—Poly Sulphide—Synthesis and Cure Reactions, Unsaturatd Polyesters, ISRO-Polyol—Synthesis and Cure Reactions, Polyether Polyols, PBAA (Poly Butadiene-Acrylic Acid Copolymer)—Synthesis-Characteristics, PBAN (Poly Butadiene-Acrylic Acid-Acrylonitrile Terpolymer)—Synthesis and Cure Reactions, LTPB or HEF-20 (Lactone Terminated Polybutadiene) Synthesis and Cure Reactions, CTPB (Carboxyl Terminated Polybutadiene)—Synthesis and Cure Reactions, HTNR (Hydroxyl Terminated Natural Rubber)—Synthesis and Cure Reactions, HTPB (Hydroxyl Terminated Poly Butadiene)—Synthesis (Free-Radical and Anionic Methods) and Cure Reactions—Micro Structure-M.wt Distribution, PVC Plastisol and Double Base Binders, Plasticizers, Metallic Fuel—Metal Powders, Metal Wires, Metal Hydrides, Cross Linkers/Curing Agents, Burning Rate Modifiers, Bonding Agents, Process Aids, Stabilizers or Anti Oxidants, Questions and References.

CHAPTER 3: Composite Solid Propellants—Energetic Aspects 56–90

Energetic Computations—Specific Impulse, Heat of Formation, First Law of Thermo-Dynamics, Calorimetric Measurements, Thermo-Chemical Calculations, Hess's Law, Relationship between Heat of Combustion and Heat of Formation, Oxidizer—Fuel Energy Balance, Relationship between Bond Energies and Heat of Formation, Calculations of Heat of Combustion of Ethane, Cyclopropane, Cyclopropene, etc. from Bond Energies, Favourable Bonds, Contribution of Metallic Fuels, Effect of Binder Type on I_{sp}, Effect of Binder-Al—AP Effect on I_{sp}, Nozzle Design—Assumptions and Parameters, Propellant Grain and Grain Configuration, Failure Modes, Performance Efficiency of a Rocket Motor, Combustion

Processes in Rocket Propellants—Double Base Propellants—Composite Propellants, Combustion Models—CMDB Propellants, Igniter and Igniter Propellants—Igniter Requirements, Pyrogen and Pyrotechnic Igniters and Comparative Properties of Motor Propellant and Igniter Propellant, Questions and References.

CHAPTER 4: Propellant Processing Technology 91–119

Solid Propellants, Advantages, Parts of a Composite Solid Propellant Rocket Motor, Propellant Processing—Motor Cases—Composite Cases—Glass Fibre Reinforced Epoxy Case and Kevlar Fibre Reinforced Epoxy Cases, Polar Winding Machine, Metallic Cases-15 CDV 6 Steel and Maraging Steel, Difference between Carbon Steel and Maraging Steel, Motor Case Properties, Motor Insulation—Types of Insulation—Requirements, Rubbers Used, Fillers and Additives, Types of Insulation—Sheeted Insulator, Castable Insulator, Sprayable Insulator, Trowelable Insulator, Reactive Insulator (Insu-liner), Typical Insulator Properties, Motor Lining and Liner Systems—Liner Features, Theory of Liner Bonding, Liner Composition and Application, Cured Liner Properties (Typical), Interface Properties, Propellant Processing—Process Flow Chart—Raw Material Preparation, Mixing Operations—Horizontal Mixer, Vertical Change Can Mixer and Continuous Mixer, Process Parameters—Mixing Sequence, Mixing Duration, Process Temperature, Humidity Control, Pot Life-Zero Flow Time (ZFT)—Penetrometric Pot Life (PPT), Propellant Casting—Vacuum Casting, Multiple Pressure Casting or Bottom Casting and Bayonet Casting or Top Casting, Motor Curing, De-coring and Trimming, Inhibition of the Grain, Non-Destructive Testing (NDT) and Motor Static Testing, Questions and References.

CHAPTER 5: Propellant Characterization and Rheology 120–154

Mechanical and Physical Characterization—Tensile Strength, Elongation, Modulus of Elasticity, Shore-a-Hardness, Density, Visco-Elastic Behaviour—Stress-Strain Curves, Stress Relaxation & Relaxation Modulus, Creep Compliance, Shear Modulus, Failure Envelope, Dynamic Mechanical Properties—Master Relaxation Modulus, Dilation in Tension—De-wetting Strain, Poisson's Ratio, Dilation in Compression—Bulk Modulus, Interface Properties—Peel Test, Tensile Bond Test, Shear Bond Test, Thermal Characterization—Specific Heat, Coefficient of Thermal Expansion, Thermal Conductivity, Glass Transition, Calorific Value, Thermal Analysis—Thermo Gravimetric Analysis (TGA)—Differential Thermal Analysis (DTA) and Differential Scanning Calorimetry (DSC), Rheological Characterization—Rheology—Viscosity (η), Coefficient of Viscosity, Methods of Determining Viscosity—Oswald's Viscometer, Brookfield Viscometer with Helipath Stand, Contrave's Rheometer, Herschel Bulklay Equation, Classification of Liquids Based on Rheology—Newtonian Liquids, Non-Newtonian Liquids—Pseudo Plastic, Dilatants and Thixotropic, Burn Rate Characterization—Burn Rate Determination—Acoustic Emission Technique, Crawford Bomb Method, Ultrasonic Method, Burn Rate Law, Temperature Sensitivity, Non-Destructive Testing of Propellants—Visual Inspection, X-ray Radiography (XR), Real Time Radiography (RTR), Computer Aided Tomography (CAT), Neutron Radiography (NR), Ultrasonic Testing (UT), Acoustic Emission Testing (AET) and Optical Holography, Questions and References.

CHAPTER 6: **Internal Ballistics of Rockets** 155–171

Solid Propellant Burn Rate, Motor Burning Pattern—Progressive Burning Motor, Neutral Burning Motor, Regressive Burning Motor, Plateau and Mesa Burning Propellants, Ultrasonic Burn Rate Determination, Pressure-Time Curves, Ignition Delay, Ignition Rise Time, Pressure—Thrust Build Up, Web Burn Time, Action Time, Thrust (F) and Total Impulse (I), Total Thrust (F) or Total Impulse (I), Specific Impulse (I_{sp}), Exhaust Velocity (V_e), Thrust Coefficient (C_F), Characteristic Velocity (C^*), Mass Flow Rate or Mass Burning Rate (M_b), Mass Discharge Rate (M_D), Discharge Coefficient (C_D), Equilibrium Pressure, Idialised P–T Curve, Importance of K & Factors Affecting Equilibrium Pressure, Erosive Burning, Factors Affecting Erosive Burning & Validation of Erosive Burning, Chuffing, Factors Affecting Chuffing and Resonance Burning, Questions and References.

CHAPTER 7: **Advanced Solid Propellants** 172–196

Advanced Oxidizers—Characteristics, Ammonium Dinitramide (ADN)—Synthesis and Performance, Hydrazinium Nitro Formate (HNF), Synthesis of HNF, Hexanitro Hexaazaiso Wurtzitane (HNIW or CL-20)—Synthesis and Properties, 1,3,3-Trinitro-Azetidine (TNAZ)—Synthesis, Nitrocubanes-Synthesis, Advanced Binders-comparative Properties, Glycidyl Azide Polymer (GAP)—Synthesis and Properties, Polyoxetanes—Bis-azidomethyl Oxetane (BAMO)—Synthesis, Poly-nitrato Methyl Methyl Oxetane (Poly NIMMO) or PLN—Synthesis, BAMO-THF Copolymer, PGN (Poly Glycidyl Nitrate)—Synthesis and Properties, Fluorinated Polymers—Synthesis and Properties, Energetic Plasticizers—Bis (2,2-dinitropropyl formal (BDNPF), Bis 2,2-dinitropropyl Formal (BDNPA), Azide Terminated Glycidyl Azide Plasticizer (GAPA), Ethylene Glycol Bis Azidoacetate (EGBAA), Diethylene Glycol Bis Azidoacetate (DEGBAA), Trimethylol Nitromethane Tris Azidoacetate (TMNTA) and Penta Erythritol Tetrakis (Azidoacetate) (PETKAA), Advanced Solid Propellants—Specific Impulse of Advanced Propellants, Cyclic Polynitrogen Compounds, Octa Azacubane-Explosive Properties, Questions and References.

CHAPTER 8: **Safety, Quality and Reliability in Solid Propellants** 197–222

Safety, Hazards Classification, Safety Tests-impact Test, Friction Test, Auto-Ignition Test, Shock or Detonation Test, Safety Approach—Safety during Propellant Processing, Cardinal Principles, Hazard Analysis, Ageing Characterization and Shelf Life, Environmental Effects on Propellants and Storage Conditions, Disposal of Waste and Aged Propellants, Origin of Propellant Waste-Propellant Processing, Machining, Batch Rejection and Aged Propellants, Handling of Propellant Wastes—Methods of Waste Disposal—Chemical Methods, Molten Salt Oxidation, Bio-Degradation, Reclamation/Recycling Method, Waste Burning Pit Method, Un-Instrumented Motor Test and as Commercial Explosives, Quality—Process Control, Quality Control, Reliability, Product Assurance and Margin of Safety, Questions and References.

CHAPTER 9: Homogeneous Propellants 223–259

Single Base, Double Base, Triple Base and Colloidal Propellants, Composite Modified Double Base, Nitramine Propellants, Ingredients—Nitrocellulose, Nitroglycerine, Nitroguanidine, RDX, HMX, Non-explosive Plasticizers, Stabilizers, Darkening Agents, Burn Rate Modifiers, Flash Suppressors, Plateaunizing Agents, Methods of Handling Nitrocellulose, Water Wet Nitrocellulose, Alcohol Dehydrated Nitrocellulose, Globular Nitrocellulose, Testing of Nitrocellulose, Explosive Properties, Processing of Homogeneous Propellants-Single Base, Double Base and Nitramine Propellants—Solvent Extrusion Process, Solventless Process, Mechanical Properties, Processing of Cast Homogeneous Propellants, Characterization of the Casting Powder, Interstitial Casting Process, Slurry Cast Process, CMDB Processing, Theoretical Performance, Combustion Products, Ballistic Properties, Burning Mechanism, Questions and References.

APPENDIX: The Evolution of Solid Propellants in ISRO 260–301

Glossary 302–308

CHAPTER 1

Propellants—An Introduction

1. HISTORY OF ROCKETRY

The history of rocketry is as ancient as the spirit of man and its exact origin cannot be authentically stated. Fire is the origin of weapon development and firepots containing flammable materials were thrown in defense even during 1000 B.C. Chinese are the leaders who have contributed to both theoretical and practical development of rocketry. In 1232 A.D. in the battle of Kai-fung-fu, the Chinese used a queer weapon, called "Fire Arrow", a device looked more like an arrow but with some incendiary (Black powder) filled pack attached to the arrow to impart better speed and range. These fire arrows were launched in salvos from arrays of boxes or cylinders to improve its power. Muratori, an Italian, was the first to use the word "Rochetta" which means in English 'rocket' in 1379 A.D. In 1500 A.D., a Chinese, named Wan Hoo, attempted a manned flight with rockets and lost his life in doing so. This can be considered as the forerunner of the modern concept of clustering of rockets used in space exploration. The Arabs emulated the Chinese and used solid propellant rockets with a war head to frighten the enemy horses with a glittering shower. The centuries that followed witnessed intense activity in Europe especially in Italy, Germany and England. The French troops, for example, used rockets extensively in the wars.

A climax was reached probably in India when Hyder Ali and his son Tippu Sultan employed 200 mm long rockets in the Mysore war in the eighteen century. Both of them with rocket contingent having 1200 rocketeers and 5000 rocketeers respectively used the rockets effectively against the British army in the battle at Srirangapattinam. These rockets travelled ferociously spitting fire all the way and landing violently in the midst of battle ground, demoralizing the British cavalry. These rockets could reach a distance of about one kilometer in a curve and hence, could be thrown by people covered by a line of infantry. The war office in London was flooded with reports about the strangeness of Tippu's rockets and Woolwich Arsenal was asked to find an antidote to this weapon.

Two rockets, belonging to Tippu's forces were captured during the fourth Mysore war in the siege of Srirangapattinam in 1799 by the companies of East India

Company. They are now in the museum of Artillery at Woolwich, London. In 1804, William Congreve developed a variety of superior rockets based on Tippu's rockets for various purposes and used them successfully in battles against the French and Germans including in the battle of Waterloo to defeat Napolene Bonaparte in 1815. William Hale improved the performance of rockets by spin stabilization, which were used in Mexican war during 1846–48. Also, Russians fired their first rocket in 1817 and established a factory to manufacture them in 1826 at St. Petersburg. In the second half of the 19th century, the propulsive power of rockets was used for other than military operations. The first two-stage rocket was developed in 1855 for transport of heavier cord and in rescue line applications.

The limitations of black powder as blasting explosive in 1846 led to the discovery of nitroglycerine. The Swedish inventor Alfred Nobel developed a process for manufacturing nitroglycerine and its absorption by 'Kieselguhr' for safe handling and transportation of nitroglycerine. This mixture of nitroglycerine and kieselguhr is known as dynamite. Along with nitroglycerine, nitration of cellulose was attempted by many workers like Schonbem and Bottger. In 1875, Alfred Nobel discovered that mixing of nitroglycerine and nitrocellulose produce a gel which can be used to produce blasting gelatin. Later in 1888, the first smokeless powder, called 'cordite', consisting of nitrocellulose, nitroglycerine, camphor and benzene was discovered. Cordite in various forms had remained as the main propellant of British force till 1930. Britishers established a cordite factory at Aravankadu, India to manufacture various types of cordites. The introduction of smokeless nitrocellulose and the double base propellant marked a significant advancement in propellant history.

A group of solid propellants, called composite propellants, came into existence as a result of increased R&D activities during World War II. The first composite propellant appeared somewhere around 1945. Since then, composite propellants have played a major role in propellant field. Though, propellants were used mostly for military applications, the advent of Sputnik and Explorer satellites opened the way for greater use of propellants for space. There have been limited applications of propellants for industrial use viz. oil well performing gun, industrial cannon for quarries, Jet Assisted Take Off (JATO) rockets.

In independent India, experiments with solid propellant rockets started with the flight of RH75 sounding rocket using double base propellants in November 21st 1967 and with the composite propellant in February 21st 1969. The first solid propellant satellite launch vehicle SLV-3 was successfully flight tested on July 18th, 1980 by putting a 40 kg Rohini satellite into a near earth orbit. This was followed by PSLV flights which had made 25 various successful missions and the successful GSLV flight in 2014.

2. WHAT ARE PROPELLANTS?

The propellant propels the rocket and hence the name. It is a concentrated source of energy behind the rockets, missiles and launch vehicles. As it starts burning, the potential energy it contains gets converted to kinetic energy to do the work. When domestic coke burns, it gives small amount of hot gases but when propellant burns, it gives out large volumes of hot gases flowing at high speeds. The hot gases ejected through the nozzle produce forward thrust to the propulsion unit as desired. This is similar to the boat reeling backwards when a row of swimmers dive into water from a small boat. The gas molecules acting as divers throw back the rocket boat in accordance with Newton's third law of motion. This is how the rocket flies using a rocket motor.

The difference between a good propellant and bad propellant mean the rocket set out for moon reaching its destination or exploding on the launch pad. The propellant is to a rocket what petrol is to a car. Both are suppliers of energy but with a difference. Petrol, diesel or any other fuel needs oxygen to burn while propellant does not need oxygen from external source as it contains its own oxygen in the form of chemical oxidizer. A propellant can burn even in vacuum. Thus, the propellant is not a fuel as it is loosely said. The fuel is only one part of the propellant and that along with the oxidizer makes the propellant a complete system, capable of burning without external oxygen source. The absence of oxygen in outer space demands that the propellant systems in rockets and spacecrafts be necessarily comprised of two components, a fuel and an oxidizer.

Solid propellants decompose by deflagration process. The propellants perform their work by slow liberation of energy, characterized by high temperature gases pushing against the surrounding air. High explosives on the other hand perform their work by their sudden shattering, as in the case of rock blasting.

Rocket propulsion systems can be classified according to the type of energy source—chemical, nuclear, or solar, the basic function—booster stage, sustainer, attitude control, orbit station keeping, the type of vehicle—aircraft assisted take off, missile, space vehicles, size, type of propellant, type of construction or number of propulsion stages used. Another way of classification is by the method of producing the thrust. A thermodynamic expansion of gas is used in the majority of rocket propulsion concepts. The internal energy of the gas is converted into kinetic energy of the exhaust flow and the thrust is produced by the gas pressure on the surface exposed to the gas. The same technique and principle are used for jet propulsion, rocket propulsion, nuclear propulsion, laser propulsion, solar—thermal propulsion and some electrical propulsion. Totally different methods of producing thrust are used in other types of electric propulsion. These electric systems use magnetic or electric fields to accelerate electrically charged molecules or atoms at very low densities.

In chemical rocket propulsion, the energy from a high-pressure combustion reaction of propellant chemicals—fuel and oxidizer—permits the heating of reaction product gases to very high temperatures of 2500 to 4500°C. These gases are subsequently expanded in a nozzle and accelerated to high velocities (1800 to 4500 m/sec). According to the physical state of the propellant, we have liquid, solid and hybrid propulsion devices.

3. ROCKET MOTOR

It is essential to know how a rocket motor, the propulsive unit of a solid rocket, works. It essentially consists of a propellant (L) containing in a combustion chamber A (made of steel light alloy or fibre reinforced plastic) capable of withstanding a certain internal pressure, a head–end (B) where igniter (D) is assembled and a nozzle (C) as shown in the Figure 1. The different parts of a rocket motor are marked in the figure. The propellant is loaded inside the chamber where it burns on ignition by the igniter located at the head end of the motor. The gases that are produced by the combustion of propellant are first converged in the convergent portion (E) of the nozzle and released to the atmosphere through the divergent portion (F) of the nozzle. In this process, the gases reach supersonic speeds and generate a reaction force that makes the rocket fly. The higher the speed of the gases, the faster the rockets fly. Thus, a rocket works on action and reaction principle—New's third law: 'To every action, there is an equal and opposite reaction'.

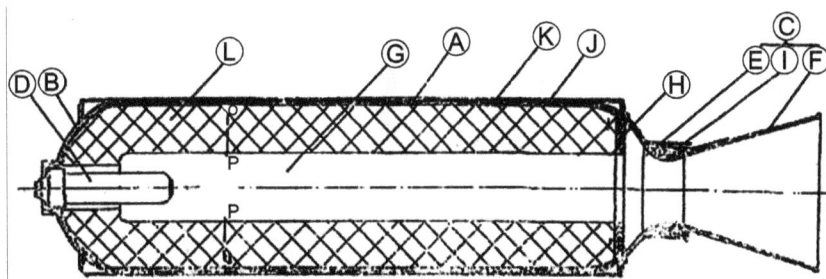

Fig. 1: Parts of a Rocket Motor

A – Chamber, B – Head End, C – Nozzle, D – Igniter
E – Nozzle Convrgent Portion, F – Nozzle Divergent Portion
G – Port, H – Inhibitor, I – Nozzle Throat Insert
J – Lining, K – Insulation, L – Propellant and P-Q – Web

4. PROPELLANTS AND EXPLOSIVE

The conversion of the chemicals in the propellant into gases moving out at high speeds and high temperatures of the order of 3000°C is known as propellant burning. The propellant has to burn in a smooth and orderly fashion giving out the gas and energy in a disciplined way over the desired duration of burning time. However, the burning can sometime be erratic, accompanied with deafening sound, shattering impact, etc. When this happens, the propellant is said to behave as an explosive. A rocket propellant can behave like a high explosive with similar effect, but when this happens, the propellant scientist is unhappy. The major difference between propellant and explosive is in the nature of burning. A propellant burns in a controlled manner and generates the gases which push the rocket forward. That is, the propellant burns in a controlled manner or deflagrates while burning, whereas an explosive burns all on a sudden and releases the hot gases at once making the pressure to buildup in the chamber, leading to explosion. Explosives undergo decomposition by a detonation process with detonation velocity ranging from 1000 to 10,000 m/s. A high velocity detonation wave propagates through the body of the explosive in all directions. Examples of propellants are composite propellants based on HTPB, CTPB or PBAN binders and that of explosives are RDX, HMX, etc.

5. ROCKET PROPULSION

Rocket propulsion is essentially governed by the basic principles of mechanics, thermodynamics and chemistry. The propulsive force is obtained by ejecting the propellant at high velocity. The definitions and basic relations of the propulsive force, the exhaust velocity and the efficiencies of creating and converting the energy and other basic parameters are discussed for easier understanding of the science and technology of solid rocket propellants.

The total impulse I_t is the thrust force F (which can vary with time) integrated over the burning time t,

$$I_t = \int F \, dt.$$

For constant thrust, this reduces to,

$$I_t = Ft$$

It is proportional to the total energy released by all the propellants in a propulsion system. The specific impulse (I_{sp}) is the total impulse per unit weight of propellant. It is an important figure of merit of the performance of a rocket propulsion system, similar to kilometers per litre concept used in automobiles. A higher number means a better performance. If the total mass flow rate of propellant is $\mathring{m}$ and the standard acceleration of gravity at sea level (g_0) is 9.8066 m/sec^2, then

$$I_{sp} = \int F \, dt / g_0 \int \mathring{m} \, dt$$

For constant thrust and propellant flow, this equation can be simplified as,

$$I_{sp} = I_t / m_p g_0 = I_t / w$$

where w is the weight of the propellant and m_p is the mass of the propellant.

Specific impulse (I_{sp}) is an index of energy. In propellant development, the goal is to increase the specific impulse. To carry a certain payload (from copper chaffs to few men) to a certain destination, the advanced and sophisticated rocket is the one which is the lightest and with maximum fuel efficiency (I_{sp}). The specific impulse is the impulse developed per unit weight of the propellant as given above. The total impulse is the product of the specific impulse and the propellant weight. It is also thrust times time, if the thrust is constant. If not constant, the total impulse is the integral of the thrust with respect to time over the burning time of the propellant. The unit of specific impulse is total impulse/weight or kg weight.sec/kg or seconds.

The specific impulse ranges from 175 sec to 250 sec for solid propellants and from 230 sec to 440 sec for liquid propellants. Specific impulse changes with operating pressure of the motor/engine and the altitude at which these operate. Hence, the specific impulse is so many sec at 70 kg/cm^2 pressure and at sea level conditions. Specific impulse gets better in vacuum (from say 240 sec at sea level to 275 sec in vacuum).

In a rocket nozzle, the actual exhaust velocity is not uniform over the entire thrust magnitude and hence, the velocity profile is difficult to measure accurately. For convenience, a uniform axial velocity C is assumed. The effective exhaust velocity C is the average equivalent at which propellant is ejected from the vehicle. It is defined as:

$$C = I_{sp} \, g_0 = F / \mathring{m}$$

It is given in meters per second. As C and I_{sp} differ only by an arbitrary constant, either one can be used as a measure of rocket performance.

In solid propellant rockets, it is difficult to measure the propellant flow rate accurately. Hence, the specific impulse is calculated from the total impulse and the propellant weight (difference in weight between initial and final motor weight). The total impulse is in turn is obtained from the integral of the measured thrust with time.

The Mass Ratio (MR) of a vehicle or a particular stage is the ratio of the final mass m_f (after the consumption of all usable propellant in the vehicle or stage) to m_0 (before rocket operation).

$MR = m_f / m_0$

This applies to a single stage or a multistage vehicle. The final vehicle mass (m_f) includes all the components that are not useful propellant and may include guidance and navigation devices, payloads, flight control systems, communication systems, power supplies, all propulsion hardware and residual or unusable propellant. The values of MR range from 10% for unmanned launch vehicles to about 60% for some tactical missiles. This fraction is important in analyzing the flight performance of vehicles.

The propellant mass fraction 'ε' indicates the fraction of propellant mass m_p in the initial mass m_0, which can be applied to a vehicle, stage or rocket propulsion system. m_f is the inert propulsion mass.

$\varepsilon = m_p / m_0 = (m_0 - m_f)/m_0 = m_p/(m_p + m_f)$

The thrust is the force produced by a rocket propulsion system acting upon a vehicle. It is the reaction experienced by its structure due to ejection of matter at high velocity. It is the same phenomenon that pushes a small boat backward when the row of swimmers on it diving into water or makes a gun recoil. The forward momentum of the bullet and the powder charge is equal to the recoil or rearward momentum of the gun barrel. A balloon works in the same way. The air inside a balloon pushes back so that the inward and outward forces are balanced. When the mouth of the balloon is opened, air escapes through mouth and the balloon is propelled (flies) in the other direction. Rocket propulsion differs from these devices primarily in the relative magnitude of the accelerated masses and velocities. In rocket propulsion, relatively small masses, which are carried within the vehicle and ejected at high velocities, are involved.

The thrust, due to a change in momentum, is given by,

$F = \{dm/dt\}\, v_2 = m^{\circledR} v_2 = w^{\circledR}/g_0\, v_2$

where F is the force, dm is the change of momentum and t is time in seconds, $m^{\circledR}$ is mass flow rate, $w^{\circledR}$ is the weight flow rate, v_2 is the gas velocity leaving the rocket. This force is the total propulsion force when the nozzle exit pressure is ambient pressure. Because of a fixed nozzle geometry and changes in ambient pressure due to variation in altitude, there can be imbalance of external environment or atmospheric pressure (p_3) and the local pressure (p_2) of the hot gas jet at the exit plane of the nozzle. Hence, for a steadily operating rocket propulsion system moving through a homogenous atmosphere, the total thrust is given by:

$$F = m^{\circledR}v_2 + (p_2 - p_3)\,A_2$$

where A_2 is the nozzle exit cross-sectional area. That is, the reaction force is the sum of momentum thrust and pressure thrust. If the exhaust pressure is less than the surrounding fluid pressure, the pressure thrust is negative. This condition is undesirable and hence, usually the rocket nozzle is designed so that the exhaust pressure is equal to or slightly higher than the ambient fluid pressure. The rocket nozzle design for which $p_2 = p_3$ is defined as the rocket nozzle with optimum expansion ratio. The change in pressure-thrust due to altitude changes amounts to 10 to 30% of the overall thrust. The variation in thrust and specific impulse is of the order of 10 to 30%.

Solid propellant decomposes by deflagration process and when ignited, it burns and releases hot gases. Parts of heat from the gases are received by the freshly exposed propellant surface within and in turn, decompose. This process continues in a self sustained manner until all the propellant burns out. The propellant burning is in parallel layers. The burning surface of a solid propellant grain recedes in a direction essentially perpendicular to the surface. The rate of regression, expressed in cm/sec or mm/sec, is the burning rate "r". For a given propellant, the burning rate depends on the pressure. However, there is a class of propellant, called Platonized propellant, which has almost constant burning rate over a pressure range.

Figure 2 given below shows the change of the grain geometry for a two dimensional grain with a central cylindrical cavity with five slots. Success in a rocket

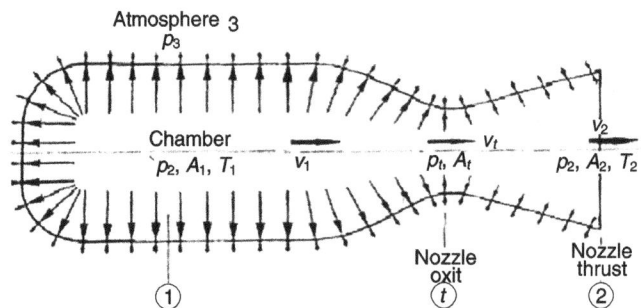

Fig. 2: Pressure Imbalance on Chamber and Nozzle

motor design and development depends, to a major extent, on the knowledge of burning rate behavior of the propellant under motor operating conditions. Burning rate is the function of propellant composition. For composite solid propellants, it can be increased by changing propellant characteristics like: i) addition of burning rate catalyst (0.1 to 3.0%) of propellant), ii) decrease the oxidizer particle size, iii) increase the oxidizer percentage in propellant formulation, iv) increase the heat of combustion of polymeric binder or plasticizer, and v) imbedding metallic wires or staples in the propellant. The burning rate also can be increased in motors: a) by increasing the combustion chamber pressure, b) initial temperature of the propellant prior to starting the firing or testing, and c) motor motion like acceleration and spin.

The burning rate of propellant in a motor at any instant governs the mass flow rate M_b of hot gases generated and flowing from the motor.

$$M_b = A_b\, r\, \rho$$

Here r is the burning rate, ρ is the solid propellant density and A_b is the burning area.

The total mass m of the propellant burned can be calculated by integrating the above equation:

$$m = \rho \int A_b\, r\, dt$$

where A_b and r vary with time.

6. CLASSIFICATION OF PROPELLANTS

Most of the modern propulsive devices are based on the energy derived from the combustion of fuels in air. The absence of air in the outer space demands that propellant system in rockets and spacecrafts be necessarily comprised of two components—fuel and oxidizer. Depending on the physical state of the fuel and oxidizer, the propellants have been classified into three categories viz. solid, liquid and hybrid propellants. Gelled propellants and fuel rich propellants are of recent origin.

Liquid Bi-Propellants

In liquid propellants, both the oxidizer and fuel are liquids. These bi-liquid systems are further classified into earth storable and cryogenic propellants. Robert Goddard demonstrated the first successful flight of liquid propellants in 1926. Some of the liquid fuels for rockets which can be stored in standard earth environment are hydrazine—N_2O_4, Unsymmetrical dimethylhydrazine—N_2O_4, aniline—Red Fuming Nitric Acid (RFNA), etc. In the bipropellant systems, the fuel and oxidizer are stored separately and injected into the combustion chamber in the form of fine droplets. The parts of a liquid bipropellant rocket is shown in Figure 3. In self-igniting

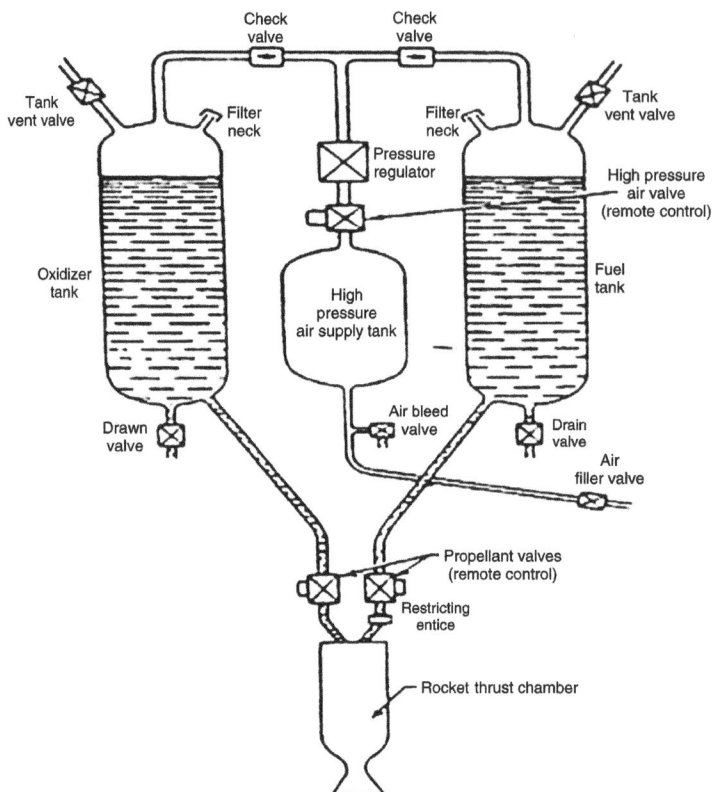

Fig. 3: Parts of a Liquid Propellant Rocket with Gas Pressure Feed System

(hypergolic) systems, ignition occurs spontaneously when the two chemicals-oxidizer and fuel—come in contact with each other. The thrust can be varied at will by controlling the rate of supply of the propellants. The bi-liquid systems, in general, give higher specific impulse than solid propellants but inferior to cryogenic propellants in specific impulse. Utilization of liquid propellant requires highly involved technology, involving numerous precision valves and regulators to get accurate metering of the two liquids. For controlling the ullage problem in zero gravity atmospheres, development of bladders, screens and other devices to ensure positive expulsion of liquids are very essential.

Liquid Monopropellants

Another class of propellant system is known as monopropellants. Usually, a liquid monopropellant has both fuel and oxidizer groups and elements in the same molecule and undergoes exothermic decomposition releasing gaseous products at high temperature. Nitromethane, methyl nitrate, hydrogen peroxide and hydrazine are some examples of monopropellants. Hydrazine based monopropellants is the workhorse liquid fuel for satellite attitude control and station keeping where the requirement of high specific impulse is not determinative. The use of mono-propellant simplifies engineering problems by reducing the number of tanks and feed lines.

Cryogenic Propellants

Cryogenic fuels are those that need very low temperature working condition for their storage in liquid form. Their boiling points are less than standard room temperature on earth. Liquid H_2 has a low boiling point of $-253°C$ and liquid O_2 of $-183°C$. They remain in gaseous form and stored under high pressure. They are highly reactive. Some typical examples are liquid hydrogen—liquid oxygen, liquid hydrogen—liquid fluorine, etc. These are cryogenic propellants.

There are systems where either fuel or oxidizer is an earth storable one while the other is a cryogenic fuel, and are called semi-cryogenic systems. An example is the well-known kerosene—liquid oxygen system used in boosters by Russians. These are less costly compared to fully cryogenic ones because of ease of operation and handling of the storable liquids.

Hybrid Propellants

In hybrid propellant systems, generally, a solid fuel and a liquid oxidizer are used. The burning of the solid fuel is different from that of solid propellants. Here the fuel vaporizes on suitable ignition and pressurizes the chamber to enable the diaphragm to burst. Oxidizer flows through an injector and mixes with the fuel vapours leading to combustion and generation of propulsive force which is better than solid propellants but less than bipropellants. The difficulty with the hybrid system is the necessity of pressurizing the oxidizer tank above the combustion chamber for positive flow. Hybrid systems have several advantages: they have variable thrust levels, the presence of voids and cracks in the solid grain is immaterial, the high temperature reaction is confined to a small area, thereby having better thermal control of the system and they are safe to handle because of the inert nature of the fuel block and has start-stop capability. The solid grains for hybrids have higher strength and resilience due to lower oxidizer content. This reduces creep and cracking of the solid fuel. The inertness of the fuel charge leads to reignition problems. This calls for separate gaseous ignition system for reignition. The cost of hybrid system is relatively low. These motors can be fired, terminated, evaluated and restarted, thereby lowering non-recurring development cost.

Some examples of experimental hybrid systems are Hydrogen peroxide (90%)—polyethylene, HTPB—LOX, etc. In spite of simplicity and advantages, rockets based on hybrid propellants have not found their due share of application. The basic problem seems to be lack of suitable hypergolic propellants, low recovery of theoretical specific impulse and uneven burn rate coupled with a decay in thrust level with time.

Fuel Rich Propellants

Fuel rich propellants for Ramjet rockets contain maximum amount of fuel (metallic or polymer) with least amount of oxidizer. Oxidizer is added only to ensure sustained combustion in the primary chamber. Oxygen from atmosphere is rammed to complete the combustion. Hence, these propellants are useful only in low altitudes or in boosters that fly below 20 to 30 kms. The metals used in these formulations are boron, magnesium, aluminum, zirconium powders and Al-Mg alloy. Polymeric fuels being tried are polycyclopentadiene, naphthalene, hydroxyl terminated poly butadiene and functionally terminated natural rubber.

Gel Propellants

Gel propellants, a variant of liquid propellants, contain metallic powders and additives in liquid propellants to make it thixotropic, thereby settling of the heavy powders is prevented. Gelled propellants remain like thick paste in stationary conditions but can flow through valves, pumps and other devices on application of shear stress. This behavior reduces the spillage during handling. The metallic powder, added to the fuel enhances the density and energy, remain uniformly dispersed on storage without settling or segregation. These gels are pushed into combustion chamber and reacted with the oxidizer as is being done in case of liquid propellants. The major disadvantage, among others, is reduction of energy output due to dilution effect of gelling agents and the susceptibility of gel's stability to temperature.

Solid Propellants

Solid propellants are of two types—double base or homogenous propellants and composite or heterogeneous propellants.

Homogeneous Propellants

They contain both oxidizer and fuel in the same molecule. Homogeneous propellants can be sub-classified as: i) Single base, ii) Double base, and iii) Triple base propellants.

Single base propellant contains Nitro Cellulose (NC) as the major ingredient and is mainly used in guns.

Double base propellant is composed of a homogenous colloidal mixture of nitroglycerine and nitrocellulose and other additives like non-explosive plasticizer, stabilizer, ballistic modifier, etc. Both NG and NC contain oxidizer and fuel elements in the same molecule and hence, are called monopropellants. Though each of these ingredients contains a fuel and oxidizer elements in the molecule, nitroglycerine is oxygen rich to the extent of >3.5% and nitrocellulose is fuel rich or oxygen deficient by 30%. Inter-diffusion of NC and NG knits the two component system into a single strong tough homogeneous grain. They are mainly made by extrusion. They are smokeless and are used for defense applications, e.g. missiles and are also used for sounding rockets. They give lower I_{sp} of the order of 200 secs compared to composite propellants which give 245 sec under standard conditions. Also, double base propellants have size limitations, case bondability problem and low density. Hence, composite propellants are preferred for launch vehicle applications. The double base propellants are discussed separately in chapter 9.

Triple base propellants contain 3 major explosive ingredients—NC, NG and nitro-guanidine. They are mainly used as gun propellants and short range rockets/ missiles.

Composite Propellants (Heterogeneous)

In composite propellants, a solid oxidizer is embedded in the matrix of a polymeric binder which also acts as fuel. Generally, solid inorganic materials such as Ammonium Perchlorate (AP), Potassium Perchlorate (KP), Ammonium Nitrate (AN), etc. are used as oxidizers. Organic polymers such as polyesters, polysulphides, polyvinylchloride, polyurethanes and polybutadienes are employed to bind the solid oxidizer powder in order to provide structural integrity to the propellant grain. At times, a metal powder like aluminum is added to the propellant mix to increase the energetics. Composite propellants are cast from a mix of solid (AP crystals and Al powder) and liquid (HTPB, PBAN) ingredients. The propellant is hardened by crosslinking or curing it in an oven, where it becomes hard and solid. In the past five decades, the composite propellants have been the most commonly used class.

They are further subdivided into the following:
1. Conventional composite propellants generally contain between 60 and 72% Ammonium Perchlorate (AP) as crystalline oxidizer, up to 22% Aluminum Powder (Al) as a metallic fuel and 8 to 16% of elastomeric binder including plasticizer and cross-linking agent.
2. A modified composite propellant where nitramines like RDX and HMX are added for getting a little more performance and density.
3. A modified composite propellant where energetic plasticizers like Nitroglycerine (NG) are added to get a little more performance.

4. High energy composite solid propellants (sometimes called elastomer modified cast double base propellants) where the elastomeric polymeric binder and plasticizer are replaced by energetic materials. These are mostly experimental propellants and a theoretical specific impulse of 270 and 275 sec is expected at standard conditions.

5. A low energy propellant with minimum smoke where Ammonium Nitrate (AN) is the crystalline oxidizer (no AP). These propellants are used in gas generators. Large amount of HMX is added sometimes to boost the performance level.

Several classifications can be confusing. The name Composite Modified Double Base Propellant (CMDB) has been used for double base propellant where some AP, Al and binder are added. Propellants can also be classified by the major manufacturing processes like cast propellant, extruded propellant or solvation process propellant. Propellants have also been classified by their principal ingredient, such as principal oxidizer ammonium perchlorate propellants, ammonium nitrate propellants or azide type propellants or their principal binder such as HTPB propellant, PBAN propellant, polyurethane propellant, etc.

Propellants can be classified by the density of smoke in the exhaust plume as smoky, reduced or minimum smoke. Most composite propellants are smoky because of the presence of Al in the composition which gets oxidized to aluminum oxide, a white smoke seen in the exhaust gas. In addition, propellants can also be classified based on safety ratings for detonation as detonable (class 1.1) or as non-detonable (class 1.3). Double base propellants and composite propellants containing HMX and RDX are examples of class 1.1.

Propellants also can be classified as toxic and nontoxic based on exhaust gases. The exhaust plume gases can be very toxic if they contain beryllium oxide particles, chlorine gas, hydrochloric acid gas, hydrofluoric acid gas or some other fluorine compounds. Eco-friendly propellants do not produce toxic gases in the exhaust plume.

In solid propellant rocket motors, the word "motor" is common to solid rockets as the word "engine" is to liquid rockets. The propellant is contained and stored in the chamber for long time storage (1 to 20 years). Motors come in different sizes and types. Solid propellant rocket motors have no moving parts except for some motors with movable nozzles and actuators for thrust control. The thrust control is with liquid propellants. Solid propellants are in greater readiness for operation as they contain oxidizer and fuel ingredients thoroughly mixed. For this reason, they are less safe to store compared to liquid propellants. Solid rocket manufacture calls for less man power compared to liquid engine manufacture. Overall, solid rockets are relatively simple, easy to apply and require little servicing. They cannot be fully checked prior to use, and required thrust cannot be varied in flight.

7. SPACE LAUNCH VEHICLES

Between the first space launch in 1957 and end of January 2014, more than 10,000 space launches have taken place in the world. Multi-step or multi-stage rocket vehicles permit higher vehicle velocities, more payload for space vehicles and improved performance for long range ballistic missiles. After the useful propellant is fully consumed in a particular stage, the remaining empty stage is dropped from the vehicle and the operation of the propulsion system of the next stage is started. The empty mass of the used stage is separated from the remainder of the vehicle, because it avoids the expenditure of additional energy for further accelerating a useless mass. As the number of stages is increased, the initial take off mass increases. The number of stages should not be too large because the physical mechanisms become more numerous, complex and heavy. The most economical number is generally between 2 and 6, depending on the mission. The payload of a multistage rocket is essentially proportional to takeoff mass, even though the payload is only a small portion of the initial mass. A single stage to orbit vehicle is attractive because of its cost and complexities of staging. However, its payload is very small. A low earth orbit can only be achieved if the propellant performance is very high.

Classifications

Space launch vehicles or space boosters can be classified as expendable or recoverable/ reusable. Other ways of classification are the type of propellant (storable, cryogenic or solid propellants), number of stages (single stage, two stage, etc.), size or mass of payloads and manned or unmanned. Space shuttle, Titan IIIC, Ariane V or PSLV or GSLV are launch vehicles used for boosting satellites into synchronous orbit/ geo-stationary orbit or into escape trajectories for planetary travel. PSLV (Polar sun synchronous satellite launch vehicle) is a IV stage vehicle with solid core and six solid strap- on motors as first stage, a liquid second stage, a solid third stage and a liquid fourth stage. The fourth stage of PSLV permit a wide variety of maneuvers, orbit changes and trajectory transfers to be accomplished with the pay load, which can be one or more satellites or spacecrafts.

Each space launch vehicle has a specific space flight objective—earth orbit or moon/mars landing. It uses between three to five stages and each with its own propulsion system and fired sequentially after the lower is expended. The number of stages depends on the specific space trajectory, the number and type of manouvers, the specific impulse of the propellant and others. The initial stage, called booster, is the largest and is operated first. This stage is separated from the ascending vehicle before the second stage is ignited. Once the propellant of a given stage is expended, the dead mass of the stage (chamber and instruments attached to the stage) is no longer useful in providing additional kinetic energy to the succeeding stages. By dropping off this useless mass, it is possible to accelerate the final stage with its useful payload to higher terminal velocity than would be attained if multiple staging were not used.

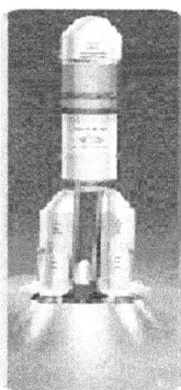

GSLV   Space Shuttle

Depending on their missions, spacecraft can be categorized as earth satellites, lunar, interplanetary and trans-solar types and as manned and unmanned spacecraft. Rocket propulsion is used for both primary (such as orbit insertion or orbit change maneuvers) and secondary propulsion functions (attitude control, spin control, momentum wheel and gyro unloading, stage separation and settling of liquids in tanks. A spacecraft usually has a series of different rocket propulsion systems, some are very small. The spacecraft is part of the launch vehicle that carries the payload. It is the only part of the launch vehicle that goes into orbit or deep space and some are designed to return to earth.

Active development or production of rocket propulsion systems are being built in approximately forty different countries compared to three countries sixty years ago. Some of them have made significant and original contributions to the state of the art (rocket propulsion) technology. The solid propellant rocket motor technology has been understood and disseminated well enough that many private companies and government laboratories are designing, developing and manufacturing solid rockets in many sizes and categories. Almost all rocket motors are used only once. The hardware that remains after propellant burning is not reusable except in the case of space shuttle solid boosters. In this case, the booster is recovered from sea, cleaned, refurbished and reloaded with propellant. Reusability makes the design more complex, but it saves the hardware cost.

The other applications of rockets include missiles, engines for research planes, assist-take-off rockets for airplanes, ejection of crew escape capsules and stores, weather sounding rockets, signal rockets, decoy rockets, spin rockets, vernier rockets, under water rockets for torpedoes and missiles, the throwing of life lines to ships and festival rockets.

Typical Questions

1. Give an account on the historical evolution of rockets?
2. What are propellants?
3. What is the difference between Propellants and Explosives? Give examples.
4. What is the basic principle in Rockets?
5. What are the different parts of a typical rocket motor?
6. Explain the following: (a) Mass ratio of a vehicle
                          (b) Propellant mass fraction
                          (c) Specific impulse and total impulse
7. How can be solid and liquid propellants classified?
8. Explain the difference between composite and double base propellants?
9. What are monopropellants and bi-propellants?
10. What is hypergolicity? Explain with suitable examples.
11. What are the merits and demerits of cryogenic propellants?
12. What are hybrids and air breathing propellants?
13. What are gel propellants? What are its advantages? Explain with suitable examples.
14. What are the different types of composite propellants?
15. What are eco-friendly or green propellants?
16. Explain the functioning of space launch vehicles?
17. What are expendable and reusable space boosters?
18. Write notes on: (a) Space shuttle (b) PSLV and (c) ISRO's Lunar and Mars missions.

REFERENCES

[1] Afroz, Javed and Debasis, Chakraborty, "Prediction of Solid Rocket Motor Nozzle Damping Coefficient using CFD Techniques," *Journal of Propulsion and Power*, Vol. 30, Issue 1, pp. 29–34, 2014.

[2] Agrawal, J.P. and Hodgson, R.D., Organic chemistry of explosives, *John Willey and Sons Ltd.*, The Atrium, Chichester, England (2007).

[3] Agrawal, J.P., High Energy Materials: Propellants, Explosives and Pyrotechnics, Wiley-VCH, Weinheim, Germany, 2010.

[4] Akhavan, J., In the Chemistry of Explosives, Royal Society of Chemistry, Cambridge, UK, 2004, 2nd edn.

[5] Allen Davenas, book on Solid rocket propulsion technology edition 1993, pp. 1–10, 35–61, 215–225.

[6] A method of manufacturing of solid rocket motors, US patent no. 6101948, Donald Lee Kenaresboro, Forest Ray Goodson, Frank Stephen Inman, Date of Patent, August, 15, 2000.

[7] Badgujar, D.M. and Mahulikar, P.P., Advances in science and technology of modern energetic materials: An overview, *J. Haz. Mat.,* 151 (2008), 289.

[8] Bottaro, J.C., Recent Advances in Explosives and Solid Propellants, Chem. Ind. 10, (1996), 249.

[9] Brown, M.E., Introduction to Thermal Analysis: Techniques and Application, New York, Kluwer Academic Publishers (2001).

[10] Cengel, Yunus; Introduction to Thermodynamics and Heat Transfer, McGraw Hill 2007.

[11] Chen, M., Sui, Y.K. and Yang, Z.G., *Initiators and Pyrotechnics*, No. 5 (2007), pp. 5–8.

[12] DeLuca, L.T., New Energetic Ingredients for Solid and Hybrid Rocket Propulsion, *Proc. 9ᵗʰ International High Energy Materials Conference*, Feb. 13–14, Trivandrum, 2014.

[13] Fischer, N., Klapötke, T.M., Matecic Musanic, S., Stierstorfer, J. and Suceska, M., TKX-50, New Trends in Research of Energetic Materials, Prat II, Czech Republic, 2013, 574–585.

[14] George, P. Sutton, "Book on Rocket Propulsion Element," Vol. 7, edition 2001, pp. 27–36, 46–84, 417–453, 474–511.

[15] Gould, R.F., Propellants Manufacture, Hazards, and Testing, Advances in Chemistry, No. 88, *American Chemical Society,* Washington, DC (1969).

[16] Knuth, W.H., Chiaverini, M.J., Sauer, A. and Gramer, D.J., "Solid-Fuel Regression Rate Behaviour of Vortex Hybrid Rocket Engines," *Journal of Propulsion and Power*, Vol. 18, No. 3, 2002, pp. 600–609.

[17] Krishnamoorthy, V.N., Energetic Materials for the New Millennium, *Proc. 3rd International High Energy Materials Conference*, Dec. 6–8, Trivandrum, 2000.

[18] Krishnan, S., Chakravarthy, S.R. and Athithan, S.K., Propellants and explosives technology, ISBN 81-7023-884-6, Allied Publishers Limited, India, 1998.

[19] Kumar, R., "Regression Rate Studies Using Wax as a Hybrid Fuel," Ph.D. thesis, Indian Institute of Technology Madras, India (2013).

[20] Lengelle, G., Duterque, J. and Trubert, J.F., Combustion of solid propellants, *ONERA*, May 2002.

[21] Maruyama, S., Ishiguro, T., Shinohara, K. and Nakagawa, I., "Study on Mechanical Characteristic of Paraffin-Based Fuel," *AIAA Paper,* 2011–5678.

[22] Meyer, Rudolf; Kohler Josef and Homburg, Axel, "Explosives", Revised Edition, 2007.

[23] Nagappa, R. and Kurup, M., Development of HTPB propellant system for ISRO solid motors, AIAA-90-2331, AIAA (1990).

[24] Nakagawa, I. and Hikone, S., "Study on the Regression Rate of the Paraffin-Based Hybrid Rocket Fuels," *Journal of Propulsion and Power*, Vol. 27, No. 6, 2011, pp. 1276–1279.

[25] Pascal. Ph, Pin. B., De Amicis. R. and Magnière Ch., "Ariane 5 MPS—the ARIANE 5 solid rocket motor," *Space Solid Propulsion,* Rome, 2000.

[26] Pastrone, D. "Approaches to Low Fuel Regression Rate in Hybrid Rocket Engines," *International Journal of Aerospace Engineering*, Vol. 2012, 2012, pp. 1–12.

[27] Selvaraj, B., Masthiraj, N. Vivek and Kumar, S.R. Dhinesh, Minimum-Signature (Smokeless) Propellant, *International Journal of Emerging Technology and Advanced Engineering,* 2, (2012).

[28] Silva, G., Rufino, S.C. and Iha, K., Green Propellants: Oxidizers, *J. Aerosp. Technol. Manag.,* 5(2), (2013), 139–144.

[29] Solid Rocket Propulsion Technology, edited by Alain Davenas, Pergamon Press, 1997.

[30] Thomas, L. Moore, "CTPB and HTPB Propellants for Extended Space missions," *AIAA,* 2002–3750.

[31] Varghese, T.L. and Ninan, K.N., "Gelled Propellants and New Energetic Materials as Propellant/Explosive Ingredients," India, 1998, pp. 419–448, In S. Krishnan S.R. Chakravarthy and S.K. Athithan (eds.). "Propellants and explosives technology," ISBN 81-7023-884-6, Allied Publishers Limited.

[32] Varghese, T.L. *et al.*, Book on *"Manual of Solid Propellant Chemicals,"* Special Publications, ISRO-TTG-SP-33-1987.

[33] Verma, Pankaaj; Bhujbal, J.G.; Ghavate, R.B.; Darekar, S.D. and Singh, R.V. Boron Viton based fuel rich propellant processing, in *15th International Seminar on New Trends in Research of Energetic Materials – 2012,* Pardubice, Czech Republic.

[34] Yakar, A.B. and Gany, A., "Hybrid Engine Design and Analysis," *AIAA Paper* 1993–2548.

[35] Yang, R., An, H. and Tan, H., *Combustion and Flame,* 135, 463 (2003).

[36] Zhai, J., Shan, Z., Li, J., Li, X., Guo, X. and Yang, R., *J Applied Polymer Science* 128 (2013), 2319–2324.

Composite Propellants—
Ingredients and their Functions

A number of common propellant ingredients for composite solid propellants are categorized by major function, such as oxidizer, fuel binder, plasticizer, curing agent, bonding agent, burn rate modifier, process-aid and so on. However, some of the ingredients have more than one function. The ingredient properties and impurities can have profound effect on the propellant characteristics. A minor change in one ingredient can cause measurable changes in ballistic and physical properties, aging or ease of manufacture. When propellant performance has tight specifications, the ingredient purity and properties must conform to tight tolerances.

Propellant ingredient properties also vary from supplier to supplier as well as from lot to lot of the same supplier. In some cases, particularly with the newly developed propellant, it is impractical or difficult to establish procurement specifications for these ingredients that will adequately control burning rate. One approach to obtaining uniformity has been to establish proprietary or sole source procurement for those ingredients that may have marked effect on burning rate. Although sole-source procurement contributes to the uniformity of propellant ingredients, competitive bids for non proprietary ingredients have significant effects on minimizing the cost of raw materials. Each manufacturing plant from which bids are accepted must be qualified to produce the ingredient. Qualification programs for selected ingredients are set up by process engineers. Selection of ingredients and suppliers to be qualified is unique for each propellant formulation and is highly dependent on the quality of the material to be purchased. Qualification programs involve processing and testing expenses that must be weighed against forecasted procurement savings and other advantages.

Reproducibility of end-item performance is highly dependent on the variation in raw material properties used in propellant formulations. Whenever possible, specifications for raw materials are developed in sufficient detail to relate the critical chemical and physical properties to burning rate in the finished propellant. Since it is difficult or impossible to specify raw material properties to ensure specific end item performance, it is generally done to establish a baseline by characterizing large lots of raw material used in a newly developed formulations. Raw materials are very often characterized by processing development motors from reserved lots of ingredients and evaluating physical and ballistic properties. For example, due to indefinable variations in the properties of AP oxidizer resulted in a variation of nearly 5% in burning rates of lots supplied by different suppliers, in spite of the fact that these materials were purchased according to identical specifications and had essentially same particle size distribution.

This chapter covers the major ingredients used in composite propellants and their role in propellants.

1. OXIDIZER

Oxidizer is the major ingredient of composite propellants and accounts for more than 70% by weight of the propellant. It has greater influence on propellant properties. An oxidizer possesses high oxidation potential and an element of high electronegative atom or group. The periodic table can be used to distinguish the oxidizer from the fuel. That is, oxidizers are highly electronegative and occupy the right hand side of neutral elements like nitrogen and fuels are highly electropositive and find on the left side of the neutral element nitrogen. Thus, fluorine is the best oxidizer, followed by oxygen and chlorine. Hydrogen is the best fuel and hydrogen/fluorine system is the best propellant system. However, fluorine is very reactive and the products of fluorine are highly corrosive. Hence, fluorine is not preferred as an oxidizer. The next best is oxygen and oxygen/hydrogen system is the best propellant combination and is used in all launch vehicles.

Usually, the oxidizer should have high oxygen content and high heat of formation. In addition, the oxidizer should have higher density, high thermal stability and low hygroscopicity. It should be safe to handle and non-metallic in nature and produce large volumes of gaseous products on reaction. The oxidizer should be compatible with other ingredients in the propellant formulation, easily available and storable without any phase transformations at the normal operating temperatures. Since oxygen is usually associated with other elements like chlorine, nitrogen or fluorine, the source of oxygen, apart from gaseous and liquid oxygen, are Chlorates (ClO_3^-), Perchlorates (ClO_4^-), Nitrates (NO_3^-) and Nitrites (NO_2^-). These decompose on heating and give oxygen to oxidize the fuel. The oxidation potential of the oxidizing group varies in the order $F^- > OF^- > NF_2^- > ClF_4^- > O > NO_3^- > ClO_4^- > NO_2^- > ClO_3^-$, etc. It is clear from the above that nitrates and perchlorates are better oxidizers compared to nitrites and chlorates. Hence, composite propellants use perchlorates and nitrates as oxidizers to oxidize the fuel, mostly, carbon containing chemicals or organic materials which act as fuels to get maximum amount of energy from the combustion reaction.

The oxidizers generally used in solid propellants satisfying the requirements are those having small bond energies with low atomic mass elements like NH_4NO_3, KNO_3, $NaNO_3$, NH_4ClO_4, $KClO_4$, $LiClO_4$, $N_2H_5ClO_4$, $N_2H_6[ClO_4]_2$, $HONH_3ClO_4$ and NO_2ClO_4. The characteristics of the oxidizers are summarized in Table 1.

Table 1: Oxidizer Characteristics

Oxidizer	Molecular Formula	M.pt/Decomp. Temp. (°C)	ΔH_f kJ/mol	Density (g/cm³)	Oxygen Balance (%)
AN	NH_4NO_3	170	−365.04	1.72	20.0
AP	NH_4ClO_4	130	−296.00	1.95	34.0
HP2	$N_2H_6(ClO_4)_2$	170	−293.30	2.20	41.0
HP	$N_2H_5ClO_4$	170	−177.80	1.94	24.0
ADN	$NH_4N(NO_2)_2$	90	−150.60	1.82	25.8
HNF	$N_2H_5C(NO_2)_3$	395	−72.00	1.92	13.1
NP	NO_2ClO_4	120	37.10	2.22	66.0
RDX	$C_3H_6N_6O_6$	204	70.63	1.82	−21.6
HMX	$C_4H_8N_8O_8$	275	74.88	1.96	−21.6

Ammonium Nitrate (AN) and Ammonium Perchlorate (AP) are generally used as oxidizer in composite propellants. AN has available oxygen content of 20% while AP has 34%. The density of AN is 1.72 g/cc while that of AP is 1.95 g/cc. The heat of formation of AN is −365 kJ/mole while that of AP is −296 kJ/mole. Though Lithium Perchlorate (LiP) has higher oxygen availability (60.1%), it is highly hygroscopic unlike AP. Nitronium perchlorate has even more higher oxygen content but is highly hygroscopic and non-compatible with organic materials used as fuels. Potassium Perchlorate (KP) though has higher oxygen content than AP, it produces solid KCl as a major combustion product which reduces the energetic or specific impulse. AP exists in two crystalline forms—orthorhombic and cubic—the phase transition occurring around 240°C. AN has five phase transformations of which the one at 32.1°C is accompanied by significant volume change, causing grain dimensional change. Hydrazinium perchlorates are highly hygroscopic and non-compatible with polymeric fuel binders. In addition, RDX and HMX are also used in solid propellants to whip the energy content and burning rate to some extent. Other oxidizers that are likely to compete with AP and AN are Ammonium Dinitramide (ADN), Hydrazinium Nitro-Formate (HNF) and CL-20 (HNIW). These are discussed in Chapter 7 on advanced solid propellants. Currently, AP is the workhorse oxidizer of solid propellants. AP is preferred over sodium and potassium perchlorates in view of the production of gaseous products on decomposition.

AP is produced from sodium perchlorate and ammonium chloride by double decomposition reaction. Sodium perchlorate itself is made by electrochemical oxidation of sodium chloride (common salt) using either platinum anodes or triple oxide coated titanium substrate lead dioxide anodes and stainless steel cathodes. Sodium chlorate is also used for making potassium chlorate used in match industry.

Electrolytic anodic oxidation,

$$NaCl \xrightarrow{\text{oxidation}} NaClO_3 \xrightarrow{\text{oxidation}} NaClO_4$$

Double decomposition,

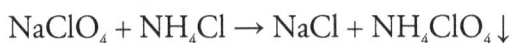

$$NaClO_4 + NH_4Cl \rightarrow NaCl + NH_4ClO_4 \downarrow$$

The sodium chloride formed remains in solution while AP crystallizes out. These crystals are separated out from the mother liquor and re-crystallized from hot water and dried in fluid bed dryers. The coarse AP is coated with small quantities (<0.5%) of anti caking agents like Tricalcium Phosphate (TCP), stearyl amine/amide (FERT FLOW), etc. to impart free flow characteristics and to prevent caking during storage. It is then classified to different particle size fractions, blended and packed in drums for further use. AP is a white crystalline solid which has orthorhombic structure at room temperature and changes to cubic structure at 240°C.

Ammonium perchlorate has almost all the good characteristics of a good oxidizer. The processibility of highly solid loaded propellant is governed by the oxidizer particle shape, size and its distribution. Generally, AP is used in propellant formulations cithcr as bimodal or trimodal distributions in ordcr to gct maximum oxidizer loading with the binder. The term "modal" refers to the number of peaks in a plot of particle size distribution. The packing density and burn rate considerations dictate proper ratio of these fractions. The minimum aggregate void ratio (volume voids/volume of aggregate particles) is preferable since such a system will only require minimum liquid binder to fill the interstices. Spherical particles give a lower void ratio under random packing due to the ease with which spheres roll over and slide past one another during compaction. Photomicrogaphs of PEPCON, USA and APEP, India show more spherical particles. The particle shape is highly dependent on the crystallizing and drying process. Rapid flash drying results in some fracturing of crystals and a rough overall shape characteristics because of fast removal of moisture. Slower drying results in nearly spherical particles with little internal cracking.

The average particle size of AP produced at the Ammonium Perchlorate Experimental Plant (APEP) at Alwaye, India is 300 μ. Fine particles of AP (<60 μ) are made by grinding it in a hammer mill to the required size at the propellant manufacturing site just before they are incorporated in to propellant mix. Sizes below

40 μ diameter are considered hazardous as these can be easily ignited and sometimes detonatable. The particle size of AP affects the burning rate of the propellant formulation and the unloading viscosity of propellant slurry. This is due to the larger surface area for the finer particles. The variation of AP particle size by adjusting coarse to fine ratio on the processibility and burning rate on a typical HTPB propellant shows that as the coarse content decreases, the unloading viscosity, tensile strength and modulus, hardness and burning rate increases and elongation decreases due to increase in surface area and better reinforcement of propellant by fine particles. Friability is another property of oxidizer crystals and is difficult to measure and not included as a specification. AP crystal friability vary significantly from supplier to supplier and sometimes from lot to lot and is important to the user because of the variations caused in particle sizes (caused by attrition while handling and mixing) and it affects the processibility and ballistic properties.

Most of the operational solid propellants have AP content of 68–70% (along with 16–20% Al powder). To achieve higher solid loadings, use of trimodal distribution of AP consisting of coarse (300 μ), fine (~40 μ) and micro fine (3 to 8 μ) has been used. Coarse AP crystals are handled as class 1.3 materials while fine and microfine grades are considered as class 1.1 high explosives and are usually manufactured on-site from coarse grades. Most propellants use a blend of oxidizer particle sizes to maximize the weight of oxidizer per unit volume of propellant with small particles filling part of the voids between larger particles. However, composite propellants based on AP produce smoke in cold or humid atmospheres due to the presence of large quantities of Hydrogen Chloride gas (HCl) in the exhaust product.

Potassium Perchlorate (KP) though has a higher oxidizer content and higher density, is not used in solid propellants as oxidizer in view of its low gas producing nature and the corrosive solid residue of potassium chloride in the exhaust product. It is made from sodium perchlorate by double decomposition with potassium chloride. KP gives higher flame temperature and higher burning rate. Hence, it is used in propellants for higher burning rates as in igniter propellants. Sodium Perchlorate (SP) is highly hygroscopic and is not preferred as oxidizer. Lithium Perchlorate (LP) has more than 60% available oxygen but is also highly hygroscopic and hence, is not preferred as oxidizer for solid propellants. Nitronium perchlorate has high oxygen content and positive heat of formation to serve as an excellent oxidizer. However, it is highly hygroscopic and very reactive with organic compounds, in addition to its toxic nature. Hence, it is not used as an oxidizer in solid propellants.

Ammonium Nitrate (AN) is another oxidizer with low oxygen content and high hygroscopicity. It is cheap, readily available and gives non toxic exhaust gases. It is used in some propellant formulations where non-smoky exhausts are required as in missiles and in gas generators. It gives lower specific impulse even with more than 80% solid loading. AN based propellant formulations have lower burning rate and are difficult to ignite. To boost the energy, other oxidizers like AP, HMX and RDX have been added to AN based formulations. AN is made from ammonia and nitric acid. Nitric acid itself is made by oxidizing ammonia. AN is also used as a fertilizer.

AN has five phase transitions occurring around –18, 32.2, 84.2, 125 and 169.6°C (melting point). Of these, the one occurring around 32.2°C is accompanied by significant volume change which can create cracks in propellants based on AN. This is severe in countries like India where the average day temperature is around 30 to 35°C. Attempts have been made to shift this phase transformation at 32°C to higher temperatures either by doping or co-crystallization. The doppants tried include Ni^{++}, Cu^{++}, and K^+. Phase stabilized AN is used in gas generators and smokeless propellants.

2. FUEL BINDERS

The fuel binder is a liquid which on solidification binds the dispersed oxidizer and metallic powder together, in addition to acting as a fuel. This fuel-oxidizer mixture on ignition reacts and is transformed into combustion gases. The modern propellants have a solid loading of 84 to 88% (including oxidizer and metallic fuel) in the formulation. This leaves 12 to 16% for the liquid. The art of propellant mixing lies in making the best use of this meager quantity of the liquid in getting a processible paste which on solidification or polymcrization adds both cncrgy and strcngth to the propellant. The binder is an elastomeric matrix that binds the oxidizer and metallic fuel to form a rubbery mass capable of withstanding the severe strains produced by thermal and mechanical stresses. The binder largely determines the propellant mechanical properties. It is also a major reducing agent and a gas producing fuel, compatible with oxidizer and other ingredients. Binders are organic polymers mostly containing carbon and hydrogen. The liquid material is based on available plastic, resinous or elastomeric materials. Both thermoplastic (those getting soft with temperature) and thermosetting (those getting solidified with temperature) have been used as fuel binders. Binders are typically cross linked polymers formed from pre-polymers and cross linkers, which provides a matrix to bind all solid ingredients (oxidizer, metallic fuel and additives) with the plasticizer and process aid to ease the processing of the uncured mix. During combustion, the binder is decomposed to give large volumes of stable gas molecules like carbon monoxide, carbon dioxide, water vapor and nitrogen per unit mass of propellant. The list of fuel binders include polystyrene, polyvinyl chloride, unsaturated polyesters, polysulphides, polyurethanes, polybutadienes, natural rubber and phenol and urea formaldehydes.

Salient Features of Binder

The binder requirements for solid propellants are somewhat stringent. Though binder is only 12 to 16% by weight of propellant, it determines the mechanical properties. The considerations related to the choice of the fuel binder are as follows:

1. Preferably the binder should be a liquid, with a workable viscosity (1000 to 10000 centipoises) at mixing temperature so that it can take maximum solid loading. It must have a molecular weight of 2500 to 5000 and minimum density (0.8 to 1.00 g/cc) for easy processibility.

2. The propellant exhaust must contain more hydrogen like light gas species to give the average molecular weight ($\overline{M}$) of the combustion products low. This means the exhaust must contain less heavy gases like nitrogen, sulphur or carbon dioxide. Hence, the binder chemical make up must contain more hydrocarbon units (CH_2) or binder must have high C/H ratio in its backbone. Also, the binder on combustion should result in products of high negative heat of formation, ΔH_f. That is, the binder should have positive heat of formation. At present, a pure hydrocarbon meets all the requirements.

3. The binder must have chemically reactive end groups like –COOH, –OH or epoxy which can be used to convert the liquid binder to a cross linked elastomer with high tensile and compressive strength during curing (good mechanical properties) with low cure shrinkage to avoid separation from motor case.

4. The binder must have low glass transition temperature (T_g) so that the propellant made from it should function satisfactorily over a temperature range of 80°C to –60°C.

5. The cross linking or curing of binder must take place at low temperatures (40–75°C) with minimum evolution of heat to avoid propellant catching fire and becoming a safety hazard.

6. The binder must have low coefficient of expansion since the rocket motors when cooled or heated, the propellant contracts or expands at quite different rates from those of the metal chamber leading to cracks.

7. For special applications, the binder requirements also include low water absorbtivity, good compatibility with high energy reactive ingredients, long storage stability and negligible vapourization loss under vacuum. Propellants for inter planetary missions must be capable of withstanding severe heating cycles as sterilization of spacecraft is done by heat. Also, for extended space missions, the binder should resist nuclear radiation, especially in Van Allen belts. The binder should also withstand high energy radiations.

8. The rocket exhaust from the combustion of the binder should attenuate radio microwave transmission as little as possible by excessive electrons due to high temperature or ionization.

9. It must be stable in intimate contact with oxidizer till ignition and must be capable of curing at low temperatures (40 to 80°C) in a reasonable time with minimum evolution of heat during curing to avoid accidental ignition of propellant and other side reactions. The viscosity buildup on addition of curing agent should be slow and gradual (long pot life) for easy casting of propellant grains of intricate geometry.
10. The binder must be capable of bonding to rubber insulation materials and metallic case.

Thermoplastic Binders

To change from castable slurry to a useful solid propellant in the rocket motor case, the binder may be thermoplastic or thermosetting. The thermoplastic materials are softened or melted with heat, mixed with oxidizer and other ingredients and cast. Then, on cooling, they become rigid and retain the shape of the mould. An example of thermoplastic binder is the asphalt type of binder used as early in 1942. The other thermoplastic binders tried are polystyrene, polyacrylates and methacrylates and polyisobutylenes. Addition polymerization type thermoplastics such as Polyvinyl Chloride (PVC) and poly vinyl acetate have also been used with plasticizers like dibutyl sebacate and dioctyl sebacate. In these combinations, called plastisols, the polymer dissolves in the plasticizer with heating and forms a rubbery solid gel. Standard techniques of the plastic industry, like molding and extrusion, were used to manufacture the propellants. Thermoplastics offer the advantages of simplicity and lack of hazard from reaction exotherm or toxic curing agents. PVC based plastisols, in particular, have been developed.

Thermosetting Binders

The thermosetting materials are initially liquids or meltable solids but after addition of curing agents and heating, they undergo a chemical change, usually formation of a cross-linked network, and become insoluble and rigid mass. Most development work today is being done in the area of thermosetting materials because of their superior physical properties and their ability to bond to rocket motor case wall. The binders in uncured state are liquids and are called more correctly as pre-polymers. These pre-polymers have reactive functional groups of known quantity on the backbone chain. When treated with chemical curing agents and or heat, a cross-linked network is formed with chemical bonds between chains. A linear pre-polymer has long unconnected chains which slip one over the other when stressed, but a lightly cross-linked polymers have inter chain connections which deform under stress and recover when the stress is removed or released. These are elastomers. If the number of cross-links are very large, the polymers become rigid just like cement and useless for making propellants except for making cartridge (non-case-bonding) propellants.

The polymers considered for use as propellant binders can be grouped according to their effect on propellant processing as:

1. Functional group containing polymers either at terminals or in the backbone chain of the polymer as given below.
2. Plastisol binders, and
3. Double base propellant binders.

Characteristics of Pre-Polymers

The characteristics of the pre-polymers are determined by the following properties:

1. The average molecular weight of the pre-polymers used as propellant binder should be in the range of 2500 to 5000. A decrease in the molecular weight of the pre-polymer reduces the viscosity of the binder, but leading to cure shrinkage. Higher molecular weight of the pre-polymer makes it viscous and renders mixing uniformly difficult, in addition to lower solid loading in the mix. The average molecular weight has further advantage of giving mixing viscosities moderately high enough to prevent settling of the oxidizer and of producing good mechanical properties after curing.

2. The functionality of the pre-polymer must be more than two. Functionality is the number of reactive functional groups (–COOH, –OH, –SH, etc.) or reactive sites (C=C, C≡C) present in the pre-polymer chain. Difunctional curing agents lead to linear coupling whereas trifunctional curing agents produce three dimensional networks. The ratio of difunctional to trifunctional molecules in the pre-polymer must be controlled to get the required amount of cross-linking and cross link density, to get the desired mechanical properties in the final product. The most effective elastomer results when the functional groups are at the chain ends.

3. Poly dispersity of the binder should be greater than 1 but less than 2. Poly dispersity gives a general idea about the distribution of different molecular weight species in the pre-polymer. As polymers are long chain molecules, their molecular weight cannot be represented by a single number as in simple molecules like CH_3OH. Polymer molecules comprise different molecular weights and hence, its molecular weight is expressed in terms of an average value. There are many averages and the two most common and experimentally verifiable methods of averaging are called number average ($\overline{M}n$) and weight average ($\overline{M}w$) molecular weights. Poly dispersity is the ratio of $\overline{M}w/\overline{M}n$. It gives an idea of the lowest and highest molecular weight species as well as distribution

pattern of intermediate species. As the molecular weight distribution becomes broader, the $\overline{M}w/\overline{M}n$ increases. Ideally it should be unity.

Starting with polysulphides, Polybutadiene—Acrylic Acid—Acrylonitrile Terpolymer (PBAN), Carboxyl Terminated Polybutadiene (CTPB) and Hydroxyl Terminated Polybutadiene (HTPB) have been used as binders for composite propellants. In addition, ISRO polyol and Lactone Terminated Poly Butadiene (LTPB or HEF-20) have been used as propellant binders in India for the first time. Currently, HTPB based propellants are mostly used in rockets, missiles and launch vehicles.

Polybutadiene pre-polymers are very widely used as binders in composite propellants, due to its higher solid loading capacity and greater fuel value compared to other binders. The most usually used pre-polymers in solid propellants are given in Table 2.

Table 2: Energetics of Polymeric Binders

S. No.	Polymeric Binders	ΔH (Cals/gm)
1.	Polyester	5320
2.	Polysulphide	5840
3.	Pedathane (NCO terminated polyether)	7040
4.	ISRO Polyol	9370
5.	HEF-20	9470
6.	PBAN	9860
7	CTPB	10360
8.	HTPB	10380

Hydroxyl Terminated Polybutadiene (HTPB), Carboxyl-Terminated Poly-butadiene (CTPB) and the Terpolymer of Butadiene, Acrylic Acid and Acrylonitrile (PBAN) are the usual polybutadiene binders used. To this list, ISRO added three more polymers viz. Lactone Terminated Polybutadiene (LTPB or HEF-20), ISRO polyol and Hydroxyl Terminated Natural Rubber (HTNR). A comparison on the energetics of different polymeric binders are given in the Table 2.

Polysulphides

Polysulphides are the first chemically cross-linked binder to find application in solid propellant. The LP-3 pre-polymer belongs to the liquid polysulphide series of the Thiokol Chemical Corporation, USA. The polysulphide LP3 is made from dichloro-diethyl formal and sodium polysulphide solution. Polysulphide resin is cured with para quinone dioxime or lead dioxide. The formation of water during the curing reaction is undesirable from the point of view of aging properties and use of metals like Aluminum or Magnesium in the propellant formulations. This reduces the specific impulse of the polysulphide propellants. Also, polysulphides have low hydrogen and high sulphur content and therefore are not as energetic as CTPB. Polysulphides, however, have excellent mechanical properties and suitable for case bonding applications. Polysuphides are used successfully in critical missions like moon exploration in the escape motors because of sure ignition and reliable performance.

Synthesis

It is prepared by the condensation reaction between sodium polysulphide and Dichloro Diethyl Formal (DDF), followed by reduction with NaHS and Na_2SO_3 and further acidification with HCl. The required molecular weight and viscosity of the pre-polymer can be obtained by adjusting the concentration of NaHS and Na_2SO_3. The functionality can be increased by adding trichloropropane in the initial polymerization stage along with dichloro diethyl formal.

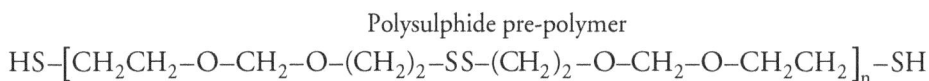

Polysulphide pre-polymer

$$HS–[CH_2CH_2–O–CH_2–O–(CH_2)_2–SS–(CH_2)_2–O–CH_2–O–CH_2CH_2]_n–SH$$

Cure Reactions

Polysulphide can be cured using lead dioxide or p-quinone dioxime. The cure reactions are shown below. The elimination reaction causes thermal shrinkage and porosity.

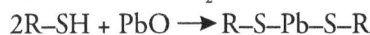

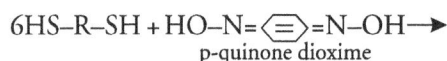

$$HS–R–SH + PbO_2 + HS–R–SH \rightarrow HS–R–SS–R–SH + H_2O + PbO$$
$$2R–SH + PbO \rightarrow R–S–Pb–S–R$$
$$\downarrow PbO_2$$
$$R–S–S–R + 2PbO$$
$$6HS–R–SH + HO–N=\langle\equiv\rangle=N–OH \rightarrow$$

p-quinone dioxime

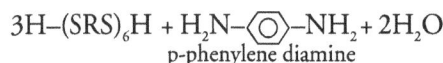

$$3H–(SRS)_6H + H_2N–\langle O\rangle–NH_2 + 2H_2O$$

p-phenylene diamine

Unsaturated Polyesters

Unsaturated polyesters cured with styrene are one of the earliest polymer systems available and considered for propellant formulation. This hydrocarbon polymer had higher fuel value than some of the polysulphide polymers. The unsaturated polyester polymers can be prepared to convenient molecular weight ranges. The unsaturated polyesters are usually prepared by condensing a polycarboxylic acid with a polyhydric alcohol, one or both of which contain olefinic linkages. Among the unsaturated carboxylic acids used to introduce the functional group as a site of cure activities are maleic, fumaric or itaconic acids. Adipic or sebacic acid was frequently used to increase the hydrocarbon fuel value. Though a number of dihydric, trihydric or polyhydric alcohols could be used, 1,2-propylene glycol or diehtylene glycol is commonly used. Many olefinic compounds can interact with these polyesters to give desired cross-linking, among which methyl acrylate or styrene is popular and easily available commercially. The addition of olefin to the double bond is catalyzed by peroxides with cobalt napthenate as catalyst. Crosslinks can be controlled by the number of unsaturated sites in the polyester pre-polymer. Theoretically, if each molecule has only two reaction sites, then infinite, linear chains could be obtained. Hence, the molecular weight distribution and average functionality of the polymer are important. The use of plasticizers could help in adjusting the average properties of the polymeric binder. The drawbacks of the unsaturated polyester system are the volatility and explosive limits of styrene, the control of peroxide cure catalyst activity, control of exotherm during mixing and cure cycles and the inability to get desired strain capability for case bonding of the solid propellants. In spite of these difficulties, a number of propellant formulations have been developed and flown in rockets. India's first composite propellant 'Mrinal' based on unsaturated polyester was flown in RH-75 rocket on 21st February, 1969.

ISRO-Polyol

ISRO-polyol, a saturated ester propellant binder developed by ISRO, is based on castor oil and has a molecular weight (about 2000), nearly bifunctional and a viscosity of 2000 cps at 30°C.

It is synthesized by the self condensation of 12-hydroxyl stearic acid, followed by condensation with Trimethylol Propane (TMP). 12-hydroxyl stearic acid is prepared by the hydrolysis of saturated castor oil,

ISRO – Polyol (IP-Pedester TM-20)

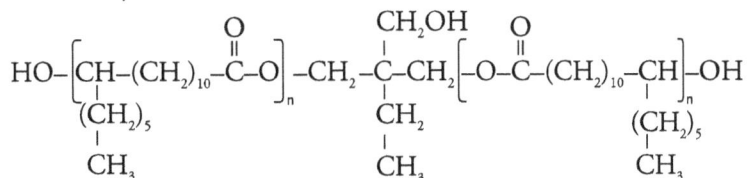

$$HO-\left[\underset{\underset{CH_3}{\overset{|}{(CH_2)_5}}}{\overset{|}{CH}}-(CH_2)_{10}-\overset{O}{\overset{\|}{C}}-O\right]_n-CH_2-\underset{\underset{CH_3}{\overset{|}{CH_2}}}{\overset{\overset{CH_2OH}{\overset{|}{C}}}{}}-CH_2-\left[O-\overset{O}{\overset{\|}{C}}-(CH_2)_{10}-\underset{\underset{CH_3}{\overset{|}{(CH_2)_5}}}{\overset{|}{CH}}\right]_n-OH$$

Cure Reactions

The basic reaction is between isocyanate group and the hydroxyl group of the polyester leading to the formation of urethane linkage. The cross-link density can be increased by adding small concentration of TMP,

$$-NCO + HO- \rightarrow -NH-COO-$$

Propellants made using ISRO Polyol has been successfully used in sounding rockets and considered as a candidate propellant for India's PSLV booster. Being a non-petroleum based binder, it is cheap and easily made from castor oil. The hexyl pendant groups impart good low temperature properties in addition to making it a self plasticized system. However, because of variations in castor oil properties, being a natural product, it showed inconsistencies in propellant mechanical properties. Hence, it was not further pursued.

Polyether Polyols

Three structurally different polyether diols are used as propellant binders. They are poly (1,2-oxypropylene) diol (PPG), poly(1,2-oxybutylene) diol (B-2000) and poly (1,4-butylene) diol. The first two diols have secondary hydroxyl group at the terminals while the third diol has more reactive primary hydroxyl termination. Of these, PPG is used widely as propellant binder due to its availability in wide range of molecular weights. The cure reaction is between the hydroxyl groups of the polyether polyol and the isocyanate forming a urethane linkage. The major advantage of polyether polyols are their easy availability, low viscosity and proper cure rate. Polyether stands between polyesters and polybutadienes in energy. However, polyether is known to absorb oxygen forming peroxides. This can be corrected by adding aromatic amine type antioxidants. Polyether binders have good aging characteristics. PPG based polyurethane propellants lose their low temperature strain capability due to moisture embrittlement after exposure to high relative humidity or moisture. This is because propylene oxide moiety absorbs moisture and in turn dissolves ammonium perchlorate. To some extent, the original property can be restored by drying the propellant.

PBAA (Poly Butadiene-Acrylic Acid Copolymer)

The first butadiene pre-polymer used in solid propellants is the liquid copolymer of butadiene and acrylic acid (PBAA),

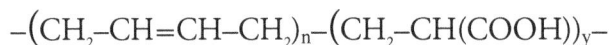

$$-(CH_2-CH=CH-CH_2)_n-(CH_2-CH(COOH))_y-$$

Synthesis

This polymer was synthesized by a free radical emulsion polymerization of butadiene and acrylic acid using AIBN as initiator and quaternary ammonium salt as emulsifier, to an average molecular weight of around 3000 and functionality of 2.

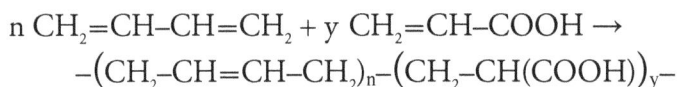

$$n\ CH_2=CH-CH=CH_2 + y\ CH_2=CH-COOH \rightarrow$$
$$-(CH_2-CH=CH-CH_2)_n-(CH_2-CH(COOH))_y-$$

Because of the method of synthesis, the carboxyl groups are distributed randomly in the chain and the number of functional groups per molecule varies widely. Hence, the pre-polymer is a mixture of nonfunctional, mono-, di-, polyfunctional molecules and exhibit a range of molecular weights. Hence, PBAA based propellants cured with epoxides and aziridines show poor reproducibility in their mechanical properties. Though the low viscosity of the pre-polymer helped in higher solid loading formulations compared to other non polybutadiene binders, the poor reproducibility in mechanical properties due to random spacing of functional carboxyl groups, wide functionality distribution of the pre-polymer and the post cure during storage led to abandoning of the PBAA pre-polymer in favor of the Terpolymer of Butadiene, Acrylic Acid and Acrylonitrile (PBAN).

PBAN (Poly Butadiene-Acrylic Acid-Acrylonitrile Terpolymer)

The drawbacks of PBAA, viz. mechanical behavior and storage characteristics were improved by using terpolymers based on butadiene, acrylic acid and acrylonitrile. The liquid PBAN pre-polymer was prepared by the same emulsion polymerization technique used for PBAA. However, the introduction of acrylonitrile group in the backbone chain probably improved the spacing of carboxyl groups, which could be the factor in the more reproducible propellant curing and mechanical properties. Also, the propellants based on PBAN showed lesser tendency to surface hardening which is due to the oxidative attack at the double bonds and is known to be suppressed by the nitrile groups as seen in nitrile rubber.

Synthesis

PBAN is synthesized by the free-radical polymerization of butadiene, acrylic acid and acrylonitrile using benzoyl peroxide or 4,4'-azobis isobutyronitrile as initiator.

Initiation

$$C_6H_5-COO-OOC-C_6H_5 \rightarrow 2C_6H_5{}^{\cdot} + 2\,CO_2$$

where the initiating radical is $C_6H_5{}^{\cdot}$ or $R^{\cdot}$.

Propagation and Termination

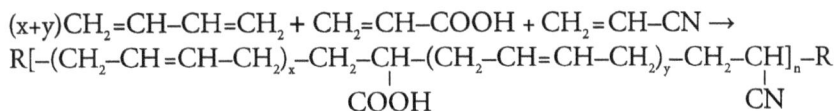

$$(x+y)CH_2=CH-CH=CH_2 + CH_2=CH-COOH + CH_2=CH-CN \rightarrow$$

$$R[-(CH_2-CH=CH-CH_2)_x-CH_2-\underset{\underset{COOH}{|}}{CH}-(CH_2-CH=CH-CH_2)_y-CH_2-\underset{\underset{CN}{|}}{CH}]_n-R$$

More solid propellants based on PBAN have been produced than from any other single pre-polymer. This is because of PBAN's low cost coupled with better thermal stability and low temperature properties of the propellants based on PBAN. The PBAN pre-polymers can be cured both by epoxides and aziridines as in the case of CTPB or PBAA. PBAN based propellants are used in Space shuttle boosters, each booster carrying more than 500 tonnes of propellant. The biggest motor ever made viz. 260" diameter motor of Aerojet General solid booster used PBAN based propellant.

Cure Reactions

PBAN

Reaction with epoxy group

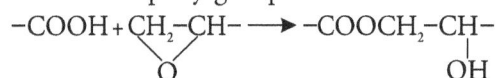

$$-COOH + CH_2-CH- \longrightarrow -COOCH_2-CH-$$
$$\underset{O}{\diagdown\diagup} \qquad\qquad\qquad \underset{OH}{|}$$

Reaction with aziridine ring

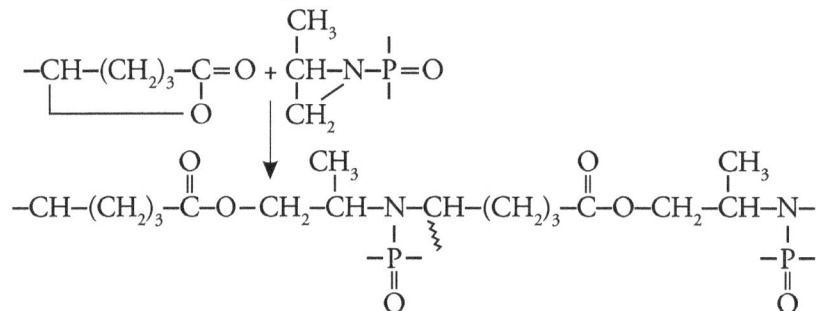

LTPB or HEF-20 (Lactone-Terminated Polybutadiene)

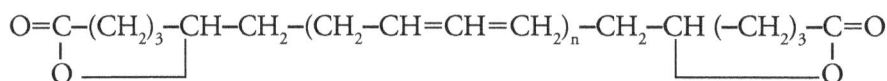

$$O=C-(CH_2)_3-CH-CH_2-(CH_2-CH=CH=CH_2)_n-CH_2-CH\,(-CH_2)_3-C=O$$

Lactone-Terminated Polybutadiene (LTPB) or High Energy Fuel-20 (HEF-20) has terminal functional groups different from those of in any of the polybutadiene pre-polymers reported above. The propellants based on this pre-polymer has been successfully used in the upper stages of India's first launch vehicle SLV-3 and ASLV in the apogee motor to put India's first developmental communication satellite, APPLE, into orbit using European Space Agency's Ariane vehicle. This pre-polymer is made by oxidative degradation of polybutadiene rubber of 0.2 to 0.3 million molecular weight. Though the polymer was expected to have carboxyl groups in the terminals, but because of acidic nature of the medium in which the oxidative degradation was carried out, the carboxyl group cyclizes by reacting with the adjacent double bonds forming lactone or internal ester.

Synthesis

It is synthesized by the oxidative degradation of high molecular weight polybutadiene rubber using per benzoic acid and periodic acid.

Step 1:

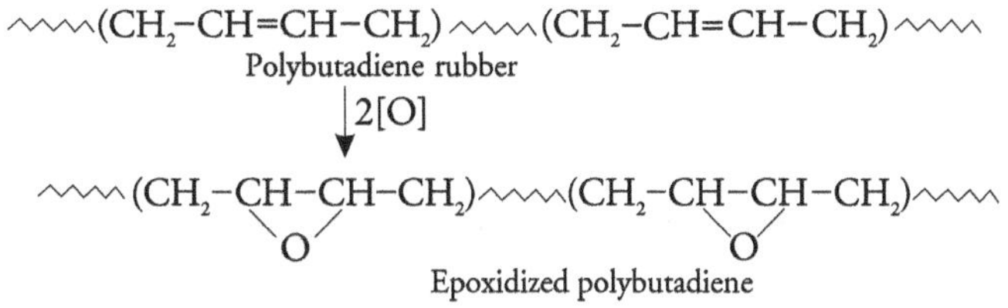

$$\wedge\wedge\wedge(CH_2-CH=CH-CH_2)\wedge\wedge\wedge(CH_2-CH=CH-CH_2)\wedge\wedge\wedge$$
Polybutadiene rubber

$$\downarrow 2[O]$$

$$\wedge\wedge\wedge(CH_2-CH-CH-CH_2)\wedge\wedge\wedge(CH_2-CH-CH-CH_2)\wedge\wedge\wedge$$

Epoxidized polybutadiene

Step 2:

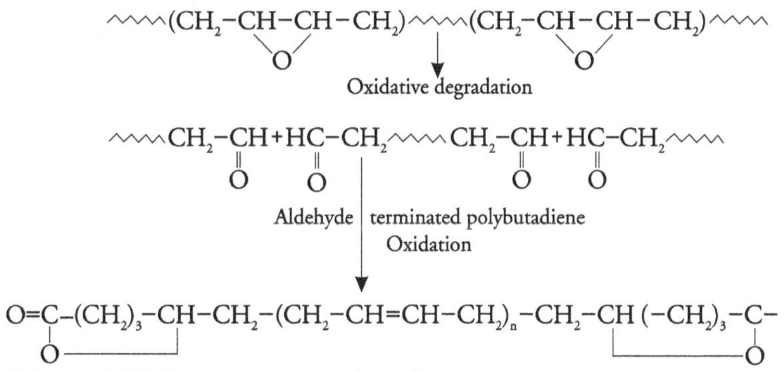

$$\wedge\wedge\wedge(CH_2-CH-CH-CH_2)\wedge\wedge\wedge(CH_2-CH-CH-CH_2)\wedge\wedge\wedge$$

Oxidative degradation

$$\wedge\wedge\wedge CH_2-CH+HC-CH_2\wedge\wedge\wedge CH_2-CH+HC-CH_2\wedge\wedge\wedge$$

Aldehyde | terminated polybutadiene
Oxidation

$$O=C-(CH_2)_3-CH-CH_2-(CH_2-CH=CH-CH_2)_n-CH_2-CH(-CH_2)_3-C-$$

HEF-20 or LTPB (Lactone terminated polybutadiene)

Cure Reactions

The lactone terminated pre-polymer reacts with epoxides and aziridines in the same way as CTPB. The propellants made out of this pre-polymer showed excellent mechanical properties. However, the poor storage stability and aging characteristics coupled with high cost of manufacturing are against the continued use of the polymer in solid propellants.

1. Reaction with epoxy group

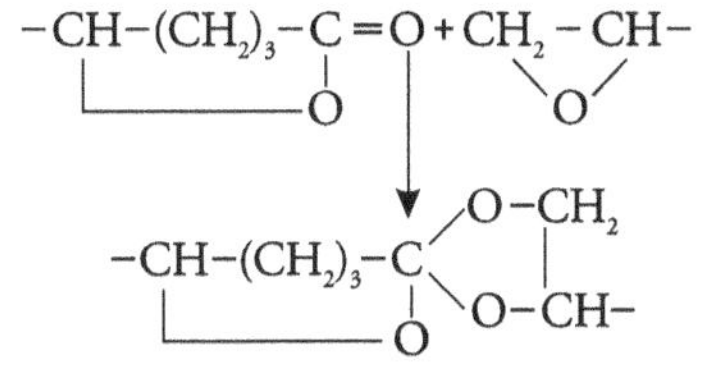

$$-CH-(CH_2)_3-C=O+CH_2-CH-$$

$$-CH-(CH_2)_3-C\begin{smallmatrix}O-CH_2\\O-CH-\end{smallmatrix}$$

2. Reaction with aziridine ring

$$-CH-(CH_2)_3-C=O + CH-N-P=O$$

(with structural formula showing CH₃, CH₂ groups and ring)

$$-CH-(CH_2)_3-\overset{O}{\overset{\|}{C}}-O-CH_2-\overset{CH_3}{\underset{\;}{CH}}-N-CH-(CH_2)_3-\overset{O}{\overset{\|}{C}}-O-CH_2-\overset{CH_3}{\underset{\;}{CH}}-N-$$

$$-P- \qquad\qquad -P-$$
$$\overset{\|}{O} \qquad\qquad \overset{\|}{O}$$

CTPB (Carboxyl Terminated Polybutadiene)

$$HOOC–R–(CH_2–CH=CH–CH_2)_n–R–COOH$$

Carboxyl terminated polybutadiene pre-polymer has carboxyl groups in the terminal positions to take full advantage of the entire length of hydrocarbon polymer chain. These butadiene polymers were synthesized by a free radical or anionic technique to an average molecular weight of 3500 to 5000 and nearly bifunctional structure. This provides improved mechanical properties for the highly loaded solid propellants, particularly at low temperatures.

Synthesis (Free-Radical Method)

There are two principal methods for preparing the free radical initiated pre-polymers. The first method uses glutaric acid peroxide as the initiator while the second method uses 4,4'-azobis 4-cyanovaleric acid as the initiator in the solution polymerization of butadiene gas.

Initiation

$$HOOC–(CH_2)_3–COO–OOC–(CH_2)_3–COOH \rightarrow 2\,HOOC–(CH_2)_3^{\bullet} + 2\,CO_2$$

or

$$HOOC–(CH_2)_2–\overset{CH_3}{\underset{CN}{C}}–N=N–\overset{CH_3}{\underset{CN}{C}}–(CH_2)_2–COOH \longrightarrow 2\,HOOC–(CH_2)_2–\overset{CH_3}{\underset{CN}{\overset{\;}{C}^{\bullet}}}+N_2$$

Thus, the initiating radical is $HOOC–R^{\bullet}$

$$CH_2=CH–CH=CH_2 + {}^{\bullet}R– COOH \rightarrow HOOC–R–CH_2–CH=CH–CH_2^{\bullet}$$

Propagation

$$HOOC–R–CH_2–CH=CH–CH_2^{\bullet} + n\,CH_2=CH–CH=CH_2 \rightarrow$$
$$HOOC–R–(CH_2–CH=CH–CH_2)_n–CH_2–CH=CH–CH_2^{\bullet}$$

Termination

$$2\,HOOC-R-(CH_2-CH=CH-CH_2)n-CH_2-CH=CH-\overset{\cdot}{CH_2} \rightarrow$$
$$HOOC-R-(CH_2-CH=CH-CH_2)_{2(n+1)}-COOH$$

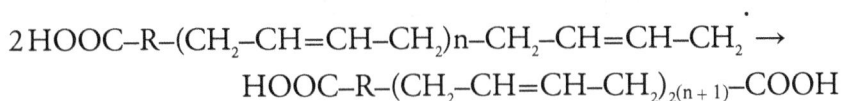

Anionic Method

The lithium initiated pre-polymers are made by organo-lithium technique as shown below:

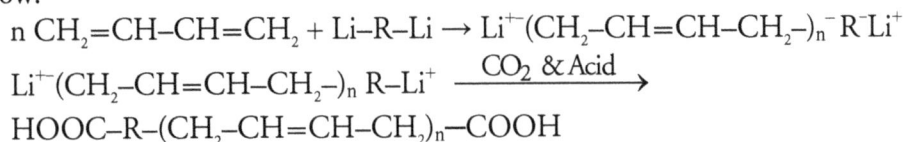

$$n\,CH_2=CH-CH=CH_2 + Li-R-Li \rightarrow Li^+(CH_2-CH=CH-CH_2-)_n\,R^-Li^+$$

$$Li^+(CH_2-CH=CH-CH_2-)_n\,R-Li^+ \xrightarrow{\;CO_2\;\&\;Acid\;}$$

$$HOOC-R-(CH_2-CH=CH-CH_2)_n-COOH$$

The anionic technique provides a pre-polymer with a narrow molecular weight distribution, but the mean molecular weight can be varied over a range. On the other hand, pre-polymers made by free radical initiation method generally exhibit a broad molecular weight range and somewhat branched structure.

Cure Reactions

There are two classes of compounds which can be used as curing agents for carboxyl terminated poly-butadienes, viz. aziridienes and epoxides. The bifunctionality of CTPB requires curing agents be polyfunctional to provide for the formation of the three dimensional network. The types of some of the cross-linkers and chain extenders used in CTPB based solid propellants are ERLA-0510 (a triepoxide based on p-amino phenol and glycidyl ether), DER-332 or GY250 (di-epoxide based on bisphenol A) and GY252 (a diepoxide based on bisphenol A plus a diluent) and aziridines like MAPO (Tris(2-methyl aziridinyl-1) phosphine oxide), and BITA (1,3,5-Tris-(2-ethylaziridinyl-1) adduct of trimesic acid.

The cure reactions of carboxyl group with epoxy and aziridine are the same as that of PBAN given above.

Almost without exception, the polyfunctional aziridines and epoxides used with CTPB undergo side reactions in the presence of ammonium perchlorate, which affects the binder network formation causing a shift in mechanical behavior and post cure reactions. Multifunctional aziridines with CTPB form chain extension and cross-linking by the formation of the amido-ester structure in the reaction with carboxyl groups. Also, aziridines and epoxies are also subject to homopolymerization and rearrangement to oxazolines, which affects the equivalents balance of the reacting groups. The polymer network formed with aziridines, however, is unstable and softens rapidly when exposed to high temperatures. This is due to the presence of three phosphorus—nitrogen bonds which cleave at high temperatures after aziridine ring reaction with carboxylic groups. In view of resultant aging behavior of cured polymers with multifunctional curing agents, a mixture of an aziridine and an epoxide is used as a practical solution for curing CTPB. In this system, the softening caused by the P–N bond cleavage in MAPO cured propellants is offset by the continued post curing of epoxides. CTPB propellants illustrate how processing order of ingredients greatly reduces processing hazards. MAPO homopolymerizes with the release of heat. Sensitive to heat, MAPO-AP combinations ignite quite readily, a characteristic that was responsible for loss of life and extensive damage to mixing facilities in two disasters in 1965.

HTNR (Hydroxyl Terminated Natural Rubber)

Synthesis

HTNR is obtained by the depolymerization of masticated natural rubber in presence of H_2O_2. The presence of –COOH group as impurity is nullified by treating with propylene oxide.

Natural rubber → Masticated → Dissolved in toluene

Toluene solution + 30–40% H_2O_2/3–4 hrs, 150°C, 200 psi → HTNR

$$HO\ (-CH_2-CH=\overset{\overset{\displaystyle CH_3}{|}}{C}-CH_2-)_n-OH$$

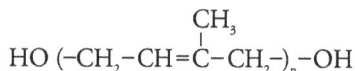

Cure Reactions

Cure reactions are the reactions of –OH and –NCO groups as given under ISRO-polyol.

HTPB (Hydroxyl Terminated Poly Butadiene)

Hydroxyl terminated polybutadienes are late comers and are the only pre-polymers specifically made for propellant binders. This pre-polymer is synthesized by two methods viz., free radical polymerization and anionic polymerization of butadiene. The free radical polymerization gives a pre-polymer with broad molecular weight distribution having a polydispersity in the range of 2 to 3 while anionic technique gives a narrow distribution of molecular weight (polydispersity of 1.1 to 1.3) as shown below. The functionality and the microstructure of hydroxyl terminated polybutadienes depend on the manufacturing process—free radical or anionic. The anionic method gives higher vinyl and cis and lower trans content compared to free radical method of manufacture. Increasing the vinyl content raises the pre-polymer viscosity and T_g and hence, vinyl content should be kept to a minimum. As with all polybutadienes, they are subject to air oxidation with subsequent hardening. Because of the primary nature of the hydroxyl group, its reactivity with isocyanate is high and hence, has short pot-life. Partial or complete hydrogenation solves this problem of oxidation and hardening, but increases the viscosity and T_g of the pre-polymer with consequent lowering of low temperature properties. Incorporation of sufficient side groups in the polymer chain can prevent the backbone chains from aligning and crystallizing. Free radical method is cheap and cost effective and gives HTPB suitable for composite propellant making.

Micro Structure and its Effects on Properties of HTPB

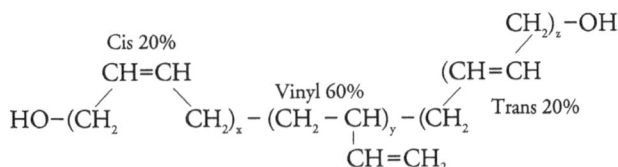

$$\text{HO-(CH}_2 \overset{\overset{\text{Cis 20\%}}{\text{CH=CH}}}{\diagup \quad \diagdown} \text{CH}_2)_x - (\text{CH}_2 - \underset{\underset{\text{CH=CH}_2}{|}}{\text{CH}})_y - (\text{CH}_2 \overset{\overset{\text{CH}_2)_z-\text{OH}}{\diagup}}{\underset{\text{Trans 20\%}}{(\text{CH=CH}}}$$

Vinyl 60%

Cis: Trans: Vinyl Ratio of Free-Radical HTPB is 20:60:20

		Viscosity (η)	T_g	*Mech. Props.*
1.	Cis (close packing not possible)	Lowers η	Decreases	More flexible
2.	Trans (Trans < vinyl)	Increases η	Increases	Less flexible
3.	Vinyl (higher inter molecular friction)	Increases η (Crystallinity increases)	Increases	Less flexible

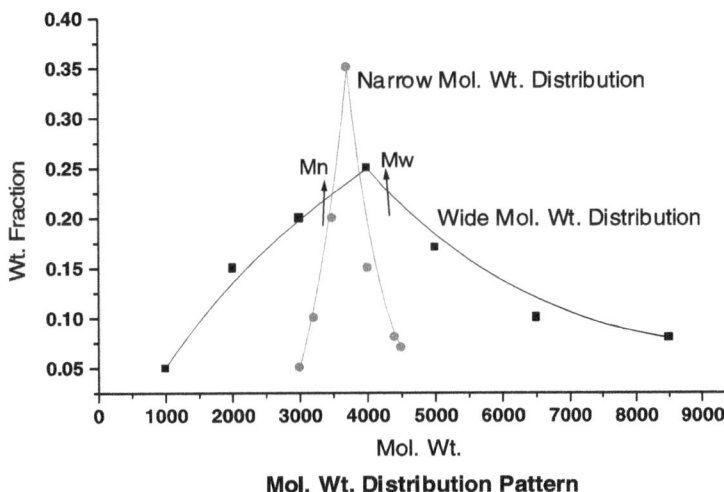

Mol. Wt. Distribution Pattern

Synthesis (Free-Radical Method)

The free radical technique of making HTPB involves polymerization of butadiene gas in benzene with hydrogen peroxide initiator in isopropanol as mutual solvent, around 110°C in a pressure reactor. The pre-polymer is washed with methanol to remove oligomers and dried under vacuum.

Initiation: $H_2O_2 \xrightarrow{\text{Heat}} 2\ \dot{O}H$

$CH_2=CH-CH=CH_2 + \dot{O}H \rightarrow HO-CH_2-CH=CH-\dot{C}H_2$

Propagation: $HO-CH_2-CH=CH-\dot{C}H_2 + n\ CH_2=CH-CH=CH_2$
$\rightarrow HO-(CH_2-CH=CH-CH_2)n-CH_2-CH=CH-\dot{C}H_2$

Termination: $2HO-(CH_2-CH=CH-CH_2)n-CH_2-CH=CH-\dot{C}H_2$
$\rightarrow HO-(CH_2-CH=CH-CH_2)_{2(n+1)}-OH$

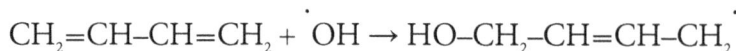

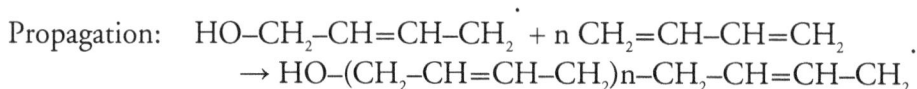

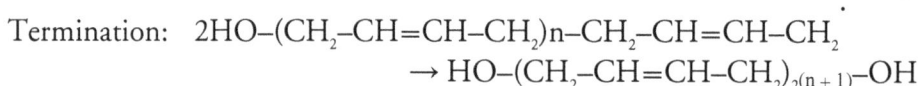

Anionic Method

The anionic polymerization of butadiene in toluene is made using organo lithium initiator. Termination, to get hydroxyl, is achieved with propylene oxide, followed by hydrolysis.

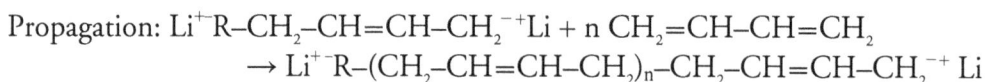

Initiation: $CH_2=CH-CH=CH_2 + Li\ (R)Li \rightarrow Li^{+-}R-CH_2-CH=CH-CH_2^{-+}Li$

Propagation: $Li^+R-CH_2-CH=CH-CH_2^{-+}Li + n\ CH_2=CH-CH=CH_2$
$\rightarrow Li^{+-}R-(CH_2-CH=CH-CH_2)_n-CH_2-CH=CH-CH_2^{-+}Li$

Termination

$Li^{+-}R-(CH_2-CH=CH-CH_2)_n-CH_2-CH=CH-CH_2^{-+}Li + 2RCHO$
$\rightarrow Li^{+-}O-CHR-R-(CH_2-CH=CH-CH_2)_n-CH_2-CH=CH-CH_2-CHR-O^{-+}Li$
$\rightarrow HO-CHR-(CH_2-CH=CH-CH_2)_n-CH_2-CH=CH-CH_2-CHR-OH$

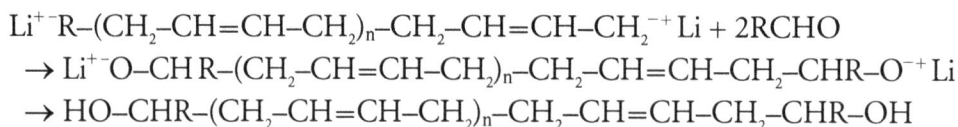

Cure Reactions

The basic reaction is the one between isocyanate group and hydroxyl group leading to the formation of urethane linkage. The cross-link density can be increased by adding small concentration of TMP.

$-NCO + HO- \rightarrow -NH-COO-$

The hydroxyl groups are reacted with multifunctional isocyanates to give urethane linkages. Therefore, this type of binder is often called a polyurethane. The isocyanate curing agents used are Toluene Diisocyanate (TDI) which is a mixture of 2,4 and 2,6 isomers of TDI in the ratio of 80:20, Hexamethylene Diisocyanate (HDI), Isophorone Diisocyanate (IPDI), or Methylene Dicyclohexyl Isocyanate (MDCI). The isocyanate is added in excess (5%) to remove traces of water and to allow for occurrence of side reactions such as isocyanate dimerization and urea, biuret and allophanate formation. During curing, the water content of all ingredients must be kept to a minimum to prevent excessive cross-linking and CO_2 bubbles. One molecule of water can destroy effectively two isocyanate groups and produce a urea linkage. The CO_2 evolved must be removed before gelation to prevent void formation. The chemistry of urethane binders is complicated by high reactivity of the isocyanates.

The structures of TDI and IPDI are given below as an example. The aromatic nature coupled with ortho-para effect makes TDI more reactive while the alicyclic nature with combinations of primary-secondary isocyanate groups make IPDI a slow reactive, yielding longer pot-life and amenability for large scale processing of HTPB propellants.

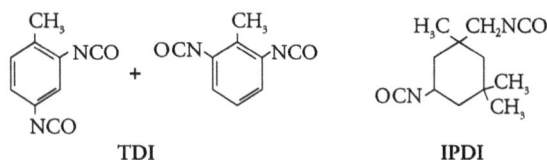

In fact, any diol/triol mixture that reacts with an isocyanate can form a network. However, to be useful as a propellant binder, it must have low cure shrinkage, low reaction exotherm, rubbery characteristics even at low temperatures, good aging stability and ease of handling during propellant manufacture. Long chain polyols with either polyester, polyether or polybutadiene backbone foot the bill. This keeps the isocyanate concentration at low levels, avoiding the problem of excessive cure shrinkage, high cure exotherm, and non-rubbery properties. Poly-(neopentylglycol azelate) (NPGA), Poly(neopentyl glycol sebacate) and poly(neopentyl glycol-6-nitro sebacate) are some useful polyester diol binders used in solid propellants. Polyesters are not desirable because of lower specific impulse compared to polyether and polybutadiene backbone diols. Also, they have comparatively higher viscosity due to wide distribution of the molecular weight in the pre-polymers. The low temperature properties are also not good.

Free radical HTPB marketed by Atlantic Richfield Co. (ARCO), USA is now made in France, Japan and India. ISRO's process (India) has been commercialized. ISRO's HTPB is being used in all the solid propellants, igniters and other auxiliary motors of PSLV and GSLV.

The hydrocarbon nature of HTPB (~98.6%) coupled with low viscosity (~60P) and low density (~0.90g/cc) makes it a good fuel binder capable of taking solid loading up to 90% without sacrificing the ease of processibility. In comparison, CTPB and PBAN which also have hydrocarbon backbones have less hydrocarbon because of end carboxyl and nitrile groups in the backbone. The hydrocarbon nature of CTPB and PBAN are 97 and 92% respectively. In addition, they have higher specific viscosity 250P for CTPB and 350P for PBAN and higher specific gravity 0.904 and 0.938 respectively. The higher viscosity is attributed to the hydrogen bonding among the carboxyl groups. These properties make CTPB and PBAN take lesser solid loading compared to HTPB for comfortable processing, thereby making these less energetic.

HTPB curing uses urethane reaction which is quantitative and goes to completion at a convenient rate giving good mechanical properties, while CTPB and PBAN use epoxides and aziridines for cure reactions with attended adverse side reactions including the one with oxidizer. The urethane linkage formed between the hydroxyl groups of HTPB and the isocyanate groups of the curing agent is more stable compared to the ester linkages formed between the carboxyl groups of CTPB and PBAN and the epoxy/aziridine groups. The –N–P– linkage formed in these cases is also easily hydrolysable hydrolytically.

The cure reaction should not be too fast or too slow as measured by slurry viscosity. This means that solid propellants containing 86 to 89% solid should be at least fluid for at least 5 to 6 hours for casting the propellant. This time, called pot life, can be increased by reducing the mixing temperature, adding pot life extenders or using the mixer bowl as the casting hopper to reduce the time of unloading of propellant slurry in to casting containers for casting.

The mechanical behavior of cross-linked polymer depends on the number of cross-links in the network introduced by the curing or cross-linking agent used for curing reaction. Chain extension which improves the strain capability is brought about by bifunctional curing agents while trifunctional curing agents introduces three dimensional network structure to prevent the plastic flow of propellants under pressure and temperature. The crosslink density (branch points/cc) is thus controlled by varying the ratio of bifunctional to trifunctional units to achieve the desired mechanical properties. Free radical technique of making the pre-polymers gives more than two functionality while the anionic technique gives two or slightly less functionality. HTPB being made by free radical technique exhibits more than two functionality (~2.3) and requires 75 to 80 equivalents of curing agent per 100 equivalents of hydroxyl groups. Compared to this, CTPB and PBAN needs 100 equivalents of epoxy/aziridine curing agents for 100 equivalents of carboxyl groups for curing. In addition, because of side reactions of epoxy/aziridines including homopolymerization, generally more than 100 equivalents are essential for the cure of CTPB and PBAN.

The completeness of the cure of a propellant is determined at a particular temperature by following the mechanical properties of the propellant, say stress, with days of cure when it attains a constant value. The HTPB propellant is fully cured in 5 days at 60°C or 8 days at 50°C while PBAN propellant is approaching cure at the end of 12 days at 75°C. The cure reactions can be catalysed to go fast and completed by using metal salts or metal complexes in small quantities. For example, ferric acetyl acetonate catalyses both urethane reaction as well as carboxyl/epoxy cure reactions. The propellants based on these pre-polymers can be cured at temperature ranging from 50 to 80°C in a number of days as indicated above. In fact, urethane propellants like the one based on HTPB can also be cured at ambient temperature (25–30°C) to the desired level in 25 to 30 days.

Plastisol Binders

The plastisol propellants are solidified through swelling of the polymer in a non volatile liquid, which has been selected to be a plasticizer for the resin at elevated temperatures, a typical example being Polyvinyl Chloride (PVC). The other polymers studied as propellant binder include cellulose acetate and nitrocellulose. Solvation or curing is accomplished by heating to a temperature at which the resin particles dissolve rapidly in the plasticizer to form a gel which on returning to room temperature has the characteristics of a rubbery solid.

Polyvinyl Chloride

It is industrially made by the emulsion or suspension polymerization of vinyl chloride.

$$CH_2=CH–Cl \quad \rightarrow \quad –CH_2–CH(Cl)–CH_2–CH(Cl)–CH_2–CH(Cl)–$$

The resin used in the manufacture of platisol propellants must be dispersion grade. The resin particles should be spherical or nearly spherical, preferably of 30 micron or less maximum diameter, free from porosity. This will help in the formation of a smooth, creamy plastisol when mixed with approximately equal amount of plasticizer. Further, the plastisol (the resin and plasticizer) must be capable of being heavily loaded with oxidizer and other solid additives to get a useful propellant composition. In the plastisol propellant process, it is essential that the resin particles should not solvate too rapidly at the processing temperature, since rapid increase of viscosity of the propellant mix interferes with the mixing and casting operations. There must be enough pot life of the propellant. The resin—plasticizer system has a dominating influence on pot-life.

Basically, PVC plastisol propellants consist of binder, oxidizer and metallic fuels. Minor ingredients (less than 2% of the total) consist of wetting agents, stabilizer, opacifier and burning rate modifier. Several plasticizers for PVC are used in propellants. Dibutyl sebacate, dioctyl sebacate and dioctyl adipate are found to be good. The plasticizer has a most important effect on the physical properties of the cured propellant and the variation of these properties with temperature. Long chain aliphatic plasticizers are preferred over aromatic plasticizers because they impart good low temperature flexibility. The wetting agent (glycerol mono oleate, pentaerythritol dioleate, or dioctyl sodium sulfosuccinate) in propellant formulation facilitate the mixing of ingredients and reduce the viscosity of the propellant mix so that it is more easily cast. Stabilizers (0.25 to 0.50% of barium ricinoleate, basic lead sulphate, barium and cadmium soaps) are added to the formulation to retard the decomposition of PVC during cure in the temperature range 150 to 180°C.

PVC plastisol propellants are cured in the temperature range of 150 to 180°C. Only a few minutes' residence time at the curing temperature is essential for completing solvation of PVC in plasticizer. Since propellant has low thermal conductivity, it takes long time to reach the temperature. During this period, the PVC closest to the heating source is exposed to this temperature for long time and hence, may decompose evolving HCl gas. The stabilizer added to the formulation absorbs the gas. The de-chlorination creates a double bond which can further crosslink, reducing the mechanical properties, mainly flexibility.

Viscosity of plastisol propellant is in the range of 500 to 1000 poise. The limit of low end viscosity depends on the tendency of ingredients like oxidizer/metallic fuel to separate from the uncured plastisol during processing or storage of the plastisol slurry before solidification. The upper limit depends on the viscosity of the slurry that can be processed through the casting fixtures in a reasonable time. Too high a viscosity may lead to defects like the propellant folding over on itself to form a void which may remain in the cured grain even after cure.

PVC plastisol propellants cure at very fast rate once the binder has reached the curing temperature. Full strength is developed at 150°C and above within five minutes. As no chemical reactions are taking place during curing of PVC plastisol propellant, the heat required to raise the temperature to curing temperature has to be supplied from outside. The heat is transferred by conduction through the propellant grain from an oven or external source. In bigger diameter grains like 560 mm, the heating of the mandrel which has perforations, is also required, in addition to outside heating, to reduce the time of curing. Prolonged heating destroys the elastomeric properties of the propellant. The change is more rapid above 150°C and the extent depends on temperature and the time propellant is exposed to high temperature. Because of higher contraction of the propellant grain compared to that of metal moulds, the cured propellant shrinks and debonds from the mold when the propellant attains the room temperature. This fact makes the plastisol propellants unsuitable for case bonding techniques to increase the mass ratio. Hence, plastisol propellants are used mostly as free standing grains in missiles and sounding rockets.

3. PLASTICIZERS

Plasticizers are low molecular weight, non-volatile, non reactive liquids added to propellants or polymers to improve processibility, flexibility and mechanical properties. The plasticizer effect is due to a reduction in cohesive forces of attraction between polymer chains. Plasticizers, thus, act as intermolecular lubricant in propellant formulations. The small plasticizer molecules compared to polymer chains, penetrate into polymer matrix and establish polar attractive forces between the chain segments and plasticizer. These attractive forces reduce the cohesive forces between the polymer chains and increase the segment mobility, thereby reducing the T_g values. Plasticizer helps the free movement of the binder molecules without undergoing any chemical reaction.

The major functions of a plasticizer are: i) to improve the processibility of propellant slurry by reducing the mix viscosity for longer potlife, and ii) to improve the mechanical properties such as flexibility, low temperature properties, etc. Propellants with more than 80% solid loading require plasticizer to get good processibility. Generally, 20 to 35% of the polymer by weight is used as plasticizer. Higher concentrations of plasticizer use in propellants lead to migration problems during storage. Propellants for deep space programs and sterilisable propellants do not use plasticizers in their formulations. Many plasticizers such as ornate (standard oil of California), spindle oil, Savanol (a hydrocarbon oil), dibutyl phthalate, dioctyl sebacate, dioctyl adipate, tricresyl phosphate etc. are used generally in propellant formulations. In addition, energetic plasticizers containing azide groups, difluoro nitramino and nitrate ester groups have been used to increase the energetic of solid propellants. Structural similarity with energetic polymer should facilitate incorporation; however, one of the most common problems has been exudation-migration of the low molecular weight plasticizer to and from the surface of the formulation. A promising approach has been to increase the structural similarity, and hence, miscibility, by using low molecular weight oligomers of the polymer matrix as the plasticizer. Branched polymers or polymers with long pendent groups can act as internal plasticizer as they increase the interchain free volume, resulting in lowering of viscosity and T_g as well as improving flexibility for the end product. A typical example is ISRO Polyol (cited under fuel binders) with $CH_3-(CH_2)_5$ pendent group acting as internal plasticizer. Selecting a plasticizer for commercial applications is obviously decided as a compromise between cost, compatibility, efficiency, and migration effect.

Requirements

The essential requirements of a propellant plasticizer is that it should be a low molecular weight, high boiling, non-volatile liquid compatible with other propellant ingredients and having high heat of combustion. When using plasticizers, it is important to choose a material that is sufficiently soluble in the gel structure formed by the polymer and yet sufficiently non-volatile. When using plasticizers in propellants, one must consider the entire rocket system. The plasticizer may be more soluble in the inert liner surrounding the propellant, and migration may take place to the liner degrading the interface properties of liner/propellant/insulator. This will cause disastrous results in the performance of propellants. If the plasticizer has significant vapor pressure, the plasticizer volatilizes from the propellant, thereby changing the physical properties of the propellant. In spite of all these problems, the plasticizer is useful in tuning the propellant formulations to get the desired properties.

4. METALLIC FUEL

Most of the modern high energy solid propellants contain very fine (say 10 to 30 micron) metallic fuels such as aluminum. These metallic fuels increase the chemical energy of solid propellants not only through the highly exothermic reaction with oxidizer/oxygen but also exclude water vapor from the exhaust product and increase its hydrogen content. Because of the high density of metals, the propellant density also increases. Presence of aluminum also suppresses combustion instability of the propellant. Generally, 18 to 20% of aluminum is used in propellant formulations. Increase of aluminum concentration increases the specific impulse up to 20% and thereafter no increase in impulse due to incomplete combustion. Metallic oxides also reduce the efficient conversion of chemical energy to propulsive energy by rocket nozzle. Among other metallic fuels used like magnesium, boron and beryllium, magnesium is highly reactive and can pose safety hazards during processing. With boron, the exhaust products are under oxidized because of the difficulty of its combustion and hence, the full potential is not available. Beryllium is very satisfactory in improving the performance of propellants but the exhaust products are extremely toxic. Other metallic fuels reported in literature are zirconium, titanium and nickel. Metallic fuels and their properties are given in Table 3.

Table 3: Metallic Fuels and their Properties

$$4\,Al + 3\,O_2 \;\rightarrow\; 2\,Al_2O_3 + \Delta H \text{ (Exothermic)}$$

Metallic Fuels	ΔH (cal/g)	Remarks
Zn	2600	Low ΔH
Mg	6000	High reactivity with air and water
Al	7000	Less reactive, moderate ΔH, low cost and easy availability
Be	15000	Poor combustion, high cost and highly toxic
B	16000	Poor combustion and high cost

Metal Wires

Metallic wires and staples of silver, copper, tungsten, etc. have been used to increase the burning rate of solid propellants enormously. This has been demonstrated in PVC propellants. In this technique, metal wires are embedded in the propellant grain before cure. Then, when the propellant is burned, the wires extend from the unburned propellant into the flame zone. The metal wires have considerably higher thermal diffusivity than the gases between the unburned propellant surface and the flame, so they provide paths for rapid heat transfer to the burning surface. As a result, the propagation of the burning surface along each wire is much faster than normal burning rate. The burning surface adjacent to each wire recesses to form a cone with the wire at its apex. This increases the burning surface area and hence, the burning rate of the propellant. A five-fold increase in effective burning rate is possible. The rate of propagation of burning surfaces around each wire is uniform. Higher the thermal diffusivity or lower the melting point, higher the burning rate. End burning grains having axially oriented wires is one way to increase the burning rate. A burning rate of 50 to 100 mm/sec at 70 kg/cm^2 has been achieved. The degree of increase that can be obtained with wires of various metals depends on the thermal diffusivity and melting point of the metals.

The metallic staples which are employed as burning rate additives in solid propellants are comprised of any metal or metal alloy of high heat conductivity and high melting point such as copper, aluminium, iron, stainless steel, brass, zirconium, silver and many others. These staples, which may be of any cross-section such as circular, rectangular, square or many sided, can vary in length up to 500 mils. However, all the staples in a given matrix will be of same length and have a maximum side to side dimensions of 20 mils. The solid rocket propellant composition is mixed in the usual manner and the staples added at some point in the process prior to casting of the propellant. The dispersions of the staples in the mixture there of will result in a random uneven orientation of the staples in the mixture. The staples containing propellants are oriented by forcing the propellant through the screen when casting the rocket motor. When metal staples oriented along the axis of the motor casting, the staples will be normal to the burning surface of an end-burning propellant grain. This will increase the burning rate to the maximum. Ferromagnetic metal (iron, cobalt, nickel) staples can be oriented by the application of external magnetic field.

Metal Hydrides

Hydrides are good gas producers and that too light gas like hydrogen. Different hydrides like aluminum hydride have been tried as metallic fuels. Their hydrogen content improves both the potential energy and the working fluid. However, there are problems like reactivity of hydrides with other ingredients, oxygen and moisture, in addition to their lower densities. To reduce the reactivity, encapsulating technique has been tried with partial success. Boron forms a family of hydrides called boranes and dihydroboranes. All of them are toxic and some are unstable. However, their high heat of combustion per unit weight led to extensive studies as fuels.

Fuel rich propellants are solid propellants containing generally a higher percentage 20–40% metallic or polymeric fuel. Boron, magnesium, aluminum and the alloys of magnesium and aluminum are the metallic fuels tried. These fuels burn in atmospheric air that is rammed into the combustion chamber. This gives additional energy or specific impulse.

5. CROSS LINKERS/CURING AGENTS

Bridging formed between linear polymer chains leading to a three dimensional network structure are known as cross-linking. The cross links between polymer molecules can be either through regular covalent bonds (chemical cross-linking) or through secondary valence type links such as hydrogen bonds (physical cross-linking). Cross-linking could be made to occur during polymerization. Thus, styrene when polymerized with small amounts of divinyl benzene results in a cross-linked polymer.

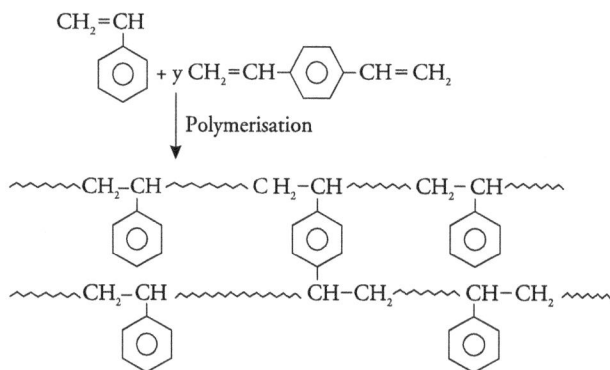

Similarly, when a condensation reaction between a difunctional monomer and a trifunctional monomer could lead to cross linking during a poly-condensation reaction as given below:

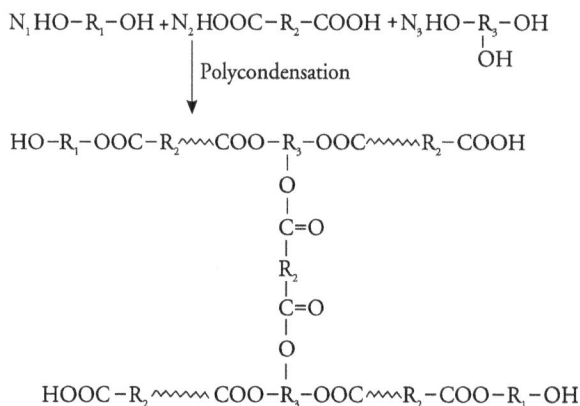

Vulcanization of rubber is also a way of introducing cross linking in polymers. For example, sulphur or sulphur related compounds react with rubber molecules to effect cross linking as follows.

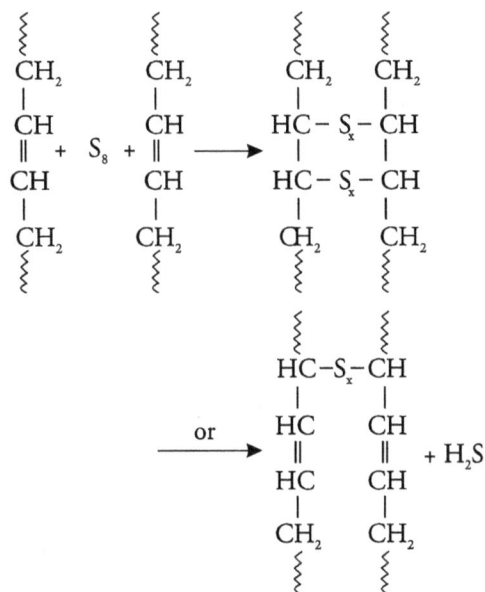

$$
\begin{array}{ll}
\begin{array}{l}
CH_2 \\
| \\
CH \\
\| \\
CH \\
| \\
CH_2
\end{array}
+ S_8 +
\begin{array}{l}
CH_2 \\
| \\
CH \\
\| \\
CH \\
| \\
CH_2
\end{array}
\longrightarrow
\begin{array}{l}
CH_2 \quad CH_2 \\
| \qquad | \\
HC-S_x-CH \\
| \qquad | \\
HC-S_x-CH \\
| \qquad | \\
CH_2 \quad CH_2
\end{array}
\end{array}
$$

$$
\xrightarrow{\text{or}}
\begin{array}{l}
HC-S_x-CH \\
| \qquad | \\
HC \qquad CH \\
\| \qquad \| \\
HC \qquad CH \\
| \qquad | \\
CH_2 \quad CH_2
\end{array}
+ H_2S
$$

Cross linking in certain polymers can also be effected by cure reaction. These uncross-inked linear polymers contain either reactive functional groups or double bonds in their molecules as in polymeric binders used in solid propellant formulations. They are usually not of a very high molecular weight and are liquid resins known as pre-polymers. When these pre-polymers are reacted with low molecular weight or polymeric substances containing appropriate functional groups capable of reacting with active groups of the pre-polymer, curing takes place resulting in a cross-linked solid polymer. The low molecular weight or polymeric material used to bring about the cure reaction is called the curative or curing agent. The functionality of both pre-polymer and the curing agent plays an important role in determining whether the reaction between them will result in cross linking or not. For cross linking to take place, both the pre-polymer and curative should not only have bi-functionality but one or both of them should have some tri-functional molecules also. Unless such molecules are present in the system, no cross-linking takes place and only linear coupling of the pre-polymer molecules through curative molecules results. The amount or degree of cross-linking depends upon the ratio of the bi-functional to the tri-functional molecules present in the system. It is the ingredient that causes the binder to solidify and become hard. Some important curing agents for pre-polymers used as binders are given in Table 4.

Table 4: Pre-Polymers and their Curing Agents

	Pre-Polymers	*Curing Agents*
1.	Unsaturated polyesters	Styrene, methyl methacrylate or any unsaturated compound
2.	Polysulphides	Lead dioxide, p-quinone dioxime or epoxides
3.	Epoxides	Thiol terminated poly sulphide, diamines, carboxyl-terminated pre-polymers
4.	Carboxyl containing pre-polymers	Epoxides, aziridines
5.	Di-hydroxyl or tri-hydroxyl polymers	Diisocyanates

The structure of different curing agents and cure reactions of individual binders are given under that particular binder in the text. Cross-linking agents are used in small amounts depending upon the reactive groups present in the pre-polymer, a minor change in the percentage of curing agents will have a major effect on the propellant physical properties, manufacturability, and curing. It is used only with composite propellants.

6. BURNING RATE MODIFIERS

Burn rate requirements are specific to the design of particular propellant grain configuration to realize the required thrust-time profile. Additives for altering burning rate are of interest to propellant formulator to meet the ballistic needs. The burning rate modifiers or ballistic modifiers are blended in carefully controlled quantities into rocket motor propellant during production. The selection of additives which can be used in small quantities is sought after to increase or decrease burning rate, to decrease the sensitivity of burning rate to pressure (pressure index 'n') or temperature, or to accomplish some combination of these changes. The ideal additives would have negligible effect on fluidity of propellant slurry, physical properties of cured propellant, thermal stability, storage stability or safety characteristics. Common burn rate modifiers are copper chromite ($CuCr_2O_4$), ferric oxide (Fe_2O_3), copper(II) oxide (CuO), ferrocene derivatives like butyl ferrocene, catocene, etc. (burn rate accelerators) and Lithium Fluoride (LiF), ammonium fluoride, ammonium sulphate, ammonium fluoroborate, nitroguanidine, oxamide, etc. (burn rate retardants). These chemicals either accelerate or retard the burning rates by accelerating or decelerating the decomposition of the oxidizer or lowering the decomposition temperature. The inorganic catalysts do not contribute to the combustion energy, but consume energy when they are heated to the combustion temperature.

In composite propellant, the fuel and oxidizer are in separate phases and give oxidizer rich and fuel rich gas streams on propellant combustion. After ignition, the propellant surface pyrolyses, resulting in the breakdown of the binder. The breakdown product of the binder mainly consists of chain scission and chain unzipping to give volatile products and monomers. The oxidizer can sublime endothermically as a monopropellant. At high pressures, monopropellant oxidizer decomposes. The more critical process is the mixing and exothermic reaction of the oxidizer vapors and with binder pyrolysis products. The heat resulting from these reactions is transferred to the propellant surface and continues the cycle. The conductive heat feed-back from the flame to the propellant surface is pressure sensitive and that is why the burn rate is pressure sensitive.

Composite solid propellants are generally fuel rich because of practical difficulties in loading the optimum amount of ammonium perchlorate. For the hydrocarbon-ammonium perchlorate combination, 90.5% by weight of oxidizer would be required for complete combustion to N_2, HCl, H_2O and some CO_2. However, in practice 84 to 88% loading is used. Ammonium perchlorate decomposes thermally into:

$$NH_4ClO_4 \rightleftharpoons NH_3 + HClO_4 \rightarrow NO + Cl_2 + N_2O + HCl + H_2O + N + O_2$$
$$\text{Reversible} \qquad\qquad\qquad\qquad \text{non-reversible}$$

Small amounts of catalysts or burn rate modifiers can significantly affect the ammonium perchlorate decomposition rate. When 0.2% by weight of copper chromite ($CuCr_2O_4$) or ferric oxide (Fe_2O_3) is blended with ammonium perchlorate, the decomposition temperature is greatly reduced as seen from Differential Thermal Analysis. Propellant burn rates can, therefore, be increased by addition of such materials. But, the effect of such catalysts is not clear due to the variations and inconsistency with concentration, type of catalyst and pressure.

Another explanation for the burning rate increase by specific additives is due to decomposition beginning at specific surface sites, like crystal imperfections. The reaction continues for a short way into the crystal and then a new site is required. These sites may be poisoned by combustion products. The catalyst may supply new points for decomposition. Many metal salts (salts of Mn, Fe, Cr. Cu, and Hg) are also catalysts for the oxidation reactions of hydrocarbons. They induce free radical decomposition of hydroperoxides, which are formed by the hydrocarbon on contact with oxidizer. Other mechanisms include chemical catalysis of gas—phase reactions and increased heat conduction from flame to solid surface. It is known that transitional metal ions are effective election acceptors and they promote the decomposition of perchlorate radicals, thus enhancing the burn rate. Ammonium compounds (retardants) retard the forward reaction and hence, reduce the burn rate:

$$NH_4ClO_4 \rightleftharpoons NH_3 + HClO_4$$

Another way to increase the burning rate is the use of small amounts of metal wires in propellants. Silver wires of 0.125 mm are introduced in solid propellant normal to the burning surface. Once the combustion starts, the tip of the wire in the high temperature flame zone conducts heat into propellant causing localized temperature gradient. Since the burning rate is temperature dependent, the propellant touching the wire burns faster than the rest. As the burning progresses, a conically shaped indentation penetrates even deeper into the more slowly receding normal burning surface, eliminating an ever widening circular area of this surface. When such wires are present in a regular pattern, a stable burning condition is reached in which all of the conical indentations intersect, leaving no normal surface. The wire charged technique has been used successfully to produce five times the normal burning rate in plastisol propellants. Metal staples have also been used to increase the burning rate especially in end burning motors.

7. BONDING AGENTS

A strong bond between binder and oxidizer particles is essential for getting high tensile strength. Bonding agents are used to improve this bonding. These are low molecular weight compounds and have functional groups which either react with oxidizer or have comparatively higher polarity than the rest of ingredients so that there exist a secondary ion-polar attraction between oxidizer and bonding agent. The bonding agents also contain two or more functional groups, like hydroxyl, which react with curing agent. Usually 0.1 to 0.3% of the bonding agents are used. Standard bonding agents used are alkanol-amine-amide type, MAPO-carboxylic acid adducts, Trimethylol Propane (TMP), etc. These bonding agents improve the mechanical properties of the propellant but do not affect its low temperature properties. For HTPB based propellant an adduct of MAPO, tartaric acid and adipic acid is used as bonding agent by some manufactures.

8. PROCESS AIDS

Process aids or wetting agents in the propellant composition is used to facilitate the mixing of ingredients and to reduce the viscosity of the propellant slurry so that it can be more easily cast. The process aids improve the wetting of the oxidizers by liquid materials during the mixing operations. Some of the common process aids are silicone oil, lecithin, detergents like sodium lauryl sulphate, etc. 0.05 to 0.1% is the usual quantity used in propellant formulations.

9. STABILIZERS OR ANTI OXIDANTS

Stabilizers are additives added to propellant formulations to inhibit or reduce degradation of polymeric fuels like HTPB or PBAN. These chemicals protect the polymeric fuel from thermal and photo degradation. Since the butadiene binders are unsaturated ones and they have to be protected from cross-linking during storage due to thermal or photo degradation. They are added in 0.1 to 0.2% and are generally phenolic compounds or aromatic amine like phenyl beta napthyl amine, butyrated hydroxyl toluene, etc.

Typical Questions

1. What are the qualities of a good oxidizer?
2. How is AP preferred over other conventional oxidizers in composite propellants?
3. How is ammonium perchlorate manufactured?
4. Explain the propellant specifications for AP?
5. Explain unimodal and bimodal distributions of AP?
6. What are the drawbacks of using AN in composite propellants? How can we overcome it?
7. Explain the salient features of propellant binders?
8. What are thermoplastic and thermosetting binders?
9. Explain the characteristic features of HTPB resin?
10. Explain the synthesis of free-radical and anionic HTPB pre-polymers?
11. Explain the micro structures and molecular weight distributions of free-radical and anionic HTPB pre-polymers?
12. Explain the features of CTPB and LTPB?
13. How is polysulphide prepared? Explain its merits and demerits?
14. Explain the basic components of HTPB composite solid propellant and explain their functions?
15. What is a ter-polymer? How is it prepared?
16. How is ISRO polyol different from poly ether polyols?
17. Explain the synthesis of HTNR?
18. What are Plastisol binders? Explain with suitable examples?
19. Give the chemical structures of MAPO, TDI and IPDI? Explain their cure reactions with pre-polymers?
20. What are the requirements of a good plasticizer in solid propellants?
21. Why Al is used in most of the stage motor propellants?
22. Explain the burning of metal wire embedded solid propellant?
23. Explain cross-linking reactions with suitable examples?
24. What are ballistic modifiers? How does it improve the ballistic performance of a motor?
25. What are the functions of bonding/processing aids in solid propellants?
26. Explain the following:
    (a) Curing (b) Pot life (c) Shelf life (d) Plasticizers (e) Ballistic modifiers (f) Eco-friendly Propellants.

REFERENCES

[1] Agarwal, J.P., High Energy Materials: Propellants, Explosives and Pyrotechnics, J.P., Wiley-VCH, 2010.

[2] Anuj, A. Varghese, K. Muralidharana and Krishnamurthy, V.N., Kinetics and Model free Prediction of Nanocatalyst Influenced Thermal Decomposition of Ammonium nitrate and Ammonium nitrate based Composite Solid Propellant, *Proceedings of the 9th International High Energy Materials Conference and Exhibits HEMCE 2014.*

[3] Bernard, Gondouin, Solid Rocket Propulsion Technology. Ed. Davenas Alain. Oxford OX3, England: Pergamon Press 1993.

[4] Bhagavan, S.S., Varghese, T.L. *et al.*, "Design and Analysis of Experiments for Selection of Solid Propellant Formulations," *Proc. 3rd International Conference on High Energy Materials*, Trivandrum, 2000.

[5] Boldyrev, V.V., Themal Decomposition of Ammonium Perchlorate, *Termochemica Acta*, 1–36, 2006.

[6] Bottaro, J.C., Recent Advances in Explosives and Solid Propellants, Chem. Ind. 10, 1996, 249.

[7] Prasad, Devi Vara, C.H.; Nair, C.P. Reghunathan and Ninan, K.N., *A method for Processing a low OH Value Hydroxyl Terminated Polybutadiene based Solid Propellant for Space Boosters,* Indian patent (applied, 2013).

[8] Gould, R.F., Propellants Manufacture, Hazards and Testing, Advances in Chemistry, No. 88, American Chemical Society, Washington, DC, 1969.

[9] Gowariker, V.R., Viswanathan, N.V. and Sreedhar, Jayadev, Book on *'Polymer Science'*, published by Wiley Eastern Limited, New Delhi, 1987.

[10] Hayri, Yaman, Veli, Celik, Ercan, Degrimenci, Experimental Investigation of Factors Affecting the Burning Rate of Solid Rrocket Propellant, *Fuel*, 115, 2014.

[11] Kannan, M.P., Thermal Decomposition of Ammonium Perchlorate, *J. Thermal Anal.* 32, 1987, 1219–1227.

[12] Knaresboro, Donald L., Goodson, Forrest R. and Inman, Frank S., Method of manufacturing solid rocket motors, US patent no. US 6101948 (A), United technologies corporation, US, 2000.

[13] Krishnamurthy, V.N., "Some Issues in the Development and Production of HTPB Propellants," *Proc. 2nd International High Energy Materials Conference,* IIT, Madras, 1998.

[14] Krishnan, S., Chakravarthy, S.R. and Athithan, S.K., Propellants and Explosives Technology, ISBN 81-7023-884-6, Allied Publishers Limited, India, 1998.

[15] Kshirsagar, D.R., Sudhir, Mehilal and Singh, P.P., Bhattacharya B., *International Journal of Energetic Materials and Chemical Propulsion*, 12, 463–474, 2013.

[16] Landcrs, L.C. and Sanley, C.B., Propellant development for the Advanced solid rocket motor, in *27th Joint Propulsion Conference June 24–26,/Sacramento,* CA. AIAA/SAE/ASME (1991), 91-2074.

[17] Lengelle, G., Duterque, J. and Trubert, J.F., Combustion of Solid Propellants, *ONERA*, May 2002.

[18] Makato, Kohga, Burning Characteristic and Thermo Chemical Behavior of AP/HTPB Composite Propellant Using Coarse and Fine AP Particle, *Propellant Explosive Pyrotechnics*, 36, 2011.

[19] Mathew, Suresh, Varghese, T.L. and Ninan, K.N., A Composite Propellant Containing Active Copper Oxide and a Process for Preparing the same, Patent No. 241/MAS/97.

[20] Miya, Hiroshi and Tanaka, Shinichiro, "Nitramine propellant," *Jpn Kokai Tokkyo Koho,* 2006, JP2006151791 A 20060615.

[21] Moore, Thomas L., CTPB and HTPB Propellants for Extended Space Missions," AIAA 2002–3750.

[22] Muravyev, N., Frolov, Yu., Pivkina Alla., Monogarov, K., Ivanov, D. and Ordzhonikidze, Olga., "Particle Size and Mixing Technology Influence on Combustion of HMX/Al Compositions," *Proceedings of the International Pyrotechnics Seminar, 36th* (2009), 43–55.

[23] Muthiah, R.M., Varghese, T.L., Rao, S.S., Ninan, K.K. and Krishnamurthy, V.N., Realization of an Eco-friendly Solid Propellant based on HTPB-HMX-APsystem for Launch Vechile Application, *Propellants, Explosives, Pyrotechnics,* Vol. 23, 1998, pp. 90–93.

[24] Nagappa, R. and Kurup, M., Development of HTPB Propellant System for ISRO Solid Motors, AIAA-90-2331, AIAA (1990).

[25] Ninan, K.N., Balagandadharan, V.P. and Katherine, K.B., Studies on the Functionality Distribution of HTPB and Correlation with Mechanical Properties, *Journal of Polymer,* Vol. 32 (1991), pp. 628–635.

[26] Ninan, K.N., Balagandadharan, V.P., Devi, K. Ambika and Katherine, K.B., "Functionality Distribution and Crosslink Density of HTPB," *Polymer International,* Vol. 31 (1993), p. 285.

[27] Pande, S.M., Sadavarte, V.S., Bhowmik, D., Gaikwad, D.D. and Singh, H., *Propellants Explos. Pyrotech,* 37, 707, 2012.

[28] Patil, Prajakta R., Krishnamurthy, V.N. and Joshi, Satyawati S., Effect of Nano-Copper Oxide and Copper Chromite on the Thermal Decomposition of Ammonium Perchlorate, Propellants, Explosives, Pyrotechnics, 33(4), 2008.

[29] Reshmi, S., Varghese, T.L. and Ninan, K.N., A Slow Burning Composite Propellant Composition based on Hydroxyl Terminated Polybutadiene and a Process for making the same, Patent application, No. 467/CHE/2007.

[30] Selvaraj, B., Masthiraj, N. Vivek and Kumar, S.R. Dhinesh, Minimum-Signature (Smokeless) Propellant, *International Journal of Emerging Technology and Advanced Engineering,* 2, 2012.

[31] Silva, G., Rufino, S.C. and Iha, K., Green Propellants: Oxidizers, *J. Aerosp. Technol. Manag,* 5(2), 2013, 139–144.

[32] Singh, Manohar, Kananga, B.K. and Bansal, T.K., "Kinetic Studies on Curing of HTPB Polymer Based Polyurethane Networks," *Journal of Applied Polymer Science,* Vol. 85, 2002, 842–846.

[33] Solid Rocket Propulsion Technology, edited by Alain Davenas, Pergamon Press, 1997.

[34] Sutton, George P., 'Book on Rocket Propulsion Element," *7th Ed., Wiley,* New York, 2001.

[35] Varghese, T.L., Prabhakaran, N., Rao, S.S. and and Ninan, K.N., "Development of a New Generation High Pot life Hydroxyl Terminated Polybutadiene Propellant," *Proc. Second International HEMCE and Exhibit,* IIT Madras, 1998.

[36] Varghese, T.L., Chemical Propellants, IITP-2002, Area Specific Intensive Course Module III, General Lectures, Book Published by VSSC, ISRO, August 2002.

[37] Varghese, T.L., *et al.,* Book on Manual of Solid Propellant Chemicals, Special Publications, ISRO-TTG-SP-33-1987.

[38] Varghese, T.L., Rao, S.S. and Ninan, K.N., A Composite Solid Propellant Composition based on HTPB based Composite Solid Propellant with Active Ferric Oxide, Patent No. 1091/Mas/1999.

CHAPTER 3

Composite Solid Propellants—
Energetic Aspects

1. INTRODUCTION

There are a number of potential materials viz. polymeric binders, metallic fuels and oxidizers for use in solid propellants. A major consideration is whether the ingredients produce the necessary energy when the combustion takes place? The initial decision must be based on knowledge or at least an estimate of the chemistry and thermodynamics of the system. The preferred use of hydrocarbons for binder materials is a result of the need to have gaseous combustion products like CO_2, CO, H_2, N_2, etc. of low molecular weight. A polymeric binder with high ratio of hydrogen to carbon and positive heat of formation having high value is the best binder.

Newton's third law states that for every action, there is an equal and opposite reaction. Propellants have the major objective of imparting motion to an object. The energy source most useful to rocket propulsion is chemical combustion. Any rocket consists of three parts—combustion chamber, throat and nozzle. In the rocket combustion chamber, the propellant chemicals—oxidizer, polymeric binder and the metallic fuel—are burnt to produce high pressure gases at high temperature ($2000°$ to $4000°K$). These hot product gases accelerate and expand as they pass through the throat and nozzle to produce the action. The rocket moving in the opposite direction of the exhaust gases is the reaction. Since the gas temperatures are almost twice that of the melting point of steel, it is necessary to insulate all the parts of rocket motor that is exposed to hot gases. Rocket movement results from the thrust generated due to pressure difference between the combustion chamber and the surroundings. This leads to the conclusion that rockets operate best in vacuum where pressure imbalance are greatest and escaping gas molecules achieve their highest energy.

To achieve the necessary continual high speed flow of gas molecules, controlled combustion must occur in the combustion chamber. Solid propellants, like the ones used in PSLV, Space Shuttle or Arian rockets, the fuel and oxidizer are stored directly, in the combustion chamber in a premixed solid state. Motor performance is dictated by a number of factors, including fuel type, fuel-oxidizer ratio, grain or motor size and grain configuration. The criterion for the selection of a propellant for a particular mission depends on the energy content, the physico-chemical and operating factors. Some are, of course, highly specific to a particular type of mission for example, heat sterilizability in interplanetary landers.

2. ENERGETIC COMPUTATIONS

Specific Impulse

A high chemical energy and large volume of gaseous products with low mean molecular weight gaseous products are the major requirements for high performance. This energy of propellant system is commonly referred to as specific impulse (I_{sp}). As discussed in chapter-1, it is defined as amount of thrust generated per unit mass flow rate of the propellant. This is mathematically represented as the ratio of the thrust (F) produced to the mass of propellant (m_p) expelled per unit time (t).

$$I_{sp} = F/m_p \times t \qquad \dots (1)$$

The specific impulse is related to the combustion energy as follows:

$$I_{sp} = (2J/g) \, [H_c - H_e]^{1/2} \qquad \dots (2)$$

where J is the mechanical equivalent of heat, g is acceleration to gravity, H_c and H_e are the enthalpies of combustion products at chamber (T_c) and exit (T_e) conditions respectively.

For achieving maximum I_{sp}, the combustion energy has to be maximum. Specific impulse is also related to combustion chamber temperature (T_c), mean molecular weight of combustion gases ($\bar{M}$), the pressures at the chamber (P_c) and exit plane (P_e), the specific heat ratio $\gamma(C_p/C_v)$ and universal gas constant (R) as follows:

$$I_{sp} = [(2J/g) \, (\gamma \, RT_c/(\gamma - 1) \, \bar{M}) \, (1 - P_e/P_c) \, \gamma - 1/\gamma)]^{1/2} \qquad \dots (3)$$

The first term $[2J/g \, (\gamma RT_c/(\gamma - 1) \, \bar{M}]$ signifies the total heat content of the propellant gases in the chamber and is strictly a property of the propellant. The second term represents the fraction of heat content converted into mechanical energy. Thus, it is clear that high combustion chamber temperature with low mean molecular weight of gaseous products results in high specific impulse. The I_{sp} can be calculated theoretically using computer programmes like NASA-SP 273.

By combining the equations 2 and 3, one can get a relationship between the amount of energy released by a given propellant combination and is proportional to the square root of combustion temperature (T_c) and inversely proportional to mean molecular weight of combustion gases ($\bar{M}$), i.e.,

$$\Delta H = k \sqrt{(T_c/\bar{M})} \quad \text{or} \quad I = \sqrt{\Delta H} \qquad \dots (4)$$

Heat of Formation

The energy content of a propellant system is primarily dependent on the thermo-chemistry involved. In a general chemical reaction,

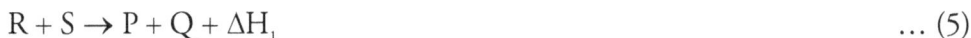

$$R + S \rightarrow P + Q + \Delta H_1 \qquad \qquad \dots (5)$$

the compounds R and S are reactants and P and Q are products and ΔH_1 is the enthalpy of the reaction. By convention, if the heat is released by this reaction, it is called exothermic and the value of heat released is given a negative sign. If heat is absorbed in the process, the reaction is called endothermic and positive sign is used for the enthalpy. The enthalpy is also called the heat of reaction. Consider a reaction,

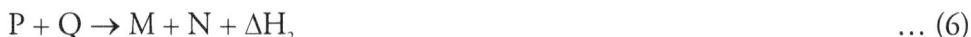

$$P + Q \rightarrow M + N + \Delta H_2 \qquad \qquad \dots (6)$$

We can add or subtract two reactions to give a third reaction. The final enthalpy is dependent only on the initial reactants and final products involved and is independent of path to achieve the desired products. This law is called Hess's law. By addng equations (5) and (6) we get,

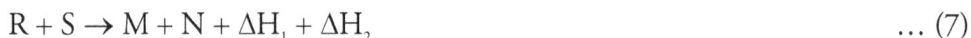

$$R + S \rightarrow M + N + \Delta H_1 + \Delta H_2 \qquad \qquad \dots (7)$$

The standard state of any element is its normal condition at ambient temperature (25°C or 298.15°K) and pressure (1 atm). If a reaction in a thermochemical equation consist only of elements in their standard states, and one mole of one product at the same temperature and pressure is involved, the enthalpy is called the standard heat of formation of the product, since the heat of formation of any element in its standard state is zero. For example,

$$C(s) + O_2(g) \rightarrow CO_2 \qquad \Delta H = 94.05 \text{ kcal}$$

$$H_2(g) + \tfrac{1}{2} O_2(g) \rightarrow H_2O \qquad \Delta H = 57.80 \text{ kcal.}$$

First Law of Thermodynamics

The first law of thermodynamics states that energy may neither be created nor destroyed in a reaction and hence, the heat of any reaction must be the difference between the heats of formation of the products and those of reactants.

A simple rule used to estimate the energy release or enthalpy (ΔH) from a chemical reaction between propellant ingredients is equal to the differences in the heat of formation (ΔH_f) of products and reactants,

$$\Delta H(T) = \sum n\Delta H_f(T) \text{ products} - \sum n\Delta H_f(T) \text{ reactants} \qquad \dots (8)$$

where ΔH_f is the standard state molar heat of formation and n is the number of moles of each product or reactant at temperature (T).

Thus, standard heats of formation of compounds are useful for calculating heat of reaction under conditions of standard temperature and pressure. Because of the law of conservation of energy, ΔH is the difference between the standard enthalpies of formation of all reactants and products. Using the convention, a large heat release from a reaction is a large negative heat of reaction. This can be achieved by forming products with large negative heats of formation from reactants with small negative or positive heats of formation. The more negative the value of ΔH_f, the more stable the compound is.

Enthalpy changes occur when a compound is formed from elements. In enthalpy measurements, the ΔH_f of a free element in its standard state is arbitrarily set to zero. For example, for the diatomic molecules $H_2(g)$, $N_2(g)$, $O_2(g)$, $Cl_2(g)$, the enthalpy is zero. Similarly, the enthalpy of carbon in the form of graphite is also zero. ΔH_f values for common substances are available in Hand books of Physics and Chemistry. A reaction in which heat is released (negative ΔH) is termed exothermic. In endothermic reaction, heat is absorbed, and ΔH is positive.

Calorimetric Measurements

Heat released or absorbed during chemical reactions can be measured by calorimetry. In calorimetry, the heat released by the system is equal to the heat absorbed by the surroundings. The insulated device that measures the absorption or release of heat in chemical or physical processes is called calorimeter. Calorimetry experiments are performed at constant volume using a device called a bomb calorimeter. It is a closed system ie, the mass of the system is constant. In the bomb calorimeter, a sample is burned in a constant-volume chamber in the presence of oxygen at high pressure. The heat that is released warms the water surrounding the chamber. By measuring the temperature increase of the water, it is possible to calculate the quantity of heat released during the combustion reactions.

Thus, it is evident that to maximize the enthalpy of propellant reaction (ΔH), the heat of formation differences between products and reactants has to be large as seen from equation (8). In addition, the molecular weight of the exhaust products has to be low. This means the heats of formation of the products has to have large negative values compared to the reactants. This implies that the propellant reactants should have small negative heat of formation or small positive heat of formation. For example, hydrogen peroxide was once used in rocket application and can be catalytically decomposed in the following manner:

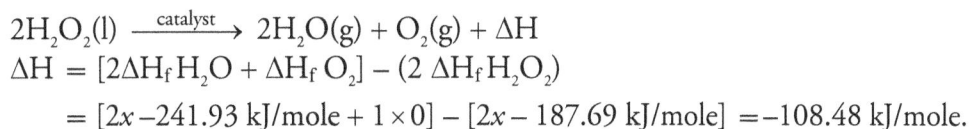

$$2H_2O_2(l) \xrightarrow{\text{catalyst}} 2H_2O(g) + O_2(g) + \Delta H$$
$$\Delta H = [2\Delta H_f\,H_2O + \Delta H_f\,O_2] - (2\,\Delta H_f\,H_2O_2)$$
$$= [2x - 241.93 \text{ kJ/mole} + 1 \times 0] - [2x - 187.69 \text{ kJ/mole}] = -108.48 \text{ kJ/mole}.$$

The catalytic decomposition of hydrogen peroxide give out 108.48 kJ/mole of energy.

Another rapid method to determine the heat of reaction of a fuel is to determine its heat of combustion. For rocket considerations, one must realize that the heat of combustion is that per unit weight of fuel and oxidizer, since the oxidizer has to be carried in the rocket system. Normally, heat of combustion is determined in gaseous oxygen. The combustion energy of elements is defined as the negative heat of formation of un-dissociated product of combustion divided by molecular weight of the product. A periodicity is seen when the heats of combustion of elements are compared in the periodic table. The magnitude of energy peaks diminishes with the atomic number. The first group elements have higher heats of combustion compared to second group elements. Beryllium has highest heat of combustion in the first period and Aluminum has maximum heat of combustion in the second period. Thus, the requirements of high combustion energy and low molecular weight for propellant selection are achieved with lighter elements—H, Be, Al, B and Mg.

The basic principle of rocket propulsion consists in making use of the energy content of certain energetic chemicals in order to generate mechanical energy. This is done by combustion, of fuels and oxidizer to generate large amounts of heat. During combustion, these compounds are converted to combustion products, mostly gaseous products. The heat generated during combustion is directly used to heat up the combustion products in the combustion chamber and the thermal energy of the combustion products is converted into kinetic energy in the nozzle. Thus, heat of combustion is directly relevant for propulsion. Heat of combustion values of many rocket propellant ingredients are available in literature. It can also be determined by experiments in bomb calorimeter. Heats of formation are also essential for rocket performance calculations.

Thermochemical Calculations

Thermochemical calculations are based on two thermochemical laws. The first law states that the molar heat (enthalpy, energy) of decomposition of chemical compound into its elements is equal to negative of molar heat of formation of the compounds from the elements,

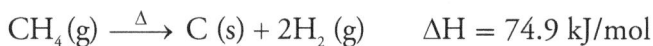

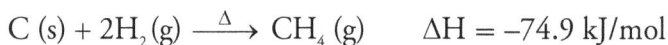

$$CH_4 \text{ (g)} \xrightarrow{\Delta} C \text{ (s)} + 2H_2 \text{ (g)} \qquad \Delta H = 74.9 \text{ kJ/mol}$$

$$C \text{ (s)} + 2H_2 \text{ (g)} \xrightarrow{\Delta} CH_4 \text{ (g)} \qquad \Delta H = -74.9 \text{ kJ/mol}$$

Second Law: Hess's Law

The second law, called Hess's law, states that heat of a chemical reaction can be obtained by the sum of the heats of reaction of several subsequent partial reactions starting with the same reactants and leading to same products.

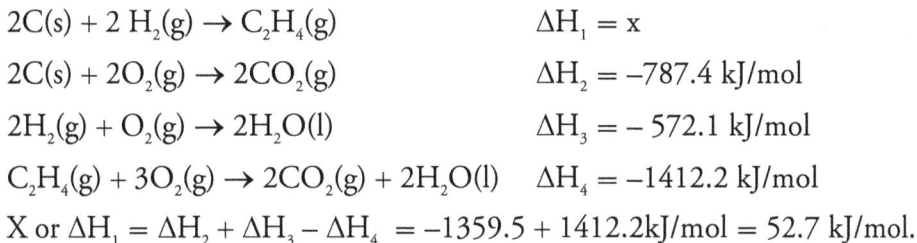

$$2C(s) + 2 H_2(g) \rightarrow C_2H_4(g) \qquad \Delta H_1 = x$$

$$2C(s) + 2O_2(g) \rightarrow 2CO_2(g) \qquad \Delta H_2 = -787.4 \text{ kJ/mol}$$

$$2H_2(g) + O_2(g) \rightarrow 2H_2O(l) \qquad \Delta H_3 = -572.1 \text{ kJ/mol}$$

$$C_2H_4(g) + 3O_2(g) \rightarrow 2CO_2(g) + 2H_2O(l) \quad \Delta H_4 = -1412.2 \text{ kJ/mol}$$

$$X \text{ or } \Delta H_1 = \Delta H_2 + \Delta H_3 - \Delta H_4 = -1359.5 + 1412.2 \text{kJ/mol} = 52.7 \text{ kJ/mol}.$$

If the process is conducted at constant pressure, the heat of formation is called enthalpy of formation (ΔH_f), if it is conducted at constant volume, ΔU_f, the energy of formation is used. Since CO_2 and H_2O are generally the common products, their heat of formation values are key values for thermo chemical calculations of combustion reactions.

Heat of Combustion and Heat of Formation Relationship:

Consider the general reaction

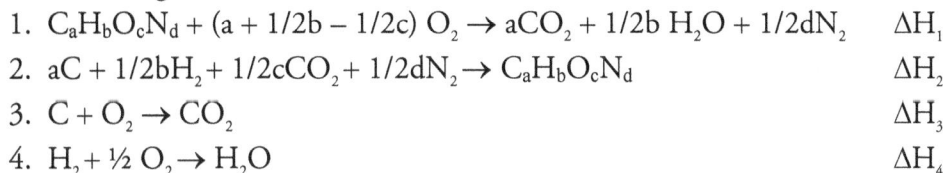

1. $C_aH_bO_cN_d + (a + 1/2b - 1/2c) O_2 \rightarrow aCO_2 + 1/2b H_2O + 1/2dN_2 \qquad \Delta H_1$
2. $aC + 1/2bH_2 + 1/2cCO_2 + 1/2dN_2 \rightarrow C_aH_bO_cN_d \qquad \Delta H_2$
3. $C + O_2 \rightarrow CO_2 \qquad \Delta H_3$
4. $H_2 + \frac{1}{2} O_2 \rightarrow H_2O \qquad \Delta H_4$

Since reaction (1) is a combustion reaction, $\Delta H_1 = \Delta HC_1$ (heat of combustion) and ΔH_2 is the reaction (2) and is equal to heat of formation as it is formed from its elements. Reactions (3) and (4) are formation reactions from its elements. Hence it follows:

$$\Delta H_1 = -\Delta H_2 + a \, \Delta H_3 + \tfrac{1}{2} b \, \Delta H_4$$

or

$$\Delta H_c = -\Delta H_f + a \, \Delta H_c \, CO_2 + \tfrac{1}{2} b \, \Delta H_c H_2O$$

Since, $\Delta H_c C = \Delta H_f CO_2$ and $\Delta H_c H_2 = \Delta H_f H_2O$, it follows:

$$\Delta H_c = -\Delta H_f + a \, \Delta H_f CO_2 + 1/2 \, b \, \Delta H_f H_2O$$

for the heat of combustion per mole of the compound $C_aH_bO_cN_d$. The molecular mass of the compound $C_aH_bO_cN_d$ is:

$$M = \sum a_i A_i = aA_C + bA_H + cA_O + dA_N = 12a + b + 16c + 14d$$

Where A_i = atomic masses of elements.

For the heat of combustion per unit mass ΔH_g is equal to,

$$\Delta H_g = [-\Delta H_f + a\, \Delta H_f\, CO_2 + \tfrac{1}{2} b\, \Delta H_f H_2O]/[12a + b + 16c + 14d]$$

For the heats of combustion per unit volume, ΔH_g has to be multiplied by the density of the fuel,

$$\Delta H_v = \Delta H_g\, \rho$$

Hence, the set of coefficients a, b, c, d are also important apart from high heat of formation to get high specific impulse.

Heat of Formation of HTPB

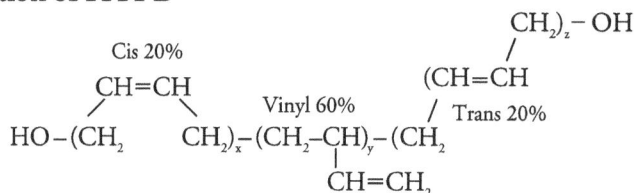

$$HO-(CH_2 \quad CH_2)_x-(CH_2-CH)_y-(CH_2 \quad (CH=CH \quad (CH_2)_z-OH$$

Cis 20% CH=CH Vinyl 60% Trans 20% CH=CH$_2$

Hydroxyl terminated polybutadiene (HTPB) is used as binder in solid propellants and has cis, trans and vinyl micro structures. The experimentally measured heat of combustion $(\Delta Hc) = -116299$ kJ/mol and average molecular weight of the polymer Mn = ~2500,

$$\Delta H_f = -\Delta H_c + a\Delta H_f CO_2 + \tfrac{1}{2}\, b\Delta H_f\, H_2O$$
$$= (116299 - 200 \times 393.7 - 151 \times 286.1)$$

$$\Delta H_f = -5642.1 \text{ kj/mol.}$$

An additional consideration to attain high performance is the extent of dissociation of the product species. As rocket combustion chamber temperature ranges from 3000 to 5000 K, many products dissociate taking away energies in atomic and free radicals. For example,

$$H_2 \rightleftharpoons 2H^{\bullet} \quad \Delta H = 300 \text{ kJ/mol}$$

$$H_2O \rightleftharpoons H^{\bullet} + {}^{\bullet}OH \quad \Delta H = 18 \text{ kJ/mol}$$

The thermal stability of the primary combustion products is therefore as important as the combustion energy. Pressure has significant effect on dissociation. Low pressure increases the degree of dissociation.

Oxidizer-Fuel Energy Balance

The general requirements for a chemical propellant system are a reaction between a high energy oxidizer and a high energy fuel to give a strongly bound, low molecular products. Thus, in an ideal system, both the oxidizer and fuel would have large positive heat of formation and would undergo highly exothermic reaction to give stable products at high combustion temperatures.

The calculation of energy release in the system by usual thermo-chemical methods is tedious, often requiring intricate balancing of chemical equations. The problem becomes more involved when the fuel contains oxidizer and oxidizer contain fuel elements. A simple method involved for the calculation of energy release is the chemical valences of fuel and oxidizer elements present in the propellant composition. The balancing of the combustion equations are required to calculate the stoichiometric composition of fuel-oxidizer mixtures. Also, when the compositions have multiple components, the balancing becomes more involved. In the chemical valence approach, it is assumed that the fuel-oxidizer mixture undergo complete combustion, resulting in balancing of total oxidizing and the reducing valances i.e.,

$$\sum n_o v_o = (-1) \sum n_r v_r$$

where n and v are the number and valence of the element. The subscript o and r refer to the oxidizing (O, Cl, etc.) and the reducing (C, H, etc.) elements respectively. Nitrogen is considered neutral and has no valence. In order to distinguish, the valence of fuel elements is considered as negative and that of the oxidizer element as positive. The right hand side of the equation is multiplied by (−1) to convert the negative sign given to fuel valences.

To illustrate the point, consider the balancing of the equation between methanol and nitric acid,

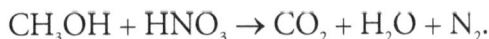

$$CH_3OH + HNO_3 \rightarrow CO_2 + H_2O + N_2.$$

Here, the total valency of the fuel (CH_3OH) = −4 − 4 + 2 = −6 and the total valency of oxidizer (HNO_3) = −1 + 0 + 6 = 5. The oxidizing and the reducing valences of methanol-nitric acid mixture could be made equal by using these numbers as coefficients and thereby balancing the equation as follows:

$$5CH_3OH + 6\ HNO_3 \rightarrow 5CO_2 + 13H_2O + 3N_2$$

This method could be applied to any fuel-oxidizer system containing C, H, O, N, and Cl atoms. Consider a mixture of ammonium perchlorate and glucose combustion. The total valences of NH_4ClO_4 is −4 + 1 + 8 = 5 and the total valences of glucose ($C_6H_{12}O_6$) is −24 − 12 + 12 = −24. Hence, ammonium perchlorate and glucose should be taken in the ratio of 24:5. The balancing equation is,

$$24\ NH_4ClO_4 + 5\ C_6H_{12}O_6 \rightarrow 30\ CO_2 + 24\ HCl + 12\ N_2 + 66\ H_2O$$

To determine whether the contents of combustible mixture is fuel rich or fuel lean, the criterion used is equivalence ratio (Ø), defined as,

$$Ø = Øs/Øm$$

where $Øs$ is the stoichiometric ratio and $Øm$ is the mixture ratio (fuel/oxidizer). A mixture is fuel rich if $Ø < 1$ and fuel lean if $Ø > 1$ and stoichiometrically balanced at $Ø = 1$. This is applicable only for 'pure' fuel-oxidizer systems as the mixture ratio does not take into account the intramolecular 'fuel' and the 'oxidizer' elements present in the oxidizer and fuel molecules, respectively.

In a generalized approach, all the oxidizing and the reducing elements in the composition are treated in a similar manner irrespective of whether they are present in the oxidizer or the fuel components. The new elemental stoichiometric coefficient $Øe$ of a composition is defined as:

$$Øe = p/r = \sum n_o v_o / (-1) \sum n_r v_r$$

where p and r represent the total number of the oxidizing and the reducing elements in the mixture composition. In other words $Øe$ is simply a ratio of the summed up valences of the oxidizing and the fuel elements of the mixture. A mixture is said to be fuel rich if $Øe < 1$, and fuel lean or oxygen rich if $Øe > 1$ and stoichiometrically balanced at $Øe = 1$. Consider for example a combustible mixture of an oxidizer (ammonium perchlorate) and a fuel polymethyl methacrylate ($C_5H_8O_2$) in the ratio of 95:5 by weight. Specific formula of 95 g of ammonium perchlorate:

$$= N1 \times 95/117.5 \ H4 \times 95/117.5 \ Cl \ 1 \times 95/117.5 \ O4 \times 95/117.5$$

$$= N_{0.8085} . H_{3.2340} . Cl_{0.8085} . O_{3.2340}$$

Similarly, the formula for 5 g of methyl metacrylate is

$$= C \ 5 \times 5/100 . H \ 8 \times 5/100 . O \ 2 \times 5/100$$

$$= C_{0.25} . H_{0.4} . O_{0.1}$$

$$Øe = [0.8085 \times 1 + 3.2340 \times 2 + 0.1 \times 2]/(-1) \ (0.25 \times -4)$$
$$+ (0.4 \times -1) + (3.2340 \times -1)$$

$$= 7.4765/4.6340$$

$$= 1.6134.$$

The value of $Øe$ is > 1, indicating the mixture is fuel lean.

Heat of Combustion of Stiochiometric Mixtures

As defined earlier, at stiochiometric composition, $Øe = p/r = 1$ or $p = r = P_o$, where P_o can be calculated as described earlier. Consider the stoichiometric mixture of glucose and ammonium perchlorate as given by the equation,

$$24 \ NH_4ClO_4 + 5 \ C_6H_{12}O_6 \rightarrow 30CO_2 + 24 \ HCl + 66 \ H_2O + 12 \ N_2$$

$Øe$ for the mixture can be calculated simply by using the specific formula. From the above equation, at stoichiometry, the weight ratio of ammonium perchlorate to glucose is 75.8/24.2. Specific formula for 75.8 g of ammonium perchlorate is,

$$= N1 \times 75.8/117.5 \cdot H4 \times 75.8/117.5 \cdot Cl1 \times 75.8/117.5 \cdot O4 \times 75.8/117.5$$

$$= N_{0.6451} H_{2.5804} Cl_{0.6451} O_{2.5804}$$

Similarly the specific formula of 24.2 g of glucose is,

$$= C6 \times 24.2/156 \ H12 \times 24.2/156 \ O6 \times 24.2/156$$

$$= C_{0.8064} \cdot H_{1.6128} \cdot O_{0.8064}$$

The $Øe$ for the mixture containing ammonium perchlorate and glucose in the ratio 75.8:24.2 by using the specific formula as follows:

$$Øe = [0.6451 \times 1 + 2.5804 \times 2 + 0.8064 \times 2]/(-1)$$
$$[2.5804 \times (-1) + 0.8604 \times (-4) + 1.6128 \times (-1)]$$

$$= 7.419/7.419 = 1.$$

Hence, $P_o = 7.419$.

It is expected that depending on the oxidizer-fuel used, various stoichiometrically balance mixtures will have different P_o values. The magnitude of P_o is linearly related to the heat of combustion of the various stoichiometric fuel-oxidizer mixtures, when P_o values of mixtures of various solid oxidizers and organic fuels, having C, H, O, and N atoms, are plotted against their heats of combustion (Q_c). The linear relationship for ammonium perchlorate - organic fuel could be expressed as,

$$Q_c = 4.395 - 0.439 \ P_o$$

From the above relationship, knowing P_o of any ammonium perchlorate-organic fuel mixture, its Q_c can be calculated. Even for metalised systems, such as ammonium perchlorate—aluminum – organic fuel, a linear relationship is found to exist between Q_c and P_o.

The plots of Q_c against P_o shows that the value of Q_c decreases as the P_o value is increased. To understand the observation further, the total heats of formation of the product and reactants with P_o were examined. It is seen that P_o exhibits linear relationships individually with total heats of formation of the products $(\sum\Delta H_f)_p$ as well as reactants $(\sum\Delta H_f)_r$. It is also seen that both $(\sum\Delta H_f)_p$ and $(\sum\Delta H_f)_r$ values increase as P_o values are increased. Since Q_c is the difference between the total heats of formation of the products and reactants, the slopes of the plots $(\sum\Delta H_f)_p$ versus P_o and $(\sum\Delta H_f)_r$ versus P_o decide the slope of the plot between Q_c and P_o. In all ammonium perchlorate oxidizer—organic fuel based systems, it is found that the slope of $(\sum\Delta H_f)_r$ versus P_o plot is much steeper than $(\sum\Delta H_f)_p$ versus P_o plot, and hence Q_c versus P_o showed a downward slope.

As the products of various combustion reactions in ammonium perchlorate—organic fuel are the same viz. CO_2, H_2O, N_2 and HCl, the slopes of $(\sum \Delta H_f)_p$ versus P_o plots of various systems may not vary much. On the other hand, the slope of $(\sum \Delta H_f)_r$ versus P_o curve may vary considerably depending on the reactants used. It therefore, seems clear that the slope of Q_c versus P_o plot is decided almost entirely by the slope of $(\sum \Delta H_f)_r$ versus P_o. This shows the significance of heats of formation of reactants in evaluating the heat of combustion. Although the heats of formation of reactants vary considerably, the relationship between Q_c and P_o is linear. The reason for such a behavior may be due to two factors: i) In the combustion reaction, $(\sum \Delta H_f)_p >> (\sum \Delta H_f)_r$ and ii) the stoichiometric mixture with ammonium perchlorate needs 10 to 25% of any hydrocarbon fuel and hence, the effect of heat of formation of the fuel is very much minimized. This relation between P_o and heats of combustion had been used to calculate the heats of combustion of various fossil fuels.

So far only oxygen balanced systems were considered. The evaluation of energetics of a fuel rich system is not as straight forward. It has been shown that the heat of explosion of various organic nitro-explosives show a trend when plotted against oxygen balance of the explosives. In this class of explosives, the maximum power is obtained when its composition is oxygen balanced. The power of such an explosive is lower when the composition is oxygen-rich or oxygen-lean. Hence, oxygen balance could be related to performance of the explosive. The energetic of a fuel-rich explosive may be dependent upon the term (R – P)/Mw, where Mw is the molecular weight of the explosive, 'R and P total reducing and oxidizing valencies'. Thus, the expression [(R – P)/R × P]/Mw will represent the energy potential of an explosive. This parameter also provides a simple relationship with detonation velocity, impact sensitivity of explosives.

Rocket propulsion systems usually do not operate with the proportion of their oxidizer and fuel in the stoichiometric mixture ratio. Instead, they usually operate fuel –rich because this allows lightweight molecules such as hydrogen to remain unreacted. This reduces the average molecular mass of reaction products, which in turn increase the specific impulse. For this reason, rockets using H_2 and O_2 propellants use the best operating mixture ratio for high performance rocket engines typically between 4.5 and 6.0, not at stoichiometric value of 8.0. For solid propellants, the fuel richness helps in achieving better mechanical properties and lower unloading viscosity for the propellant slurry.

Relationship between Bond Energies and Heat of Formation

The large quantities of heat is liberated when hydrogen atoms combine to form hydrogen molecules. This release suggests that the product is more stable than the reactants. The covalent bond in hydrogen molecule is so strong that it requires

435 kJ of energy to dissociate one mole of hydrogen molecules to hydrogen atoms. This energy, which is required to break the bond between two covalently bonded atoms, is known as bond dissociation energy. A carbon—carbon single bond has bond dissociation energy of about 347 kJ. Compounds only with C–C and C–H single bonds such as methane, ethane etc is quite unreactive chemically, because of the high dissociation energies of the bonds present in the molecules.

In many instances, it is not possible or convenient to determine the heat of formation of a compound experimentally. The compound may not exist or not available and this is quite common in propellant research where heat of formation is required for performance evaluation before synthesis. The proper use of bond energies provides a method of estimating the heat of formation of a compound.

The heat of reaction of a compound depends mainly on the bond energies between the atoms of the compound. This implies that for a good propellant, the propellant ingredients or reactants should consist of molecules with small bond energies, while the combustion products should consist of molecules with large bond energies. Bond energies of covalent bonds between atoms of various elements are available in literature and hand books of physics and chemistry, one can choose molecules based on bond energies. Knowing the bond energies and the atoms present, one can calculate the heat of formation. For example, the heat of formation of ethane and the heat of combustion of ethane is calculated as follows:

In ethane, there is one C–C and 6 C–H bonds.

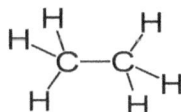

Hence, $\Delta H_f = [\Delta H \, C\text{–}C + 6 \, \Delta H C\text{–}H]$ where ΔH is the bond dissociation energy.

Hence, heat of formation of ethane is $\Delta H_f = [-347.5 - 6 \times 414.5] = -2834.5$ kJ/mol.

(a) *Heat of Combustion of Ethane* ΔH_c

$\Delta H_c = - [(\Delta H \, C\text{–}C + 6 \times \Delta H \, C\text{–}H) - 3DH_2 - 2L_c + 2 \times \Delta H_f CO_2 + 3 \times \Delta H_f H_2O]$ (where DH_2 is the dissociation energy of hydrogen, and L_c is the latent heat of sublimation of carbon)

$= 2834.5 - 3 \times 432.9 - 2 \times 711.8 - 2 \times 393.7 - 3 \times 286.1$

$= 2834.5 - 1298.7 - 1423.6 - 787.4 - 858.3$

$= - 1533.5$ kJ/mol.

There is no strain energy or resonance energy for ethane.

(b) *Heat of Combustion of Cyclopropane, C_3H_6*

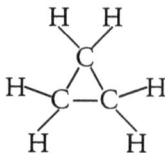

Cyclopropane has 3 C–C and 6 C–H bonds.

$$\Delta H_c = -(3 \times \Delta H\, C\text{–}C + 6 \times \Delta H\, C\text{–}H) - 3 \times DH_2 - 3L_c + 3 \times \Delta H_f CO_2$$
$$+\, 3 \times \Delta H_f H_2O$$
$$= -(-3 \times 347.5 - 6 \times 414.5) - 3 \times 432.9 - 3 \times 711.8 - 3 \times 286.1 - 3 \times 393.7$$
$$= -1944.0\ kJ/mol.$$

The measured heat of combustion is 2092.7. The difference 148.7 kJ/mol is the strain energy.

(c) *Heat of Combustion of Cyclopropene, C_3H_4*

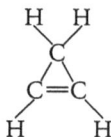

Cyclopropene has 2 × C–C bonds, 1 × C=C bond and 4 × C–H bonds.

ΔHc of cyclopropene,

$$= -(-2 \times \Delta H\, C\text{–}C - \Delta HC = C - 4 \times \Delta HC\text{–}H) - 2 \times DH_2 - 3 \times L_c$$
$$+\, 3\Delta H_f CO_2 + 2 \times \Delta H_f H_2O$$
$$\Delta Hc = 2 \times 347.5 + 611.0 + 4 \times 414.5 - 2 \times 432.9 - 3 \times 711.8 - 3 \times 393.7 - 2 \times 286.1$$
$$= 1790.5\ kJ/mol$$

The measured heat of combustion for cyclopropene is –2030 kJ/mol. Hence, the strain energy is 239.5 kJ/mol.

(d) *Heat of Combustion of Benzene, C_6H_6*

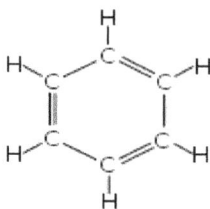

Benzene has 3 C–C bonds, 3 C=C bonds and 6 C–H bonds.

$$\Delta HC = -(3 \times \Delta HC\text{–}C + 3 \times \Delta HC = C + 6 \times \Delta HC\text{–}H) - 3 \times DH_2 - 6\,L_c$$
$$+ 6 \times \Delta H_fCO_2 + 3 \times \Delta H_fH_2O$$
$$= 3 \times 347.5 + 3 \times 611.0 + 6 \times 414.5 - 3 \times 432.9 - 6 \times 711.8 - 6 \times 393.7$$
$$- 3 \times 286.1$$
$$= 1042.5 + 1833.0 + 2487.0 - 1298.7 - 4270.8 - 2362.2 - 858.3$$
$$= -3427.5 \text{ kJ/mol.}$$

The measured heat of combustion is –3269.8 kJ/mol. The difference is the resonance energy 157.7 kJ/mol.

(e) *Heat of Combustion of Cubane,* C_8H_8

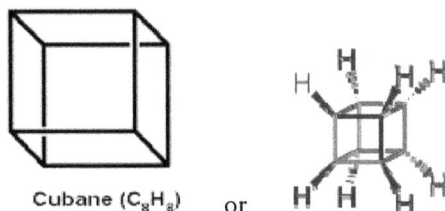

Cubane (C_8H_8)   or

Cubane has 12 C–C bonds and 8 C–H bonds.

$$\Delta HC = (12\,\Delta HC\text{–}C + 8\,\Delta HC\text{–}H) - 4DH_2 - 8L_c + 8\,\Delta HfCO_2 + 4\,\Delta HfH_2O$$
$$= 4170.0 + 3316.0 - 1731.6 - 5694.4 - 3149.6 - 1144.4$$
$$= -4234.0 \text{ kJ/mol.}$$

(f) *Heat Release in Propellant Combustion*

The energy release $\Delta H(T)$ from a chemical reaction between propellant ingredients is given by the net heat of formation of products and the reactants, at a temperature (T) is given by the equation:

$$\Delta H(T) = \{\sum \Delta H_f(T) \text{ products} - \sum \Delta H_f(T) \text{ reactants}\}.$$

This implies that for getting maximum combustion energy, the heat of formation of products are to be large compared to those of reactants. This means, the reactants must have large positive heat of formation and the products with large negative heat of formation must be present in the propellant ingredients. As heat of formation is dependent on the bonds present in various compounds, the reactants should have small bond energies and the products have large bond energies. The favourable bonds in propellant ingredients and the favourable bonds in products are as follows:

Favourable bonds in propellant ingredients,

N–S, O–O, Cl–O, N–N, Cl–N, N–O

Favourable bonds in products,

O–H, H–Cl, N–H, C–O, H–S, C–N, C=O, N≡N, Al–O

Thus, the initial decision on propellant ingredients must be based on a knowledge or at least an estimate of the chemistry and thermodynamics of the system, the main criterion being that the reactants should have an endothermic heat of formation (ΔH_f, positive) and the combustion products should have a high exothermic heat of formation (ΔH_f, negative).

The preferred use of hydrocarbons for binder is a result of the desire for gaseous combustion products (CO_2, CO, H_2O) of relatively low molecular weight. The higher the hydrogen to carbon ratio and the more positive the heat of formation, the better is the binder. Once the decision is made to use hydrocarbon binder, none of the structural variations has significant effect on energetic. Only changes in chemical composition, such as inclusion of double or triple bonds, which have high heats of formation or introduction of nitro groups can increase the energetic. The combustion products of carbon are not among the most exothermic exhaust species possible, but the only practical binder material available today are compounds of carbon. If an element existed with a lower atomic number than carbon and formed a polymeric rubbery material with high heat of formation compounds with hydrogen, it would be the best choice.

With the highly oxygenated oxidizers like nitronium perchlorate (NO_2ClO_4), the theoretical specific impulse is greatly affected by the ratio of oxygen to carbon and hydrogen in the binder. Hence, polyesters and polyethers which carry large amount of carbon-oxygen bonds give less energy compared to hydrocarbons. With conventional propellant oxidizers like NH_4ClO_4 and NH_4NO_3, this is less of a problem. When bond energies are used for estimation of heats of formation, the effect of molecular environment on the character of a particular bond should be considered. Figure 1 shows the specific impulse comparison of various binders in

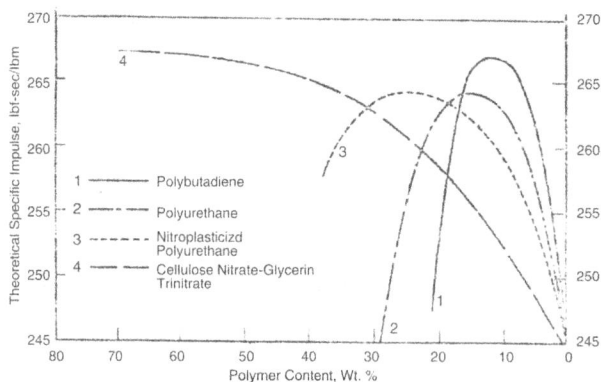

Fig. 1: Effect of Binder Type and Percentage on I_{sp} of AP-AI Propellants

propellant systems with aluminum and ammonium perchlorate. As we move from pure fuel binder polybutadiene to binders containing increasing amounts of oxygen, peak specific impulse is reached at progressively lower solid loadings.

Polybutadiene is a better fuel binder than polyurethane, showing a peak specific impulse of approximately three units higher. However, the solid loading required to reach the peak impulse imposes problems in propellant processing. When plasticizers with high oxygen content, like nitroplasticizers, are added to polyureathane binders, the peak impulse does not change but is achieved at a lower solid loading. The disadvantage of propellant systems with peak impulse at lower solid loading is lower density of propellant and consequently lower impulse/unit volume. In practice, the mechanical properties of solid propellant improves as the ratio of binder to oxidizer increases. However, in most systems, peak energetic occur at 9 to 11% by weight binder level whereas minimum acceptable physical properties are achieved at 14 to 16% binder level. The importance of reliable mechanical properties can be illustrated by showing that most operational systems like PSLV's solid motors, Space shuttle and Arian strap on boosters use 86% solid loaded propellants, sacrificing energy for mechanical properties.

Contribution of Metallic Fuel

A major advancement occurred when metallic fuel like aluminum is added to the binder-oxidizer mixture to give higher energy as well as higher density propellants without affecting the mechanical properties of the system (i.e., at allowable binder level). The situation can be understood as the hydrocarbon-oxidizer system is balanced into to give carbon monoxide, carbon dioxide and steam as combustion products. The metallic additive can be considered to be oxidized by steam and hence, does not require additional oxidizer. When aluminum is added to hydrocarbon - ammonium perchlorate system, the combustion proceeds as follows:

$$NH_4ClO_4 + -(CH_2)_x^- + 2Al \rightarrow \frac{1}{2} N_2 + CO + 5/2 H_2 + Al_2O_3 + HCl.$$

The performance of a propellant system has been shown to be proportional to $\sqrt{(T_c/ \bar{M})}$, where T_c is flame temperature and $\bar{M}$ is the average molecular weight of combustion products. Although introducing aluminum as a reactant results in the formation of higher molecular weight products (Al_2O_3), the formation of low molecular weight hydrogen gas (aluminium-steam reaction) in the combustion process offsets this. Also, the combustion of metallic aluminum fuel gives rise to very high flame temperature.

In fuel systems containing aluminium as a reactant, the performance improves as the percentage of aluminium increases to an optimum of about 22% aluminum. The peak performance does not require lower binder levels although the percentage by weight of total fuel has increased as a result of increasing aluminum level. The peak performance occurs over a flatter region of binder content. This is important because it means that improved mechanical properties can be achieved by increasing the binder content without an appreciable fall in the theoretical performance. In actual practice, peak energetic are not achieved when propellants containing optimum percentage of aluminum are burned in rocket motor. This is because of incomplete combustion as well as lag which occur in transferring energy from the metal oxide particles to the gas stream during the expansion process in the nozzle.

The specific impulse of a propellant as function of chemical parameters can be calculated according to:

$$I_{sp} = K \sqrt{n} \, T_c = K' \sqrt{(T_c / \bar{M})}$$

Where T_c is the flame temperature, n is the number of moles of gas produced per unit weight of propellant. $\bar{M}$ is the average molecular weight of gaseous combustion products and K and K' are constants.

Binder-Al-AP Effect on I_{sp}

A high specific impulse can be achieved by increasing n as well as T_c and decreasing $\bar{M}$. The binder mainly contributes to n, which in turn is favoured by maximum hydrogen content in the prepolymer. On the other hand, T_c increases with more exothermic ΔH and low heat capacities of combustion products per unit weight. ΔH becomes more exothermic with more positive heats of formation of propellant ingredients and more negative heats of formation for its combustion products. Thus, polymeric binders with more positive heats of formation are preferred. The contributors to this are unsaturation and functional groups with carbon-nitrogen and nitrogen-nitrogen bonds as in nitrile, amine oxide, hydrazine, nitramine, etc. Carbon-oxygen bonds as in carbonyl, carboxyl, hydroxyl, ether and esters, carbon-halogen bonds and carbon-sulphur bonds are not desirable because of their low energy contribution.

In combination with aluminum (high ΔH but no gaseous combustion products) and oxidizer (NH_4ClO_4), there exists a composition with maximum specific impulse. A triangular composition diagram (Figure 2) for a propellant consisting of Al/AP/polybutadiene shows lines of constant n (moles of gas/100 g of propellant) and contours of constant flame temperature. The diagonal solid line across the triangle is a line of constant chemical stoichiometry, representing all propellant compositions in which the amount of oxygen present exactly that required to form CO and Al_2O_3 (no formation of CO_2, H_2 or other oxygen compounds). As the binder content is decreased along either a line of constant oxidizer or constant Al content, flame temperature increases while the number of moles of gas decreases. In the triangular diagram (Figure 2) for the same system showing contours of constant specific impulse, the maximum specific impulse occurs near the line at some point between 15 and 10% binder (85 to 90% solids).

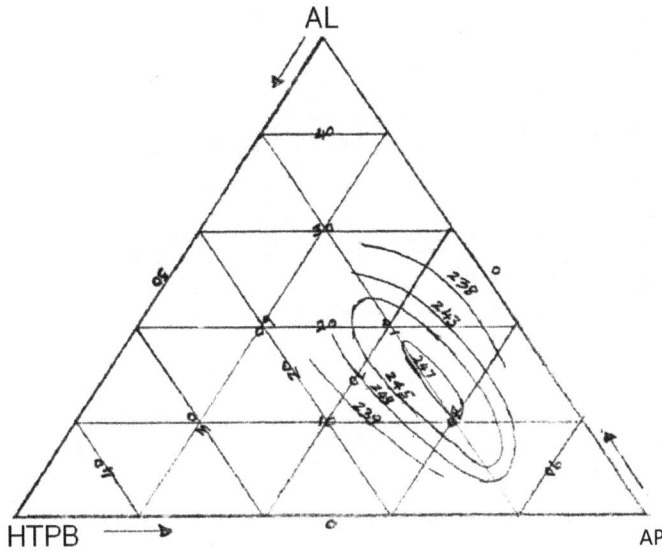

Fig. 2: Triangular Diagram of Expected I_{sp} for AP-Al-HTPB Propellants

3. NOZZLES

The purpose of Nozzle is to expand the hot propellant combustion gases from the high pressure combustion chamber to or near the external ambient pressure, thereby converting thermal energy into directed kinetic energy or thrust. The flow of combustion gases as they are expanded through the nozzle is assumed to be isentropic flow i.e. adiabatic and reversible. The nozzle of the rocket motor provides for the expansion and acceleration of the hot gases from propellant combustion in the chamber. It has to withstand the severe environment of high heat transfer and erosion. Optimum expansion ratio of combustion gases to the outside pressure becomes a prime requirement and effective exhaust velocity is considered a true measure of propulsion system performance. A rocket engine nozzle is usually a convergent-divergent De-level type and is used to expand and accelerate the combustion gases produced by burning propellants. The nozzle performance is maximized by complete exhaust expansion to ambient pressure. In space applications, the ambient pressure is continuously varying with altitude or zero, outside the earth atmosphere. Nozzle throat diameter size ranges in size from 2 mm to 160 cm, with operation durations of a fraction of a second to several minutes.

All solid rocket nozzles are ablatively cooled. The general construction of a solid rocket nozzle features steel or aluminum housings designed to carry the structural loads and ablative liners which are bonded to the housings. The ablative liners are designed to protect the metallic housings, provide the internal aerodynamic contour necessary to efficiently expand combustion gases to generate thrust and to ablate and char in a controlled manner to prevent the build-up of heat which could damage or weaken the bonding materials. The construction of nozzles ranges from simple single-piece non-movable graphite nozzles to multi-piece nozzles capable of moving to control the direction of the thrust vector. The simpler, smaller nozzles are for applications with low chamber pressure, short durations, low area ratios and or low thrust. Complex nozzles are usually for meeting requirements such as providing thrust vector controls, operating at high chamber pressures and/or higher altitudes, producing very high thrust levels and surviving longer motor burn durations.

Nozzle Design

From a performance angle, the nozzle design is to efficiently expand gas flow from the motor combustion chamber to produce thrust. Simple nozzles with non-contoured conical exit cones can be designed using basic thermodynamic relationships to determine throat area, nozzle half angle, and expansion ratio. A more complex contoured nozzle like the bell-shaped ones is used to reduce the divergence loss, slightly improve the specific impulse and reduce nozzle length and mass. Nozzle throat erosion results as a result of impingement of Al_2O_3 particles travelling at high velocity along with the combustion gases. This impact of particles mechanically removes the charred liner material. The radial throat erosion is also due to the carbon liner material reacting chemically with oxidizing species in the combustion gas flow. The nozzle throat erosion causes the throat diameter to enlarge during operation and is one of the major problems encountered in the design. A throat area increase of >5% is not acceptable for solid motor applications because it causes reduction in thrust and chamber pressure. Erosion occurs not only at the throat region but also at sections very near upstream and downstream of the throat region. The selection and application of the proper material is the key to the successful design of nozzles.

Initially nozzles were made out of a single piece of moulded polycrystalline graphite and some were supported by metallic housings. We still use them for short duration, low chamber pressure and low altitude flights and tactical missiles. For more severe conditions, a throat insert is placed into the graphite piece. This insert is a denser, better grade of graphite. Later pyrolytic graphite washers and fibre-reinforced carbon materials were used. For a period of time tungsten inserts were used because of good erosion resistance. The high strength carbon fibre and carbon matrix are used for small and medium nozzles. The orientation of fibers can be Two Dimensional (2D) or Three Dimensional (3D). For larger nozzles, layups of carbon fiber or silicon fiber cloth in a phenolic matrix or 3D or 4D carbon-carbon matrix after graphitization are used.

Assumptions

An ideal nozzle assumes one-dimensional steady isentropic flow for which the following assumptions are valid:

1. The working substance (product of combustion gases) is homogenous and invariant composition throughout the rocket chamber and nozzle.
2. The working fluids are gaseous substances and always obey the perfect gas laws.
3. There is no friction.
4. There is no heat transfer across the nozzle wall and hence the flow is adiabatic.
5. The propellant gas flow is steady and consistent. No shock and vibration.
6. Chemical equilibrium established in rocket chamber should not shift in the nozzle.
7. The gas velocity, pressure, temperature or density is uniform across any section normal to the axis.

Except for very small chambers, the loss of heat to the walls of the rocket is typically less than 1 to 2% of the total energy and hence, can be neglected. Short-term fluctuations of the steady propellant flow rate and pressure are usually less than 5% of the rated value, their effect on rocket performance is small and can be neglected. In well-designed supersonic nozzles, the conversion of thermal energy into directed kinetic energy of the exhaust gases proceeds smoothly and without normal shocks or discontinuities. Thus, the flow expansion losses are generally small.

For a given combustion chamber pressure P_c and mass flow rate $\mathring{m}$, the motor thrust can be optimized. The rocket thrust is given by:

$$F = \mathring{m}\, V_e + A_e(P_c - P_e) + F_{ext}$$

where $\mathring{m}$ is propellant mass flow rate (kg/sec), V_e is the exhaust verlocity (m/sec), A_e $(P_c - P_e)$ is the resultant force on the rocket due to the pressure difference between nozzle exit and ambient, and F_{ext} denotes the extra force in the direction of motion due to external forces like gravity or aerodynamic drag. The nozzle performance is maximized by complete exhaust expansion to ambient pressure, when $(P_e - P_a) = 0$. An under-expanding nozzle discharges the fluid at a pressure greater than ambient pressure. It has too small exit area. Shocks are generated outside the divergent section. In this case, $P_e > P_a$. In an over expanding nozzle, the fluid is discharged to a lower pressure than the ambient pressure. It has too large exit area. Shocks are generated in the divergent section of the nozzle. In this case, $P_e < P_a$.

The nozzle used in rocket motor is considered as circular for easy manufacture. Simple convergent—divergent are often found acceptable. Divergence half angle between 12 and 18° and convergence half angle between 30 and 45° usually provides the best performance. The thrust of ideal rocket nozzle is reduced by nozzle correction factor (λ) due to non-axiality of the exhaust gases.

$$V_{ex} = \lambda V_2 \text{ and } \lambda = 1 + (\cos \alpha)/2$$

Where V_{ex} is the axial component of gas exit velocity, λ is the nozzle angle correction factor, V_2 is velocity of combustion gases at exit plane.

Nozzle Design Parameters

While designing the nozzle, following points are considered for better performance:
1. Higher exhaust velocity
2. Low heat losses in the nozzle wall using suitable materials
3. Low weight
4. Low frictional losses by making smooth surface
5. Lower length of nozzle
6. Optimum expansion.

The minimum nozzle area is called the throat area. The ratio of the nozzle exit area A_e to the throat area A_t is called the nozzle area expansion and is designated by the Greek letter ε. It is an important nozzle design parameter,

$$\epsilon = A_e/A_t$$

The converging nozzle section between chamber and the nozzle throat contributes little in achieving high performance. The subsonic flow in this section can be easily turned at very low pressure drop and any radius, cone angle, wall contour curve or nozzle inlet shape is satisfactory. The throat contour also is not very critical to performance, and any radius or other curve is usually acceptable. The pressure gradients are high in these two regions and the flow will adhere to the walls. The principal difference in different nozzle configurations is found in the diverging supersonic-flow section. The wall surface throughout the nozzle should be smooth and shiny to reduce friction, radiation absorption, and convective heat transfer due to surface roughness.

When the chamber has a cross section that is larger than about four times the throat area, the chamber velocity can be neglected. However, grain design considerations lead to small void volumes or small perforations of port areas for solid propellants. Under these conditions, the chamber velocity cannot be neglected as a contribution to the performance. The gases in the chamber expand as heat is being added. The energy necessary to accelerate the gases within the chamber will cause a pressure drop and additional energy loss. The acceleration process in the chamber is adiabatic but not isentropic. This loss is maximum when the chamber diameter is equal to the nozzle diameter, which means there is no converging nozzle section. This has been called a throat less rocket motor and has been used in a few tactical missiles. This performance improvement due to inert mass savings outweighs the nozzle performance loss in these motors.

Propellant Grain and Grain Configuration

The grain is the shaped mass of processed solid propellant inside the rocket motor. The propellant material and geometrical configuration of the grain determine the motor performance characteristics. The propellant grain is cast, moulded or extruded body and its appearance and feel is similar to that of hard rubber or plastic. Once ignited, it will burn on all its exposed surfaces to form hot gases that are then exhausted through the nozzle. Most rockets have single grain. However, a few rocket motors have more than one grain inside a single case or chamber and big motors have propellant segments or segments made of same composition or different propellant compositions.

There are two ways of holding the grain in the chamber—cartridge loaded or free standing grains and case bonded grains. Cartridge—loaded grains are manufactured separately from the case, by extrusion or by casting into cylindrical mold or cartridge, and then assembled or loaded into the case. In case bonded grains, the case is used as a mould and the propellant is cast directly into the case and is bonded to the case or insulation. Free standing grains can more easily be replaced if the propellant grain is aged excessively. Cartridge loaded grains are used in some small tactical missiles and some medium sized motors. They are easier to inspect and have a lower cost. The case bonded grains give slightly higher performance, a little less inert mass (no holding device or support pads and less insulation), a better volumetric loading fraction, and are more highly stressed, and somewhat difficult and expensive to manufacture. Almost all large motors and many tactical missiles are made by case bonding technique.

Grain Requirements

A grain has to satisfy several connected requirements:
1. The flight mission determines the rocket motor requirements. This includes the total impulse, a desired thrust time curve and the tolerance on it, motor mass, ambient temperature limits during storage and flight operation, available vehicle volume and vehicle accelerations caused by vehicle forces.
2. Grain geometry is selected to fit the above requirements, taking into use of available volume and have appropriate burn surface to generate the required thrust time profile.
3. The selection of propellant is done based on performance capability, mechanical properties, ballistic properties, manufacturing characteristics, exhaust plume characteristics and aging characteristics.
4. The structural integrity of the grain including the liner/insulator must be safe to assure that the grain will not fail under the all conditions of loading, acceleration and thermal stresses.

5. As port volume increases due to burning, it should be ensured there is no combustion instability.
6. The processing of propellant grain should be simple and cost less.

The grain configuration is designed to meet most of the requirements, but sometimes the above categories are satisfied only partially. The geometry is critical in grain design. For approximately constant thrust (neutral burning), the burning surface area has to be approximately constant and for regressive burning grain the burning area will reduce during the burning time. The change of burning surface area with time has a strong influence on chamber pressure and thrust and there is a trade-off between burning rate and burning surface area. Since the density of most modern propellants are within a narrow range, there is little influence on grain design.

Grain Configurations

Many grain configurations are available for motor designers. The number of configurations decreased as methods to increase the burning rate are evolved. Current configuration designs concentrate on relatively few of them since the needs of a wide variety of solid rocket applications can be fulfilled by combining known configurations or by slightly altering a classical configuration.

Various Grain Configurations

1. End burning—case bonded—neutral burn
2. Internal burning tube—progressive burn
3. Radial grooves and tube/slots and tube—neutral burn
4. Star (conocyl)—neutral
5. Wagon wheel—neutral
6. Multi-perforated tune—progressive regressive
7. Dog bone—case bonded
8. Dendrite—case bonded.

The effect of propellant burning on surface area is readily apparent for simple geometric shapes such as rods, tubes, wedges and slots. Certain other basic surface shapes burn as follows: external burning rod—regressive, external burning wedge—regressive. Most propellant grains combine two or more of these basic surfaces to obtain the desired burning characteristics. Configurations that combine both radial and longitudinal burning as in internal-external burning tube without restricted ends are frequently called "three dimensional grains," although all grains are geometrically three-dimensional. A typical progressive, regressive and neutral burning pattern of solid motors is shown in Figure 1 of Chapter 6 "Internal ballistics of rockets."

Grain configurations can be classified according to their web fraction (bf), their Length to Diameter Ratio (L/D) and their Volumetric Fraction (Vf). Web thickness, in case of a case bonded grain is the minimum thickness of the grain from the initial burning surface to the insulated case-wall or to the intersection of another burning surface. For end burning grain, it is the grain length. Web fraction for case bonded internal burning grain is the ratio of web thickness (b) to the outer radius of the grain (b/radius of the grain). Volumetric fraction is ratio of propellant volume to the chamber volume (excluding nozzle) available for propellant, insulation and inhibitors or restrictors. The end burning grain (burning like a cigarette) has the highest volumetric loading fraction, the lowest grain cavity volume for a given impulse, relatively low burning surface area/thrust and long burning time. The internal burning tube is easy to make and is neutral burning with unrestricted ends of L/D = 2.

The star configuration is suitable for web fractions 0.3 to 0.4 and it is progressive above 0.4. Wagon wheel is better to the star shape around 0.3 web fraction. Dendrites are used in the lowest web fraction when a relatively large burning surface area is needed (high thrust and short duration). Rockets used in missiles, weather rockets and other tactical applications benefit from reducing the thrust with burn time. A high thrust is desired to give initial acceleration, but as propellant is consumed and vehicle mass is reduced, a decrease in thrust is desirable. Hence, there is a benefit to vehicle mass, flight performance and cost in having a higher initial thrust in the boost phase, followed by a lower thrust during the sustaining phase of a flight.

Any remaining unburnt propellant left after the burning of the web is known as slivers. This useless material caused a reduction in propellant mass fraction and vehicle mass ratio. The technology of grain design has reduced the slivers from 2 to 7% to less than 1%. If slivers were to occur, the designer will replace the sliver volume with lower-density insulator, which gives less of a mass ratio penalty.

Failure Modes

Solid propellant is a viscoelastic material. This viscoelastic nature is time—history dependent and the material accumulates the damage from repeated stresses, known as cumulative damage phenomenon. The most common modes of failure are: i) surface cracks formed when the surface strain is excessive. They open up new additional burning surfaces and this in turn causes the chamber pressure as well as thrust to increase. With many cracks or deep cracks, the case becomes over-pressurised and will fail. At a high strain rate, deeper and more highly branched cracks are more readily formed than at a lower strain rate. The limiting strain depends on the stress level, grain geometry, temperature, propellant age, load history and sizes of flaws and voids, ii) The bond at the grain periphery is broken and an unbounded area or gap can form next to liner, insulator or case. As grain surface regresses, a part of the unbounded area will be exposed to the hot, high pressure combustion gases and suddenly the burning area is increased by the unbounded area.

Other failure modes like excessively high ambient grain temperature causing a large reduction in the physical strength properties, ultimately result in grain cracks and/or debonding. Air bubbles, porosity or non uniform density can reduce the propellant strength to cause failure. Other failure modes are excessive deformations of the grain (slump) and involuntary ignition due to the heat absorbed by the viscoelastic propellant from excessive mechanical vibration.

Performance Efficiency of a Rocket Motor

Performance efficiency of a rocket motor is defined as follows:

$$\eta = I_{sp} \text{ (measured)}/I_{sp} \text{ (theoretical)} = 1 - \sum L$$

Here, η is the efficiency. The two specific impulses values must be for the same expansion conditions and I_{sp} (theoretical) will be calculated for equilibrium flow conditions. The term $\sum L$ is the sum of all losses as fractions of theoretical performance. The propellant test motor hardware is massive so that it can be used over and over again. Hence, low performance efficiencies are attributed to excessive heat losses. The major possible performance losses in rocket motors are the following:

1. *CL*, combustion loss is the fractional loss due to incomplete combustion or incomplete reaction between propellant ingredients.
2. *EL*, equilibrium loss is due to kinetic lag behind shifting equilibrium during expansion, including deviation from ideal gas behavior.
3. *HL*, heat loss is due to friction, heat transfer to chamber and nozzle hardware and energy used up in the ablation of insulation and nozzle material.
4. *VL*, vector loss is the reduction in the axial component of exhaust gas thrust because of divergent angle of nozzle exit cone. The vector loss is calculated directly from geometry.
5. *TPL*, two phase losses is due to presence of metallic oxides (Al_2O_3) in the exhaust in metalized propellants. The two phase losses are directly proportional to mass of condensed (solid or liquid) phase in the exhaust. The two phase loss can be reduced or eliminated from combustion of metalized propellants by producing smaller oxide particles. A ten fold size reduction would reduce the lag effects to a negligible level.

$$\sum L = CL + EL + HL + VL + TPL$$

In addition to the estimation of the losses, propellant density affects the propellant mass fraction. The only effect of density is on the volume of the motor or in the case of volume—limited systems on the mass of the propellant that can be loaded. Propellant mass fraction is important in the propellant development plans. In case, the propellant is not amenable for case bonding due to mechanical properties, such a propellant formulation must offer a large increase in specific impulse to compensate the low mass fraction (cartage loading). This calls for extra insulation thickness. Higher flame temperature calls for higher insulation thickness both in the nozzle and the combustion chamber. Higher flame temperature consumes more insulation due to ablation of the materials and hence, requires greater thickness. Similarly, motor inert mass requirement also is affected by higher flame temperature. Zirconia coated steel nozzles work for non aluminized propellants but not for metalized propellants.

Burning rate of the propellant also affects performance, by way of weight penalty. Some propellants burn efficiently only at high pressures and to get good specific impulse, it has to be operated at high pressures. This calls for thicker chamber walls, reducing the mass fraction and the reduction in efficiency. When one substitutes a fast burning high energy propellant for a slow burning propellant, the chamber pressure will be higher for the fast burning one compared to the slow burning, and also in the nozzle at each corresponding station. Hence, both chamber and nozzle will be thicker resulting in lower mass fraction. The chamber pressure can be brought back to the original level by enlarging the throat sufficiently. This, however, results in considerable increase in inert mass because the whole nozzle gets bigger. Another problem with higher burning rate is higher pressure exponent, n. The burning rate of high 'n' propellants are more sensitive to bulk grain temperature and since chamber pressure depends on burning rate, the motors containing these propellants are heavier to withstand the resulting higher pressures.

Combustion Processes in Rocket Propellants

The combustion process in rocket propulsion systems is very efficient compared to other power plants since the combustion temperatures are very high. This high temperature accelerates the rate of chemical reaction to go to nearly complete combustion. The energy released in the combustion is between 95 to 99.5%. This is difficult to improve as the combustion of propellant involves highly complex reactions in the heterogeneous mixture. The burning rate of all propellants is influenced by pressure, the initial ambient temperature of the solid propellant, the burn rate catalyst, aluminum and ammonium perchlorate particle size and distribution and to a lesser extent by other ingredients and manufacturing process variable. Experimental observations of burning propellants show complicated three-dimensional microstructures, a three dimensional flame structure, intermediate products in liquid and gaseous phase, spatially and temporally variant processes, aluminum agglomeration, formation of carbon particles and other complexities are yet to be understood and reflected in mathematical models. Some insight into combustion process can be gained by understanding the behavior of major ingredients such as ammonium perchlorate, aluminum powder and polymeric binder.

Double Base Propellants

Visual observations and measurements of flames in simple experiments like in strand burner tests give an insight into combustion processes. For double base propellants, the combustion flame structure appears to be homogeneous and one dimensional along the burning direction. When the heat from the combustion melts, decomposes and vapourizes the solid propellant at the burning surface, the resulting gases seem to be already premixed and the radiating bright flame zone can be seen where most of the chemical reactions is believed to occur and a dark zone between the bright flame and the burning surface. The bright radiating hot reaction zone seems to be detached from the combustion surface. The combustion that occurs inside the dark zone does not emit strong radiations in the visible spectrum, but does emit in the infrared region. The dark zone thickness decreases with increasing chamber pressure and higher heat transfer to the burning surface causes the burning rate to increase. The overall length of visible flame becomes shorter as the chamber pressure increases and heat release per unit volume near the surface also increases.

In the combustion zone or the bright, thin fizz zone directly over the burning surface of the double base propellant, some burning and heat release occurs. Beneath is a zone of liquefied bubbling propellant which is thought to be very thin and has been called the foam or degradation zone. Here, the temperature is high enough for the propellant molecules to vapourise and break up or degrade into smaller molecules, such as NO_2, aldehydes or NO which leave the foaming surface. Below this, is the solid propellant, but the layer next to the surface has been heated by conduction within the solid propellant.

Burning rate catalysts seem to affect the primary combustion zone rather than the processes in the condensed phase. They catalyse the reaction at or near the surface, increase or decrease the heat input into the surface, and change the amount of propellant that is burned.

Composite Propellants

The flame structure of a typical composite propellant (AP/Al/HTPB) looks very different. Here, the luminous flame seems to be attached to the burning surface, even at low pressures. There is no dark zone. The oxidizer-rich decomposed gases from the AP diffuse into fuel-rich decomposed gases from the fuel ingredients, and vice versa. Some solid particles like Al, AP crystals, small pieces of binder or combinations of these break loose from the propellant surface and the particles continue to react and degrade while in the gas flow. The burning gas contains liquid particles of hot aluminum oxides, which radiate intensively. The propellant material and the burning surface are not homogeneous. The flame structure is unsteady, three dimensional, not truly axisymmetrical, and complex. Thus, the flame structure of AP composite propellants is dependent on the physical and chemical properties of ingredients mixed within propellants, such as concentration of binder, size of AP and the presence of burning rate catalysts. The temperature increases smoothly from initial propellant temperature (T_o) to burning surface temperature (T_s) and rapidly from T_s to flame temperature (T_f).

Combustion Models

Of the number of combustion models for AP based composite propellants, the Granular Diffusion Flame Model (GDF) of Summerfield is the oldest. It is based on the assumption that fuel and oxidizer gasify at the propellant surface. These gases leave the surface in pockets that diffuse together. This model, though popular, cannot explain the variation in burning rates with pressure, solid loading and oxidizer particle size. As per the model, the burning rate can be calculated by the equation $1/r = a + bp^{2/3}$, where a and b are constants referred as chemical reaction and diffusion time parameters respectively. Hermance proposed a heterogeneous model at the burning surface between oxidizer and binder, producing increased surface area. The vapour phase flame, which transfers the heat to the condensed phase, gets closer to the surface as the AP particle size is reduced, leading to higher heat transfer and to increased burning rate.

The Beckstead, Derr and Price proposed a multiple flame model, called BDP model in early seventies. According to this model, a complex interaction between oxidizer mono-propellant and fuel decomposition products and another between the oxidizer-rich decomposed gases from AP and the fuel rich combustion products from the first diffusion flame takes place above the oxidizer-binder interface. As per BDP model, burning rate is dependent on particle size of oxidizer due to primary diffusion flame. For very small particles, primary diffusion flame reduces and kinetic aspect becomes dominant. For large particles, the reverse phenomenon is dominating.

CMDB Propellants

The flame structure and burning rates of composite modified cast double base propellants (CMDB) with AP and Al approach those of composite propellant, particularly when AP content is high. There is no dark zone and flame structure is unsteady. The flame structure for double base propellant with nitramine shows a thin dark zone and a luminous degradation zone on the burning surface. The dark zone decreases with increase in pressure. The decomposed gases of HMX/RDX are essentially neutral when decomposed as pure ingredients. In this CMDB propellant containing RDX, the degradation products of RDX interdiffuse with the gas from double base matrix just above burning surface before the RDX monopropellant flamelets appear. Thus, an essentially homogeneous premixed gas flame is formed from the solid propellant which is heterogeneous. The burning of this propellant decreases with increase in RDX and unaffected by the change in RDX particle size.

4. IGNITER AND IGNITER PROPELLANTS

Non-spontaneously ignitable propellants need to be activated by absorbing energy prior to combustion initiation. The energy is supplied by ignition system. Igniter ensures motor ignition in a controlled and reproducible manner such that the ignition sequence is completed within the specified interval and without leading to excessive motor pressure spikes. Once ignition has begun, the flame is self supporting. In composite propellants based on AP, endothermic sublimation of AP is followed by exothermic decomposition which leads to onset of ignition and it occurs either in the gaseous phase in front of the propellant surface or through gas-solid reaction at the surface of the solid propellant. Hence, the ignition is dependent on:

1. the rate of heat flux,
2. gaseous composition in front of propellant surface,
3. gas pressure above the propellant surface,
4. surface conditions and presence of AP particles, and
5. gas velocity across the propellant surface.

A typical motor pressure-time history is shown in Figure 4 of Chapter 6, "Internal ballistics of rockets. The ignition transient has three phases. The ignition lag time (phase I) is the period from initiation of igniter to the onset of motor propellant ignition, flame spreading interval (phase II) covers the time required from first propellant ignition until the grain surface is completely ignited and chamber filling interval (phase III) is the time required to reach the equilibrium chamber pressure.

The normal ignition of a motor is characterized by:

1. Ignition peak not to exceed the Maximum Expected Operating Pressure (MEOP) of the motor,
2. Ignition delay, the time lapse from the firing pulse to the chamber pressure attaining 10% of motor pressure, should be within acceptable limit,
3. Ignition interval, time from firing pulse to the time of attaining 90% of the motor peak pressure, should meet the time specification fixed from mission considerations, and
4. Rate of pressure rise (dp/dt) in the motor chamber during ignition transient should not be large to induce a shock which could damage the grain.

Igniter Requirements

In addition, the general requirement of an igniter includes:

1. Rate of heat generation is more than the rate of loss of heat to the surroundings.
2. Total heat flux of igniter should be sufficient enough to raise the surface temperature of propellant from ambient to its auto ignition temperature.
3. Should give desired performance under all specified conditions.
4. Should be compatible with motor and its propellant.
5. Container dimensions should meet the igniter charge mass, size, weight and interface requirements.
6. During combustion, fragments of container should not damage the propellant or choke the nozzle.
7. Should have long storage life and easy to produce and assemble.

Constituents of an Igniter Motor

The main constituents of a solid motor igniter can be classified as: a) an initiation system, and b) an energy release system. In the initiation system, an electrical stimulus is converted to thermal output which ignites the energy release system. In the initiation system sometimes a safety mechanism is also introduced depending on the mission requirement. The energy release system supplies heat energy to ignite the propellant and induces chamber pressurization to ensure sustained, smooth combustion. The initiation system will have an electrical initiator (squib) and a relay charge which together constitute an ignition cartridge. The conventional squib uses a small nichrome wire (0.012 to 0.0125 mm diameter) attached to two electrical terminals. On passing current, the wire gets heated up, because of resistive heating, to cause deflagration of heat sensitive charge in contact with it. The electrical characteristics of a squib are stipulated by fixing 'no fire' and 'all fire' current levels. 'No fire' current is the maximum current that could be applied to the circuit without firing the unit. The 'no fire' current is to provide a safe current level which could be passed through them to check the bridge-wire continuity. 'All fire' is the current required to consistently fire the squib.

Generally the device should have adequately large margin of safety against premature initiation due to induced emf from power lines and radio frequency sources in the vicinity. The functional characteristics to firing current response could be adjusted by controlling the heat lost to the surrounding hardware components, selecting suitable material and size of wires and forming squib charge materials having lower or higher auto ignition temperature. To keep the functional delay of squibs and its dispersion to the application of 'all fire' current to a minimum, a higher current called 'recommended fire current' is always specified.

The first flash generated by a squib is picked up by a delay charge which is an intimate mixture of pyrophoric metal powders like boron, magnesium, aluminum and strong oxidizers like potassium nitrate, potassium perchlorate, etc. In igniting large quantities of propellant as in booster motors of launch vehicles, a safety mechanism (S/A) is introduced in the initiation system. This prevents ignition propagation from initiation system to energy release system either by providing a barrier between successive explosive elements of the train causing sufficient misalignment to prevent propagation. In the 'arm' position, the ignition can be reliably and reproducibly transmitted. Often this mechanism incorporates provision for interrupting the electrical circuit concurrently with the mechanical blocking of the explosive train and also provides 'safe' or 'arm' status. The mechanism works remotely by electrical pulses.

Pyrogen and Pyrotechnic Igniters

The energy release system of the igniter provides the heat necessary to ignite the motor propellant and raise it to self sustaining combustion. Based on energy release system, the igniters are classified as: a) pyrotechnic igniters, and b) pyrogen igniters. Based on the position, igniters are also classified as: a) head end, b) aft end and c) grain mounted igniters. The pyrotechnic igniter contains pelletised pyrotechnic charge as main charge in vented/perforated containers and normally used in small motors. A pyrogen igniter is basically a small rocket motor having a fast burning propellant grain as main charge loaded in a composite or insulated metallic case, provided with single or multiple nozzles. In large motors, pyrogen igniters because of better control of burning rate and surface area are used where high level of reproducibility is required.

A comparison of pyrogen and pyrotechnic igniters is listed in the Table 1.

Table 1: Comparison of Pyrogen and Pyrotechnic Igniters

	Characteristics	Pyrogen	Pyrotechnic
1.	Main charge	fast burning propellant	pyrotechnic
2.	Configuration	cast as grains	powder or pellets
3.	Action time	300 to 1000 ms	shorter burn time (100 to 150 ms)
4.	Applications	In large solid motors	for smaller motors
5.		Relative merits: Performance closely controllable, Least change in performance Parameters when monitored in Stand-alone mode and assembled to motor. Complex construction and processing Higher cost	Performance moderately controllable P–t traces depends on test chamber volume. simple construction and processing low cost

The requirements of propellants for igniters, a specialised field of propellant technology, will include the following:

Basic Requirements of Igniter Propellant

1. Fast high heat release and gas evolution per unit igniter propellant mass to allow rapid filling of grain cavity with hot gas and partial pressurization of chamber.
2. Suitable initiation and operation over a wide range of pressures and smooth burning at low pressure with no ignition over pressure surge.
3. Rapid initiation of igniter burning and low ignition delays.
4. Low sensitivity of burn rate to ambient temperature change and low burning rate pressure exponent.
5. Operation over required ambient temperature range.
6. Safety and easy to manufacture, safe to transport and handle.
7. Good aging characteristics and life.
8. Minimum moisture absorption or degradation with time.
9. Low cost of ingredients and fabrication.

Some igniters not only generate hot combustion gas, but also hot solid particles which radiate heat and impinge on the propellant surface and assist in achieving propellant burning on the exposed grain surface. There have been a large variety of different igniter propellants and their development has been largely empirical. Black powder, which was used in early motors, is no longer favoured because of difficulties in duplicating its properties. Extruded double base propellants are used, usually as a large number of cylindrical pellets. In some cases, rocket propellants that are used in the main grain are also used for the igniter grain, A common pyrotechnic igniter formulation uses 20 to 35% boron and 65 to 80% potassium nitrate with 1 to 5% binder. The binders include epoxy resins, graphite, nitrocellulose, vegetable oil, polyisobutylene etc. Another formulation used magnesium with a fluorocarbon (Teflon). It gives both hot particle and hot gases.

To maintain the igniter mass and volume at the minimum and achieve the desired mass flow rate, the surface will have to be kept the least and this is possible by having a higher burning rate,

$$\dot{m}_{ig} = S_b r \rho$$

where $\dot{m}_{ig}$ is the igniter mass flow rate, S_b is igniter grain surface area, r is the burning rate, and ρ is the density of the propellant. To achieve higher burn rate, accelerators like copper chromite, ferric oxide, etc. are used in pyrogen igniter propellants. Metallic fuel though will give higher flame temperature and calorific value, is not favoured in higher proportions as principally it reduces the gaseous content of combustion products and to some extent the oxide deposition could inhibit the motor surface. The basic difference between the motor and igniter propellant formulations can be summarized as given in Table 2.

Table 2: Comparative Properties of Motor Propellant and Igniter Propellant

	Characteristics	*Motor Propellant*	*Igniter Propellant*
1.	Burning rate	Low (8 mm/s at 70 ksc)	High (15 to 20 mm/s at 70 ksc).
2.	Oxidizer	Higher percentage of coarse	Fines in higher percentage.
3.	Metallic fuel, %	High, 18 to 20%	Low, 1 to 2%
4.	Calorific value	High	Low
5.	Combustion products	40 to 50% gases	80 to 90% gases
6.	Specific impulse	High, 245 s	Low, 210 s
7.	Burning rate modifiers	Present in small fraction	Present in large fraction

Igniter propellant grains are configured to give regressive burning to prevent excessive igniter mass flow in the motor once the motor propellant is ignited. These pyrogen igniter grains are thin webbed grains and grain web thickness greater than required for minimum burning duration is provided to obtain adequate grain strength and for processing convenience.

Typical Questions

1. What is the basic principle in Rockets?
2. What are the favourable bonds/groups in propellants and exhaust products?
3. What is Hessess' law? How is it used for computing the heat of formation of ethane?
4. How is heat of formation of cyclopropene computed?
5. How is heat of combustion of cubane computed?
6. Explain the relationship between heat of combustion and heat of formation of benzene?
7. How is heat of combustion of a fuel experimentally determined?
8. How is heat of formation of HTPB binder computed?
9. How is oxidizer-fuel ratio found from the composition? What is its influence on I_{sp}?
10. Show the effect of binder type on Isp of AP-Al-Binder propellants?
11. Explain the effects of HTPB, Al, and AP on Isp of propellants using a triangular graph?
12. How is Specific impulse of a solid propellant containing HTPB binder-AP-Al (weight ratio of 13:67:20) theoretically computed?

13. What are the different grain configurations? What are its effects on motor pressure-time curve?
14. What are the functions of a rocket nozzle?
15. What are the important parameters required for nozzle design?
16. Explain the nozzle efficiency using over expanded, optimum expanded and under expanded nozzles?
17. How is a rocket nozzle designed? What are the factors for improving the performance?
18. What are the grain configurations adopted in solid motors. Explain their merits and demerits?
19. What are the factors affecting the performance of a solid rocket?
20. Explain the combustion process in composite and double base propellants?
21. Explain the merits and demerits of pyrotechnics and pyrogens?
22. Why pyrogen igniters are preferred over pyrotechnic charges for igniting stage motors?
23. What are the basic requirements of igniter propellants? How does it differ from the requirements of main motor propellants?
24. Explain the pressure-time graph of a solid rocket test?
25. Compare and contrast the salient features of pyrogen and pyrotechnic igniter?
26. What are the main constituents of an igniter motor? What are the factors dependent on ignition?

REFERENCES

[1] Adam, Cumming, "New Directions in energetic Materials," *J. of Defence Science,* I, 319–327, 1997.

[2] Afroz, Javed and Chakraborty, Debasis, "Prediction of Solid Rocket Motor Nozzle Damping Coefficient using CFD Techniques," *Journal of Propulsion and Power,* Vol. 30, Issue 1, pp. 29–34, 2014.

[3] Agarwal, J.P., High Energy Materials Propellants, Explosives and Pyrotechnics, J.P., Wiley-VCH, 2010.

[4] Beakstead, M.W., "Solid propellant combustion mechanism and flame structure," *Pure & Applied Chemistry*, 65, 1993, 297–300.

[5] Bernstein, E.R., Overviews of Recent Research on Energetic Materials, ed. Thompson, D., Brill, T. and Shaw, R., New Jersey, World Scientific, 2004.

[6] Charles, E. Roger, Solid Propellant Grain Design and Internal Ballistics, Vol. 33, No. 5–6, October/November 2002.

[7] Gould, R.F., Propellants Manufacture, Hazards, and Testing, Advances in Chemistry, No. 88, *American Chemical Society,* Washington, DC, 1969.

[8] Green, J.M., "Flow Visualization of a Rocket Injector Spray Using Gelled Propellants Simulants," AIAA-91-2198, pp. 1–16.

[9] Keicher, Thomus; Kunglstatter, Werner; Eisele, Siefried; Wetzel, Tim and Horst Krause Isocyanate free curing GAP with Bis Proparygil succinate via 3,3 dipolar cycloaddition PEP, Vol. 34, Issue 3, 210–17, June 2009.

[10] Krishnan, S., Swami, R.D., *Prop, J. and Power,* Vol. 14(3), 1998, pp. 295–300.

[11] Lengelle, G., Duterque, J. and Trubert, J.F., Combustion of Solid Propellants, *ONERA,* May 2002.

[12] Mathew, S., Manu, S.K. and Varghese, T.L., Propellants, *Explos., Pyrotech.,* 33, 146, 2008.

[13] Miller, R.S., "Resaarch on New Energetic Materials," *Material Research Society Symposium Proceedings,* Vol. 418, 1996, pp. 3–14.

[14] Mul, J.M., Korting, P.A.O.G. and Schyer, H.F.R., A Search for New Storable High Performance Propellants, *ESA Journal,* Vol. 14, 1990. pp. 253–270.

[15] Munjal, N.L., Gupta, B.L. and Varma, M., "Preparative and mechanistic studies on UDMH-RFNA liquid propellant gels," *J. Prp. Explos. Pyrotech,* Vol. 10, 1985, pp. 111–117.

[16] Palaszewski, B. and Powell, R., "Launch Vehicle Performance using Metallised Propellants," *J. Propulsion and Power,* Vol. 10, No. 6., 1994.

[17] Pande, S.M., Sadavarte, V.S., Bhowmik, D., Gaikwad, D.D. and Singh, H., *Propellants Explos. Pyrotech.,* 37, 707, 2012.

[18] Pein, DLR, Research Centre, Germany, Article on Introduction to Analysis of Rockets, Propellants and Explosives Technology, ISBN 81-7023-884-6, by Allied Publishers Limited, India, 1998.

[19] Reshmi, S., Varghese, T.L., Ninan, K.N., *et al.,* *34th International Annual Conference of ICT,* Germany, 2003.

[20] Sadavarte, V.S., Singh, R.V. *et al.,* Advanced High Energy Rocket Propellants Containing Hexanitrohexaaza Isowurtzitane (CL-20), *Proc. 9th International High Energy Materials Conference,* Feb. 13–14, Trivandrum, 2014.

[21] Sekkar, V., Krishnamurthy, V.N. and Jain, S.K., *J. Appl. Polym. Sci.,* 66, 1795, 1997.

[22] Singh, Haridwar, "High Explosives—Past, Present and FUTURE," In S. Krishnan S.R. Chakravarthy and S.K. Athithan (eds.), "Propellants and Explosives Technology, ISBN 81-7023-884-6, Allied Publishers Limited, Chennai, 1998, pp. 245–270.

[23] Sutton, Rocket Propulsion Elements, Sixth Edition, John Wiley, New York, 1992.

[24] Varghese, T.L., *et al.,* "Developmental Studies on Metallised UDMH and Kerosene Gels," *Def. Sci. Journal,* Vol. 45, No. I, 1995, pp. 25–30.

[25] Varghese, T.L., Ninan, K.N., Krishnamurthy, V.N., *et al.,* Performance Evaluation and Experimental Studies on Metallized Gel Propellants, *Defence Sci. J.,* Vol. 49, No. 1, 77–78, 1999.

[26] Wells, W.W., "Metallised Liquid Propellants," Rocket Propulsion Lab, Space/Aeronautics, June 1966, pp. 76–82.

[27] Wong, W. and Palaszewski, B., "Cryogenic Gellant and Fuel Formulation for Metallised Gelled Propellants—Hydrocarbon and Hydrogen with Aluminium," AIAA-95-3175, 1995, pp. 1–18.

[28] Yang, R., An, H. and Tan, H., *Combustion and Flame,* 135, 463, 2003.

[29] Zhai, J., Shan, Z., Li, J., Li, X., Guo, X. and Yang, R., *J Applied Polymer Science* 128, 2013, 2319–2324.

Propellant Processing Technology

1. INTRODUCTION

Solid Propellants have been very extensively used in modern times in space launch vehicles and missiles. They are energy sources for propelling rockets, launch vehicles or missiles. They are chemical substances that provide the kinetic energy and working fluids for propulsion, on burning. Propellant is a combination of oxidizer and fuel. The oxidizer supplies oxygen for the fuel to burn. Hence, propellant is not a fuel, but a combination of fuel and oxidizer. Hence, it can burn even in vacuum or under water. Propellant gives a controlled burning and release of energy. Hence, they are known as low explosives. Specific impulse (I_{sp}), which is an index of energy, is the most relevant performance parameter for propellants. I_{sp} is defined as impulse (thrust × time) per unit weight of the propellant.

Composite solid propellant is a heterogeneous mixture of mainly 3 basic ingredients—an inorganic oxidizer (AP—Ammonium Perchlorate), a powdered metallic fuel (Al—Aluminium) and a polymeric binder (e.g. HTPB—Hydroxyl Terminated Polybutadiene). AP supplies oxygen for the fuels—Al and binder to burn. The oxidizer and metallic fuel are dispersed uniformly throughout the binder matrix which imparts the required mechanical properties to the propellant. In addition to the above, it may contain small quantities of additives: as plasticizer (process aid), curing agent (for conversion of liquid paste to solid), ballistic modifier (for burn rate control), cross-linker (for mechanical strength), antioxidant (for better ageing), etc. Composite propellants are used in launch vehicles, sounding rockets and missiles. Advantages of solid propellants are: a) simplicity, b) high reliability, c) instant readiness for flight, d) higher thrust, e) low cost, and f) storage capability.

Double base propellants contain two explosive components as major ingredients— Nitro Cellulose (NC) and Nitroglycerine (NG) and other additives like non-explosive plasticizer, stabilizer, ballistic modifier, etc. Inter -diffusion of NC and NG knits the two component system into a single strong tough grain. They are made by casting as well as extrusion process which are separately dealt with. They are used in sounding rockets and for defense applications, e.g. missiles.

2. PARTS OF A COMPOSITE SOLID PROPELLANT ROCKET MOTOR

The solid motor consists of a motor case made of high strength steel or fibre-reinforced plastic. Insulation made of silica filled nitrile rubber is bonded inside the motor case for protecting the case from high temperature of around 3000°C. The propellant is bonded to the insulation using a polymeric liner. Igniter is assembled at the rear end to ignite the motor. On firing the motor, the combustion gases are accelerated and exit through the nozzle. The parts of a typical solid propellant rocket motor is shown in Figure 1.

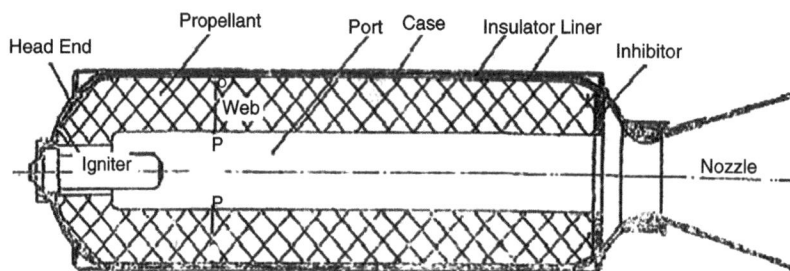

Fig. 1: Solid Propellant Rocket Motor

3. PROPELLANT PROCESSING TECHNOLOGY

Processing of solid propellants especially highly filled system with 85–90% solids in 10–15% liquid poses complex technological problems. This is due to the fact that tangible success in formulating and processing a solid propellant calls for a hard compromise on many contradictory requirements such as high I_{sp}, good mechanical properties, high density, required burn rate, easy processibility with longer pot-life, long and reliable storage life, etc.

Propellant Processing Technology involves the following steps:
1. Motor cases
    (a) Metallic Cases
    (b) Composite cases
2. Motor Insulation—Types of insulation and materials
    (a) Sheeted insulator (b) Castable insulator (c) Sprayable insulator
    (d) Trawable insulator and (e) Reactive insulator
3. Motor Lining—Liner systems
    Liner features, Theory of liner bonding, Liner composition and Liner properties
4. Propellant Processing
    Propellant processing steps—Process flow chart

    (i) Propellant mixing

        (A)   Raw material preparation

        (B)   Mixing operations: a) Horizontal mixer, b) Vertical Change Can Mixer, and c) Continuous Mixer

        (C)   Process parameters—Mixing sequence, Mixing duration, Process temperature, Humidity control, Pot Life—Zero Flow Time (ZFT)—Penetrometric Pot Life (PPT)

    (ii) Propellant casting: a) Vacuum Casting, b) Multiple Pressure Casting or Bottom casting, and c) Bayonet Casting or Top casting

    (iii) Motor Curing

    (iv) De-coring and trimming

    (v) Inhibition of the grain

  5. Non-Destructive Testing (NDT)

  6. Motor Static Testing.

Motor Case—Composite and Metallic Cases

Metallic Cases

The motor case not only contains the propellant grain, but also serves as a highly loaded pressure vessel. Besides constituting the structural body of the rocket with its nozzle, propellant grain and others, the case frequently serve also as a primary structure of the launch vehicle or missile. Booster motors and strap-on motors preferably use steel cases since the motors are cylindrical and segmented and the practical difficulty in making large composite cases. Ordinary steel contains 0.5 to 1.5% carbon while mild steel contains less than 0.5% carbon. Cast iron or pig iron contains 2.2 to 4.5% carbon. Special steels are used for rocket motor cases. Metal cases have a number of advantages compared to filament-reinforced plastic/composite (FRP) cases: a) these are rugged and can take rough handling, b) since they can withstand higher temperatures compared to FRP cases, less insulation is only required, c) they have better ageing characteristics compared to FRP cases, and d) for the same envelope, metallic cases take more propellant compared to FRP cases.

For very large and long motors, like PSLV booster or space shuttle strap-on motors, both the propellant grain and the motor case are made in segments or sections. The case segments are mechanically attached and sealed to each other at the launch site. Segments are made when an unsegmented motor would be too large and too heavy to be transported over roads or railways and often too difficult to fabricate. Multiple O-ring seals are used between the segments.

Two types of special steels used in booster and strap-on motors are:

1. 15 CDV 6 Steel and
2. Maraging steel

15 CDV 6 is a high strength low alloy steel. 15 CDV 6 is a French designation, representing C for chromium, D for molybdenum and V for vanadium and the number 15 represents 0.15% carbon, 1.5% chromium and less than 1.5% for molybdenum and vanadium.

Typical composition of 15 CDV 6:

Carbon	:	0.1 to 0.16%
Chromium	:	1.25 to 1.45%
Molibdenum	:	0.8 to 0.9%
Vanadium	:	0.2 to 0.35%
Manganese	:	0.8 to 1.0%

Uses: Used in booster motors and second stage motors of space launch vehicles.

Maraging steel is a special steel which is also known as 18 Ni 250 steel. The specialty of maraging steel is that it has 18% Nickel and has yield strength of 250 KSI (Kilo Pounds Per Square Inch). The high specific strength (yield strength/density) with high fracture toughness results in substantial reduction in the case thickness and hence, reduction in the weight of the booster. Compared to 15 CDV 6, maraging steel case requires only about half the thickness, thus effectively reducing the case weight to about half.

Typical composition of Maraging steel:

Nickel	:	18%
Cobalt	:	8%
Molibdenum	:	5%
Aluminium	:	0.1%
Titanium	:	0.4%

Other features are:

1. High fracture toughness
2. Good weldability
3. Simple heat treatment procedures
4. Good machinability
5. Negligible distortion during heat treatment.

Difference between Carbon Steel and Maraging Steel

The difference between carbon steel and maraging steel is that in maraging steel, the iron-nickel martensite phase is formed with slow cooling while in carbon steel, the martensite phase is formed with rapid cooling. Ordinary steel is hard and brittle where as maraging steel is soft and ductile. Dimensional changes are less and better machinability for maraging steel compared to carbon steel.

Uses: Maraging steel cases are used in large solid booster motors of launch vehicles like PSLV. Other booster motor casings of USA and ESA are made of D6AC and SC1092W, both containing Ni (0.5–0.6%), Mo (1.0%), Cr (1.0%), V (0.075–0.1%) and C (0.45%).

In the case of liquid motors, Ti alloy tankages are used for UDMH/MMH—N_2O_4 propellants and Ti gas bottles made of Ti-6Al-4V are used for He pressure tanks. Cryogenic propellant tankages are of AA2219 with Al-Cu alloy.

Composite Cases

Composite case is preferred for upper stage motors due to weight reduction and special design. Two types of composite cases are generally used. They are:

1. Glass fibre reinforced epoxy case—Here glass fibre dipped in epoxy resin system is wound on a suitably designed collapsible mandrel, cured well and then the collapsible mandrel is detached to get the composite case.

2. Kevlar fibre reinforced epoxy cases—Kevlar is a poly aramide, prepared by condensation between phenylene diamine and terephthalyl chloride. The aromatic nature and hydrogen bonding capability gives high temperature resistance up to 550°C and very high strength closer to the value of steel. Kevlar fibre- reinforced epoxy case is processed in a similar way as glass fibre reinforced epoxy case. Typically, the inert mass of a case made of carbon fibre is about 50% of a case made with glass fibres and around 67% of a case made of Kevlar fibres.

Both helical type filament winding machine and polar winding machine are used to give a designed contour to the composite cases made of glass fibre/Kevlar fibre-reinforced epoxy cases. Carbon fibre reinforced epoxy cases are superior giving higher strength and weight reduction. A typical polar winding machine for making composite cases is shown in Figure 2.

The forward end, aft end and cylindrical portion are wound on a preform or mould which already contains the forward and aft rings. The direction in which the bands are laid onto the mould and the tension that is applied to the bands is critical in obtaining a good case. The curing is done in an oven and may be done under pressure to assure high density and minimum voids of the composite material. Since filament-wound case wall can be porous, they must be sealed. The liner/insulator between the case and the grain can be the seal that prevents hot gases from seeping through the walls. Scratches, dents and moisture absorption can degrade the strength of the case. In some designs, the insulator is placed on the preform before winding and the case is cured simultaneously with the insulator.

Fig. 2: Polar Winding Machine

Comparative motor case properties of 15 CDV 6, Maraging steel, Glass-epoxy case and Kevlar-epoxy cases are shown in Table 1.

Uses: Composite cases are used in upper stage solid motors. Carbon fibre reinforced epoxy cases though very costly, gives much reduction in inert weight.

Table 1: Motor Case Properties

Case Material	15CDV6	Maraging Steel	Glass/Epoxy	Kevlar Epoxy
T.S. (MPa)	1660	1765	1230	1760
Elong %	8	5	–	–
Specific Strength, KN	21.7	23.1	52.5	123.7
Cost/kg	600	1200	150	700

Motor Insulation—Types of Insulation

Introduction

Thermal insulation or internal insulation is provided inside the motor case to protect the case from combustion gases whose temperature is of the order of 3000°K. The insulation materials are generally inert, non-combustible and are primarily polymeric and elastomeric in nature. Insulator is bonded to the inner side of the motor case with suitable adhesive. Insulator also acts as a structural member for transmission of load between the case and the propellant. Insulation thickness will vary depending on the motor design, configuration and the burning time or exposure time. Generally, the insulation thickness is minimum at the cylindrical portion and maximum at the head end/nozzle end portions. A simple relationship for the thickness d at any location in the motor depends on the exposure time t, the erosion rate r of the insulation material and safety factor f which can range from 1.2 to 2.0. The simple rule is that insulation depth is twice the charred depth.

$$d = t.r.f$$

Insulation also helps in reducing the thermal stresses during the processing of case bonded motors. Due to the large difference in the coefficient of linear expansion of motor case ($\sim 0.12 \times 10^{-4}/°C$) and propellant ($\sim 1.12 \times 10^{-4}/°C$), stresses and strains will be produced during the cure cycle, especially the cooling phase from curing temperature (50–70°C). This can lead to de-bond at the interface of the case bonded propellant or cracks in the grain port. The insulator, being a highly flexible elastomer and having coefficient of linear expansion ($\sim 0.71 \times 10^{-4}/°C$) between that of the case and propellant, can reduce the thermal stresses. Also, loose flaps are provided in the highly stressed regions—dome regions and cylindrical regions between the insulator and liner to overcome the de-bond/crack during the cooling phase of the motor. Loose flap is of same insulation material and is continuous and bonded to the insulator.

Requirements

1. Improved thermal properties- low thermal conductivity ($4–5 \times 10^{-4}$ cals/cm/°C) and high sp. heat (0.4–0.5 cals/g, °C). This is accomplished by filling the insulator with silicon dioxide, Kevlar fibres or ceramic particles. Asbestos is an excellent filler, but no longer used because of health hazard.
2. Low erosion rate or ablation rate (0.1–0.2 mm/s) and produce porous char after pyrolysis.
3. Good compatibility with case and propellant—good bonding.
4. Low density to reduce dead weight—every 1 kg reduction in inert weight gives about 0.8 kg improvement in payload.

Titan 3D mission failed shortly after launch due to insulator de-bond causing exposure of steel case to 8000°F.

Rubbers Used

Three types of rubbers are mainly used for rocket insulation. They are:
1. Nitrile Rubber (NBR)—It is a random copolymer of butadiene and acrylonitrile.

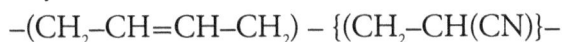

$-(CH_2-CH=CH-CH_2) - \{(CH_2-CH(CN)\}-$

2. Chloroprene rubber $-(CH_2-C(Cl)=CH-CH_2)-$

3. EPDM – Ter-polymer of ethylene-propylene-diene (cyclopentadiene) random copolymer.

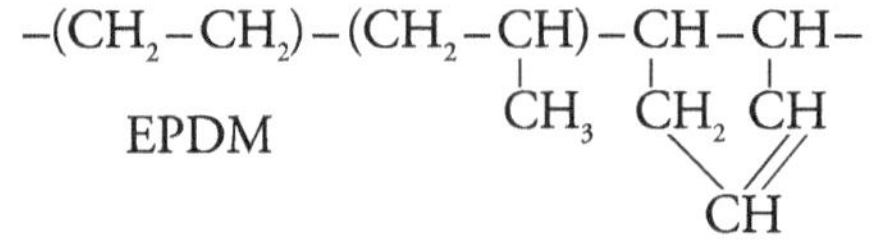

$$-(CH_2-CH_2)-(CH_2-CH)-CH-CH-$$

EPDM CH_3 CH_2 CH
                                    CH

Fillers Used

1. Silica, clay or mica powders
2. Phenolic microbaloons
3. Asbestos fibres
4. Carbon black
5. Kevlar fibres/Carbon fibres (chopped).

Other Additives

Sulphur (S) (vulcaniser), Zinc Oxide (ZnO), Stearic acid (activators), plasticizer, Mercapto Benz Thiazole (MBT) (modifier), antimony trioxide or tricresyl phosphate (flame retardants), etc.

The required rubber is masticated and then compounded with suitable fillers, vulcanizing agents, activators, modifiers, etc. calendered and finally sheeted to thickness of 1 to 3 millimeter. The required thickness in the motor case is achieved by bonding insulator sheets layer by layer. Dispersion of microbaloons or chopped fibres is a difficult task which is to be tackled to get uniform dispersion without much damage to the chops or microbaloons. In modern insulators, asbestos is avoided due to health hazards and difficulty in processing.

An external insulation is applied to the outside of the motor case, particularly in high-acceleration launch vehicles or tactical missiles. This insulation reduces the heat flow from the Air-boundary layer outside the vehicle surface (aerodyanamic heating) to the case and to the propellant. It thus prevents the FRP cases from becoming weak or propellant from becoming soft or getting ignited. This insulation must withstand oxidation, good adhesion; have structural integrity to loads imposed by the flight or launch, etc. Materials used as internal insulators are unsatisfactory because they burn in the atmosphere and generate heat. Non-pyrolysing, low thermal conductivity refractory materials are best.

Types of Insulation

1. Sheeted Insulator*

Important features of sheeted insulation are the following:

(a) Insulator is sheeted to thickness 1–3 mm and then bonded layer by layer to required thickness using adhesive.

(b) Thickness will vary at different locations of the motor depending on flame exposure time. Normally, thickness is minimum at the cylindrical portion and maximum at the head end/nozzle end.

(c) Either vulcanised or unvulcanised rubber sheets are used. In unvulcanised system, after laying, the motor is cured at elevated temperature and pressure in an autoclave. Vulcanization is the process of cross-linking rubber through sulphur linkages.

(d) Loose Flaps are laid in the highly stressed regions—dome and cylindrical regions to minimize the damages such as de-bond/crack due to thermal stresses during processing of large grains. Loose flap is continuous and bonded to insulator with epoxy or polyurethane adhesives after curing and de-coring of the propellant.

2. Castable Insulator

In castable insulator, insulator in the slurry form is cast on the inner surface of the hardware by either pressure or vacuum casting using suitable mandrel, followed by vulcanization/curing at high temperature.

3. Sprayable Insulator

In sprayable insulator, insulation compound is mixed with suitable solvent and sprayed on the inner surface of the motor. The solvent is allowed to evaporate and spraying is repeated to get required thickness, followed by vulcanization/curing at high temperature.

4. Trowelable Insulator

In trowelable insulator, insulation compound is made in the form of slurry with good thixotropic properties. It is then applied to the inner surface of the rocket case using trowels like cement plastering by mason, followed by vulcanization/curing.

5. Reactive Insulator (Insu-Liner)

In all the above 4 cases, insulator is to be lined with suitable liner to bond with the propellant. In the case of reactive insulator, no liner is required for bonding. The

---

*During laying sheets, care is to be taken to avoid wrinkles, air entrapment and de-bonding between the layers. Care should be taken to minimize the moisture content in the insulator to ensure strong interfacial bond strength.

reactive insulator itself contains reactive groups for direct bonding with propellant. Reactive insulator, in addition to rubber, filler, vulcanizing agents, etc., it contains a small percentage of the propellant binder (*e.g.,* HTPB) and compatible plasticizer. It gives direct bonding with the propellant without using a liner.

Advantages are: Liner is not required, process steps and time reduced, low cost, etc.

Typical Insulator Properties

Typical properties of two types of insulators are shown in Table 2. Rocasin is silica (~60%) filled nitrile rubber based insulator and Repin is silica filled EPDM rubber based insulator. Though Repin gives lower tensile strength and elongation (which is sufficient for rocket insulation), its lower density and lower erosion rate reduces the inert weight of the rocket. Among mechanical properties of the interface, tensile bond strength, wheel peel strength and shear bond strength with propellant and casing material are the major ones. High values of these parameters are most desirable. A tensile bond strength of 5 kg/cm^2 and peel strength of about 1 kg/cm is the minimum requirement.

Table 2: Typical Insulator Properties

Sl. No.	Insulator	Rocasin (Nitrile Rubber)	Repin (EPDM)
1.	T.S. (KSC)	150	25
2.	Elong (%)	800	70
3.	Hardness	65	55
4.	Density (g/cc)	1.19	1.0
5.	Erosion rate (mm/sec)	0.2	0.15
6.	Sp. Heat (Cals/g, °C)	0.4	0.5
7.	Thermal conductivity (Cals/cm.sec., °C)	$4\text{–}5 \times 10^{-4}$	$4\text{–}5 \times 10^{-4}$

Motor Lining and Liner Systems

Liner is a thin layer of elastomeric polymer material applied uniformly on the insulator surface for proper bonding of propellant to the insulator. Propellant-insulation interface must be strong enough to withstand the stresses and strains caused by thermal contraction of the propellant and also pressurization and acceleration loads. Hence, liner must be capable of giving good peel strength, tensile bond strength and shear bond strength to the interface so as to give cohesive failure in the interface.

Liner Features

The major requirements of a liner composition are given below:
1. Liner is sprayed uniformly over the abraded, cleaned and dried surface of the insulator.
2. Liner thickness is about 100–200 microns.
3. Liner should ensure good bond between propellant and insulator.
4. Adhesive properties of the liner should be superior so that any failure or de-bond should be cohesive in nature.
5. Liner system is usually based on the same pre-polymer used in propellants for better compatibility.
6. It should act as a diffusion barrier between insulator and propellant and avoid plasticizer migration.
7. Liner composition should be curable at or around room temperature.
8. Liner should have good flow properties and ageing characteristics

Theory of Liner Bonding

The bonding mechanism of liner may be due to polarity of liner and propellant materials or due to electric charge transfer between adherents. The wettability of substrate by propellant is the major requirement for good bonding:
1. The polar groups in the liner system form good adhesive bonds with the polar groups of the propellant binder system.
2. Wettability of the liner surface with the propellant slurry enhances good bonding. This shows that the components in the liner which can reduce surface tension or surface energy or contact angle give better wettability of liner with propellant and hence better bonding.
3. There is a critical stress below which the bond will never fail, when a deformable substrate bonded to a rigid substrate is subjected to a constant tensile stress.
4. The more the extent of hydrogen bonding, greater the interfacial bond strength. Apart from the chemical cross-linking between the liner components and the propellant binder components, the higher hydrogen bonding capability of the urethane link of the liner system with the urethane link of the propellant binder enhances further the interfacial bond strength.

Liner Composition and Application

Liner system consists of a binder filled with fillers, plasticizers and curing agents. A typical liner composition for HTPB propellant contains HTPB binder, tri methylol propane crosslinker, carbon black filler, toluene di-isocyanate curing agent

(10% excess), solvent Dichloromethane (DCM) for spraying, etc. Moisture content in liner composition should be kept minium to get higher bond strength. Liner composition is usually diluted with low volatile solvent like DCM and sprayed uniformly on to the insulator surface using a spray gun. After application, solvent evaporates, leaving a thin layer of liner. In smaller motors, lining can be done with manual brushing of the liner composition without dilution. In very large motors, centrifugal spinning and spraying of the solvent diluted liner on the insulator surface is applied using bone dry nitrogen gas pressurized nozzles. After liner application, rocket motor casing is kept rotated for uniform distribution of liner. Liner is cured to a tacky state before casting of propellant.

Cured Liner Properties (Typical):

T.S	:	20–30 ksc
Elong	:	50–70%
Modulus	:	15–25 ksc
Density	:	1.0–1.2 g/cc
Hardness (shore-A)	:	50–60

Interface Properties:

TBS	:	5–7 ksc (Cohesive)
Peel strength	:	0.8–1.5 kg/cm
Shear bond strength	:	5–7 ksc

4. PROPELLANT PROCESSING

Composite propellant processing operations are all complex and hazardous. The process converts simple chemicals into highly energetic powder packs capable of generating highest thrust generators for a given system weight. In this process, a highly viscous, non-Newtonian slurry consisting of oxidizer-fuel-additives is converted into a solid grain under carefully monitored situations. Propellant processing involves the following steps:

Propellant Processing Steps (Process flow chart)

1. Propellant mixing
   A. Raw material preparation
   B. Mixing operations using:
       (a) Horizontal mixer
       (b) Vertical Change Can Mixer and
       (c) Continuous Mixer

C. Process parameters:
> (a) Mixing sequence,
> (b) Mixing duration,
> (c) Process temperature,
> (d) Humidity control,
> (e) Pot Life-Zero Flow Time (ZFT)—Penetrometric Pot Life (PPT)

2. Propellant casting (different types)
   (a) Vacuum casting
   (b) Multiple pressure casting (Bottom casting) and
   (c) Bayonet casting (top casting)
3. Motor curing
4. De-coring and trimming
5. Inhibition of the grain
6. Non-destructive testing
7. Motor static testing

Process Flow Chart

Process Flow Chart for HTPB Propellant

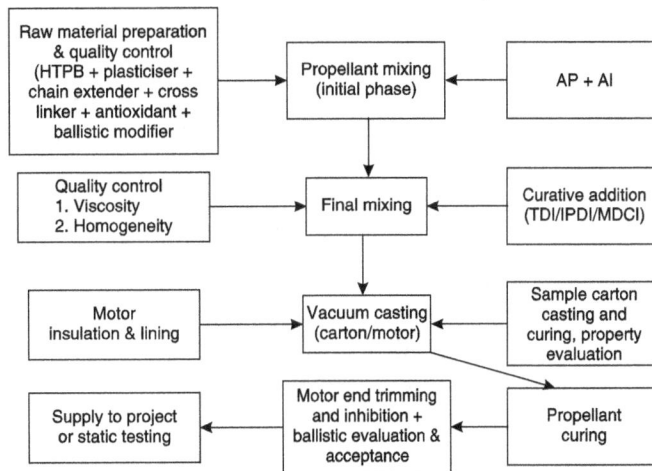

Composite propellants are usually processed by slurry casting technique. A typical process flow chart for HTPB solid motor is shown here. The process flow chart explains the various steps involved in the processing of a typical solid propellant starting from raw materials preparation to the final finished motor. The various steps involved in this process are propellant raw material preparation and quality acceptance, motor insulation and lining, propellant mixing, motor casting, curing, end trimming, inhibition, non-destructive testing, assembly and static testing.

Propellant Mixing

Mixing is a major operation in the processing of composite solid propellants especially when the solid loadings are of the order of 85–90%. The initial step involved in the mixing process is the propellant raw material preparation and quality acceptance. All the raw materials are screened before using to avoid lumps, agglomeration or any foreign materials. Moisture sensitive materials are dried, sealed and kept in air tight containers before using in propellant.

Raw Materials Preparation

The raw materials required for a typical HTPB composite propellant are HTPB binder, plasticizer, chain extender/cross linker, antioxidant, ballistic modifier, metallic fuel, oxidizer and curing agent. Plasticizer can be hydrocarbon type like liquid paraffin or ester type like dioctyl adipate, isodecyl pelargonate, etc. Diols like 1,4 butane diol or triols like glycerol, 1,1,1 trimethylol propane, etc. can be used as chain extender and/or cross linker. Phenyl beta-napthyl amine or hydroquinone can work as antioxidant. Generally used metallic fuel is aluminium powder or magnesium powder of average particle size 10–30 microns. Most of the launch vehicle propellants use Al powder as the metallic feel, on account of higher I_{sp}, higher density and low cost. The widely used oxidizer is Ammonium Perchlorate (AP) which can be unimodal, bimodal or trimodal particle size distributions. Three grades of AP, namely coarse grade of average particle size 300–500 microns, fine grade of average particle size 40–60 microns and ultra fine grade of 5–10 microns are used for this purpose. Bimodal and trimodal paricle size distributions are generally used for high solid loaded composite propellants to obtain better End of Mix (EOM) viscosity. The particle size distributions are selected in such a way that the smaller particles occupy the gap between larger particles, leading to compact packing. The commercially available AP is ground to get fine and ultra fine grade particles of AP before incorporating in the propellant mix.

The preparation of oxidizer involves the following steps: a) sieving, b) grinding, c) blending, d) drying, and e) weighing. Unimodal fine/ultra fine grade of AP is used for making fast burning propellants used in igniter propellants. The curing agents for HTPB propellant can be diisocyanates like toluene diisocyanate, isophorone diisocyanate or methylene bis cyclohexyl isocyanate. All the raw materials should pass the stringent quality control tests before using in propellant mixing.

Another step is the motor preparation, insulation laying and lining as explained earlier. Lined sample cartons are also to be prepared for making test samples.

The motor case or mould is assembled with casting fixtures such as the core (mandrel) at the central which gives the initial burning surface area, top and bottom assemblies for central alignment of the core as well as to receive the propellant slurry, depending on the casting technique selected. For the free standing block type, the propellant is cast into a mould which is stripped off once the propellant becomes a solid by curing process. For case bonded type of motors, the propellant slurry is directly fed into the prepared motor case. The motor preparation steps for case bonded motors involve operations like sand or grit blasting, degreasing, insulation laying and lining, vulcanization, abrading of the cured insulation, liner application to ensure proper bond between propellant and insulation during curing operations. Insulator surface is abraded using flexible shaft grinder. After abrading, the rubber dusts on the surface are cleaned by industrial vacuum cleaner and with a solvent like trichloroethylene. The abraded motor is then dried in an oven before liner application. The assembly of motor case with casting fixtures and propellant feed mechanisms is important and carried out as per casting technique.

Mixing Operations

Mixing is the most critical step of solid propellant processing. In this step, the liquid and solid ingredients are homogenised in a mixer, there by achieving almost complete wetting of inorganic solid ingredients by the organic liquid polymer. The mixing is carried out either in horizontal sigma mixer or in vertical change can mixer.

Mixing takes place as a result of stretching, folding, pressing, wiping, kneading, shearing and transfer of materials in all directions. The efficiency of mixing depends on the type of mixer, blade configuration, clearance between the blade and the bowl, mixing sequence, mixing time and temperature. The end of mixing results in a product with viscosity ranging from 3000 to 10000 poises still flowable but still continuously curing non-Newtonian mass. For bigger scale mixings, the mixing is carried out in two stages—premixing and final mixing. In the pre-mixing stage, all ingredients except curing agent are added and mixed to attain homogeneity. In this condition, propellant slurry can be kept in storage for few days. In the final mixing stage, curing agent is added and mixed. This kind of mixing is followed for casting large motors using number of mixings and to avoid interruption in the casting process. This calls for more number of bowls in the vertical mixer.

Horizontal Mixer

Horizontal mixer has a two blade configuration with Z-shaped blades as shown in Figure 3. They have fixed speed ratio of the blades normally with 40:60 ratio. They are meant for mixing almost fixed quantity of propellant. For example, a 50 kg capacity mixer can mix comfortably 40 to 50 kgs propellant only. Horizontal mixers are of low cost and are indigenously available in the country. However, they have the problem that the glands come in contact with propellant during mixing and hence, mixer cleaning is difficult. Hence, horizontal mixer requires frequent servicing of gland packing as part of safety clearance for propellant mixing. The mixing can be done at elevated temperature by circulating hot water through the jacket surrounding the bowl.

Fig. 3: Horozontal Mixer

Fig. 4: Vertical Change Can Mixer

Vertical Change Can Mixer

Vertical change can mixers have two blade as shown in Figure 4 or three blade configurations. Usually small mixers of 2 to 3 kgs capacity have two blade configurations while larger mixers of 50 kgs to 2500 kgs have three blade configurations. The central blade has axial motion whereas side blades have axial and circumferential motions. In the axial motion, the blades are allowed to rotate at 5–20 rpm where as in planatory motion, the maximum speed used is 5–6 rpm for propellant mixing in vertical mixers. The axial and circumferential motions in vertical mixers give more efficient mixing and mixing duration can be reduced compared to horizontal mixers. The raw materials can be added remotely from external bins to the mixer even when mixing is on. The propellant need not be unloaded and the bowl can be directly used for casting. Using a number of mixing bowls with automation in the feeding of raw materials, the mixing process can be enhanced and the production rate of big motors can be increased.

Mixing in vertical mixer is safer as the glands do not come in contact with the propellant and hence, cleaning is very easy. Variable quantities of propellant can be mixed in a vertical mixer. For example, a 50 kg capacity vertical mixer can comfortably mix 12.5 to 50 kgs of propellant. However, the cost of vertical mixer is three to four times higher than that of horizontal mixer. Vertical mixers are very widely used in the manufactures of big rockets.

Continuous Mixer

One of the recent improvement in propellant mixing technology is the introduction of continuous mixing technique which gives:

1. Continuous process with limited quantity of propellant in the mixer at a time.
2. Enhanced production rate with minimum man power

3. Better reliability and safety as the process is fully automated
4. Propellant wastage is minimum
5. The cost is very high.

NASA has used continuous mixing technology for processing Advanced Solid Rocket Motor (ASRM) of 6.5 meter diameter with 8 million kg propellant slurry. The production rate of such mixers are in the range of 8 tons per hour as against 0.6 ton per hour for vertical mixers. A typical continuous mixer is shown in Figure 5.

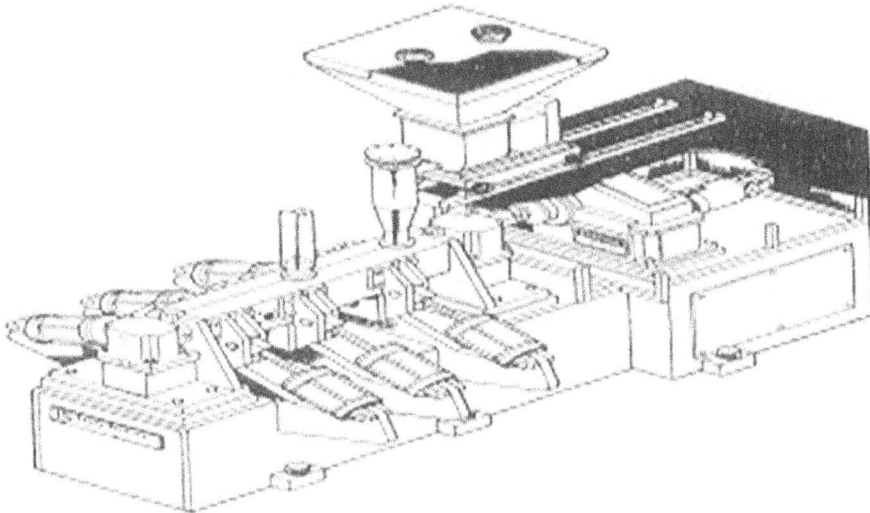

Fig. 5: Continuous Mixer

In the continuous mixing, propellant raw materials are fed into the mixer in three streams. The first stream consists of polymeric binder, metallic fuel, plasticizer, bonding and/or wetting agent, ballistic modifier, antioxidant, etc. as a pre-mix. The second stream consists of oxidizer and the third stream is with curing agent. All feedings are done at computer controlled rates. The process is fully automated, which ensures higher safety and better quality and reliability.

Process Parameters

Mixing Sequence

Mixing sequence is followed as per process flow chart designed for a particular composite propellant during its development phase. In a typical mixing sequence for HTPB propellant, the raw materials such as HTPB binder, plasticizer, chain extender and/or cross linker, antioxidant and ballistic modifier are charged to the mixer as pre-mix in the case of small size mixings (2–30 kgs) and as individually in the case of large size mixings and mixing continued for a designed period during the initial phase of mixing. This is followed by addition of metallic powder in two to three steps and continued mixing. The oxidizer coarse grade is added first in few steps, mixing continued, followed by addition of oxidizer fine grade in steps and mixing continued. This slurry is called pre-mix slurry.

Next step is final mixing. Final mixing can be done in the same day or on a later date depending on the casting requirement. Pre-mix is to be stored properly for a later date final mixing. During final mixing, curing agent is added to the mixer and mixing continued for a period of 30–60 minutes and the slurry is sent for casting. Control checks carried out for the pre-mix (without curing agent) are viscosity measurements and homogeneity checks for ensuring uniform mixing. Control checks for final mix slurry are continuous viscosity measurements to monitor the pot-life of the slurry, slurry burn rate to check the burn rate requirements, slurry density to check the overall composition and rapid cure test to check the curability of the propellant.

Mixing Duration

Mixing duration ranges from 2.5 hours to 5.0 hours depending upon the type of propellant, mix quantity, type of mixer, type of curing agent and mode of addition as manual or automated. Mixing time also affects the end properties to some extent. Larger the mixing time, better the mixing effectiveness. But longer mixing time affects the slurry pot-life and also leading to grinding of bigger oxidizer particles which can enhance the burn rate of the propellant. Hence, an optimum mixing time is followed for a particular propellant.

Process Temperature

Process temperature is also important because it affects the slurry viscosity. Higher process temperature reduces the slurry viscosity, but it enhances the cure reaction and affects pot life. The processing temperature varies from ambient to 60°C depending on the type of propellant.

Humidity Control

Humidity control during mixing is also critical as some formulations are very sensitive to moisture. Unlike CTPB and PBAN propellants, processing of polyurethane propellants requires stringent humidity control (RH < 50%) as isocyanate curing agents are highly reactive with moisture.

Pot-Life, Zero Flow Time (ZFT) and Penetrometric Pot-Life (PPT)

Pot-Life, Zero Flow Time (ZFT), and Penetrometric Pot-Life (PPT) are important process parameters and are to be closely monitored for realizing defect-free solid motors. Pot life and ZFT are found from measuring viscosity of the slurry every one hour duration at the process temperature using Brooke field viscometer with helipath stand. PPT is measured using a standard penetrometer.

Pot-Life

Pot-life is the period from the End of Mix (EOM) viscosity to the time up to which the propellant slurry is amenable for easy casting, which is arbitrarily fixed as the duration till the slurry viscosity reaches 16000 poises. Higher pot-life is essentially required for processing defect free large motors.

HTPB propellant using TDI (Toluene Diisocyanate) has a shorter pot-life of 4–5 hours. The aromatic nature coupled with ortho-para effect makes TDI more reactive and hence shorter pot-life for the propellant. Higher pot-life can be achieved using curing agents having sufficiently low reactivity over a wide range of temperature.

IPDI (Iso Phorone Diisocyanate) and MDCI (Methylene Dicyclohexyl Isocyanate) are slow reacting curing agents which can give longer pot-life of 15–20 hours for HTPB propellants. One NCO group in IPDI is primary and the other is slow reactive secondary where as in MDCI, both NCO groups are secondary.

Zero Flow Time (ZFT)

Zero Flow Time (ZFT) is the period from End of Mix (EOM) to the time up to which the slurry shows amenability to flow, which is arbitrarily fixed as the duration till the slurry viscosity reaches 50,000 poises. ZFT of HTPB-TDI propellant is 10–12 hours and that of HTPB-IPDI or MDCI is 40–50 hours. ZFT is very advantageous in the casting of large motors for better intermixing and bondability between slurry batches cast at varying intervals.

Penetrometric Pot-Life (PPT)

Penetrometric pot-life is the time up to which the propellant slurry already cast remains in a semi-cured or compressible state. PPT is measured with a standard penetrometeter as shown in Figure 6 using propellant slurry kept in a standard cup maintained at the process temperature.

The penetrometer measures the depth of penetration of a standard cone placed just touching the top surface and after a drop of 5 seconds with an accuracy of 0.1 millimeter. As the curing progresses, the yield stress of the slurry increases and the depth of penetration decreases and remains constant. From the plot of penetrometric reading (0–350 which corresponds to 0–35 mm depth) in Y-axis and time (0–80 hours) in X-axis, the time at which lowest steady state penetrometric reading is taken as PPT. The PPT of HTPB-TDI propellant is 22–24 hours and that of HTPB-IPDI or MDCI propellant is 60–70 hours. It is practically seen that the steady low penetrometric reading is around 50 (penetration depth is 5 mm) and at this state, the slurry is not compressible.

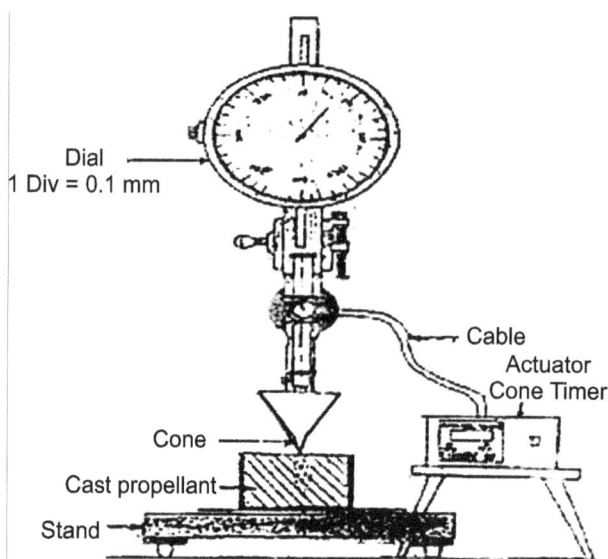

Fig. 6: Penitrometer—A General Arrangement

It is seen that for casting a 30 ton motor or segment, 13 numbers of 2.5 ton mixings are to be carried out. Here, the casting of each batch is to be completed with in the pot-life of the propellant. The safe time from the beginning of first batch to the end of second batch and so on must be with in ZFT. All the 13 batches are to be completed with in the penetrometric pot-life time. With proper follow up of pot-life, ZFT and PPT, defect free motors can be achieved. Sometimes the casting process may have to be held up either awaiting further subsequent propellant mixes or manipulation of casting tooling. A high zero flow time helps to get void free grains under this situation. If the gap between two batches casting exceed the zero flow time, the voids formed would not collapse upon release of vacuum.

Propellant Casting

Propellant casting is another important step in the processing of composite solid propellants. Here, the viscous slurry flows into the intricate parts of the motor case to fill without creating any porosity or blow holes. Internal geometry of the grain is maintained using suitable mandrels with star or cylindrical shape or any other shape as shown in Figure 7. The critical factors during casting are vacuum level, valve/slit diameter, slit area, feed rate, slurry distribution pattern, etc.

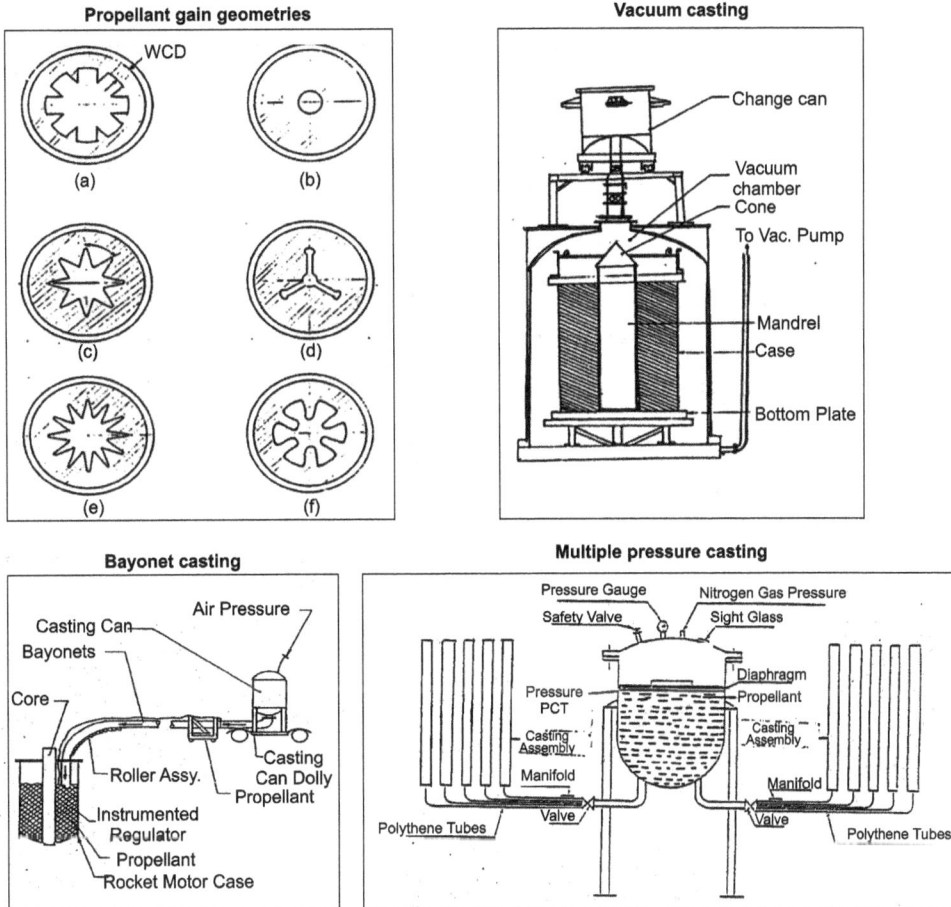

Fig. 7: Propellant Processing Techniques and Grain Geometries

Pre-Casting Operations

1. Rocket chamber cleaning—the metallic rocket chamber is first sand blasted to remove rusting/dirt, and then cleaned with suitable solvents like Dichloro Methane (DCM) or Trichloro Ethylene (TCE).

2. Chamber insulation as detailed under insulation systems.

3. Liner application on insulated chamber as detailed under liner systems.

4. Mandrel preparation—Teflon coated mandrel is cleaned with suitable solvents like DCM or TCE and then given a thin coating of silicone grease. It is then positioned and centralized inside the motor with suitable casting fixtures at top and bottom. The assembled motor is then transported to the casting station and kept inside the vacuum bell for casting operations. A cone and a splash guard are provided.

Casting Methods

The three most common methods of propellant casting and different grain geometries are shown in Figure 7. The casting methods are:

(a) Vacuum casting

(b) Multiple pressure casting or bottom casting and

(c) Bayonet casting or top casting.

Vacuum Casting

Vacuum casting is a very widely accepted casting technique in which the propellant slurry is passed through a slit plate and falls as thin strips into the rocket chamber either enclosed in a vacuum bell or serving as its own bell. The driving force for propellant flow are: a) the head of the propellant column and b) the differential pressure which is always less than one atmosphere. The differential pressure and the falling of slurry in thin strips enables proper de-aeration of the slurry leading to defect free (without any porosity or blow holes) propellant grains. The vacuum level is usually maintained at 3–7 torr residual pressure to get defect free grains. Very high vacuum of less than 1 torr residual pressure is detrimental as it removes some of the volatile components like curing agent from the slurry. Very poor vacuum of more than 10 torr residual pressure leads to porosity or blow holes in the final grain. It is practically seen that propellant slurry in the viscosity range of 4000–16000 poises can be vacuum cast comfortably to realize void free grains. Optimization of casting rate is trade between: i) the need to have rapid flow to accommodate time saving industrial requirements and to avoid significant viscosity build up due to progress of curing, and ii) the need for a fairly slow flow to permit sufficient degassing of the slurry after it has gone through the slit plate.

Vacuum casting eliminates the de-aeration as a separate time consuming step after mixing. The casting kettle with propellant is kept over the propellant feed valve assembly. Vacuum casting process produces foamy propellant at the top portion of the grain, which slumps by 25 to 30 mm when the vacuum is released. This propellant is removed mechanically after the propellant has cured to hard solid mass. Frothing indicates equilibrium between escaping traces of gas or air and the settling propellant.

Advantages:

1. It is the surest method of casting without entrapping air.

2. Propellant wastage can be minimized.

3. Very widely used for making large motors like 230 tons Ariane booster of ESA (Europian Space Agency), 139 tons and 200 tons boosters of ISRO, etc.

However, extremely viscous slurry can not be cast by this method and intermingling of propellant batches is only to some extent.

Multiple Pressure Casting or Bottom Casting

Multiple pressure casting or bottom casting is another casting method suitable for making rocket motors with thin webs and complicated geometries. In this method, the propellant slurry is first de-aerated in the mixer bowl or vacuum cast as explained above into the casting can, which acts as the pressure vessel for subsequent operation. The pressure vessel is provided with hot water jacket to keep the slurry temperature, a pressure gauge and a pressure release safety valve. The 2 outlets with control valves given at the bottom of the pressure vessel are connected to two manifolds on either side. The 5 numbers of assembled motor cases arranged in parallel on either side are connected to the 2 manifolds using flexible transparent pipes as shown in the figure given above. A pressure pad or diaphragm is placed on the slurry and the chamber is covered with an air tight lid.

When nitrogen gas pressure is applied, the slurry slowly flows through the bottom, reaches the two manifolds from which it is distributed to 10 numbers of motor cases at a time. The pressure applied ranges from 0.5 to 3 ksc. At the end of pressure casting, the control valves are closed, pressure is slowly released and the cast propellant motors are detached along with the flexible tubes and held properly to avoid any slurry leakage.

Case bonded grains with complicated configurations such as that of pyrogen ignitors, sounding rockets such as RH-75—RH-200 can be processed in 2–10 numbers in a batch. Web thickness varied from 4 mm to 25 mm with L/D ratios of 3 to 10. The grain length varied from 200 mm to 1200 mm.

Advantages:

1. Ideal method for making thin webbed motors with complicated grain geometries.
2. More numbers of motors can be cast simultaneously from a batch.
3. High viscosity propellants can be cast.
4. Unlike vacuum casting, motors with smaller annular gap of even 4 mm between the case and the mandrel can be processed by this method.

However, pressure casting has size limitations as large motors can not be made by this method.

Bayonet Casting or Top Casting

In bayonet casting, the de-aerated slurry in the casting can is gradually pumped into the motor case from top under nitrogen pressure using bayonets. The tip of the bayonet is kept always below the surface of the propellant being cast. The motor is lowered or the bayonet is raised as the propellant level rises in the motor. This method has been used successfully for making large booster motors of Aerojet's 260 inches diameter motor. For large motors, several bayonets are used and are controlled while retraction.

Advantages:

1. It is advantageous for making large motors.
2. Large vacuum bells are not required for casting.

However, it has the inherent problems such as more voids in the propellant compared to vacuum casting, more propellant wastage and the bayonet or motor has to be moved during casting.

Current manufacturing technology of making large solid propellant motors for launch vehicles and missiles is to cast them in segments of short and convenient length and later assemble them together to get the full motor. The number of cast segments varies from 3 to 7 depending on the design. India's Polar Satellite Launch Vehicle (PSLV) first stage full motor has 5 cast segments while space shuttle strap-on booster has 4 segments. The segment technology has many advantages like manufacturing facility cost minimization, handling of short segments and minimum quantity of propellant at a time.

Multiple casting is the technique in which segments corresponding to two or three motors are cast at the same time to ensure close and reproducible performance between the motors. This is essential where motors are attached to the sides of the lower stage motor, as strap-ons required to burn at the same time. The differential performance between the two motors is required to be as minimum as possible.

Some of the other techniques rarely used in propellant industry are centrifugal casting, barometric leg casting and block casting which have limitations to size and type of motor.

Motor Curing

Curing is the process of converting a liquid slurry into a solid mass. In composite propellants, curing can be a chemical or physical phenomenon. In chemical curing, the functional groups of the binder react with the functional groups of the curing agent to form a three dimensional network structure which gives the structural properties to the propellant. In HTPB propellant, the hydroxyl groups of HTPB reacts with the isocyanate groups of curing agent to form a cured network. Generally, curing is done at elevated temperature of 50–60°C using electrically heated hot air ovens or water heated ovens and the curing duration varies from 5 days for a 50 kgs block to 15–20 days for a 30 tons 3 meter diameter segment.

The variables that need control during curing process is time-temperature-pressure. The choice of time and temperature is determined generally after laboratory tests using various combinations of time-temperature sets for the propellant system ensuring completion of cure reaction and attainment of mechanical properties. The final temperature-time programme for the given size of the motor after consideration of thermal properties and making heat transfer calculations is designed such that the propellant reaches the temperature and mechanical properties in this time. After completion of curing, the grain/motor is cooled at a controlled rate to ensure minimum thermal strains on the propellant. These strains must not be severe enough to induce a failure either in propellant or in the propellant-liner bond. From safety point of view, water heated oven is preferred for curing solid propellants. For big motors, oil or water circulated through the mandrel to heat the motor uniformly is used.

Pressure curing is also adopted in composite case motors to avoid shrinkage. Normally pressure is applied on the motor propellant surface and this pushes some more propellant into the motor. This increases the propellant loading.

Only physical curing can take place in the case of PVC plastisol propellants (and also in double base plastiosl propellants). Here, PVC powder dissolves in a plasticizer like DOP (Dioctyl Phthalate) at 175°C to form a colloidal gel, which on cooling to ambient conditions gives the mechanical properties. PVC propellants are not case bondable and have size limitations and with lower specific impulse. Hence, they are not considered for launch vehicle applications.

Inhibition of the Grain

It is very difficult to cast the propellant to exact dimension because of porosity at the top of the cured propellant grain up to certain depth and dimensional changes occurring during elevated temperature curing of propellant. Extra propellant is trimmed off using special purpose cutting or trimming machine.

Loose flaps, provided on either end of case-bonded composite propellants to provide space for dimensional changes in the grain due to thermal expansion or shrinkage during propellant curing, need to be filled with suitable materials like epoxy resins, to avoid localized burning and oscillating pressure load created in gaps during burning.

The cured motor is cooled to ambient conditions gradually by stepped cooling for avoiding de-bonding, then the mandrel is removed (de-coring process), then the ends are trimmed to the required dimensions and subjected to NDT as detailed in chapter-V. For large and complicated grain designs, collapsible mandrels can be used for easy de-coring.

Inhibition is given to restrict the combustion of certain specific areas of the propellant grain, by forming a non-inflammable layer. They are usually filled polyurethanes. Short fibres (4 to 20 parts) of asbestos, carbon, glass or alumina are used to improve the char retention properties. A star shaped case bonded motor with inhibition on nozzle and head end sides gives a neutral burning curve. A tubular motor with end inhibitions gives a progressive curve.

The inhibition systems must have the following features:

1. It must be compatible with the propellant and have good interface properties.
2. Being a dead weight material, it should have low density to minimize the quantity.
3. It should have higher mechanical properties than that of propellant.
4. It should have low erosion rate of the order of 0.1 mm/sec.
5. It should have good char retention characteristics.
6. Thermal properties like low thermal conductivity and high specific heat.
7. Good ageing characteristics.
8. Good interface properties like tensile bond strength, peel strength and shear bond strength.
9. It should have good flow properties and be able to cure to a rubbery mass at room temperature.

A typical polyurethane system consisting of castor oil, fillers like asbestos and/or ferric oxide, etc. and diisocyanate as curing agent has the following properties:

Tensile strength	:	6–8 ksc
Elongation	:	30–50%
Hardness	:	Min 40 SAH
Density	:	Min 1.15 g/cc
Erosion rate	:	~0.1 mm/sec.
Thermal conductivity	:	~7.5 × 10^{-4} cals/cm sec. °C.

In this operation, the inhibition material is usually prepared to form a viscous liquid and cast over prepared propellant surface. It is cast layer by layer to attain the desired dimensions and geometry.

In some tactical missiles, the composite propellant (cartridge loaded type) is inhibited on both faces and outside with cloth fabric impregnated with resin. The fabric is continuously wound on the propellant surface to the required thickness. Here the inhibitor acts as an insulator also.

5. NON-DESTRUCTIVE TESTING (NDT)

Refer Chapter-V characterization and rheology.

6. MOTOR STATIC TESTING

The inhibited motor is assembled with head end and nozzle ends. Suitable igniter is fixed at head end. Nozzle throat of graphite or carbon phenolic is inserted at the convergent portion of the nozzle. Both thrust and pressure pickups are connected and the motor is static tested. Motor test details are presented in chapter-VI Internal ballistics.

Typical Questions

1. What is Maraging Steel? How is it preferred over 15CDV6?
2. What are composite cases? How is Kevlar-epoxy is preferred over Glass-epoxy?
3. Why composite cases are preferred for upper stage motors while metallic cases for boosters?
4. What is stainless steel?
5. What are the functions of an insulator?
6. Give the salient features of Rocasin insulator?
7. What are the different insulation methods?
8. What is vulcanization?
9. What is loose flap? How does it function?
10. What is EPDM? What are its merits over nitrile rubber?
11. What is reactive insulator? How does it work?
12. Explain the essential thermal properties of a good insulator?
13. What are the important casting parameters?
14. What are the in-process quality checks to realize defect free motors?
15. What is the role of liner in rocket motor? What are the postulates for improving interfacial bonds?
16. What is the liner used in HTPB motor? Explain the components and methods of application?
17. What are the important interface properties?
18. How do we ensure good bond between propellant and insulator?
19. What is the preferred liner thickness in HTPB motor? How do we measure it?
20. Why do we inhibit certain areas in a motor?
21. Explain an inhibition composition for HTPB motor?
22. Explain the process of HTPB motor with suitable flow diagram.
23. What are the actions taking place during mixing?
24. Explain the merits and demerits of vertical mixer over horizontal mixer.
25. What are the factors affecting mixing efficiency?

26. What are the advantages of continuous mixing?
27. What are Pot-life, ZFT&PPT? How can we improve the pot life of HTPB propellant?
28. What are the advantages of vacuum casting?
29. What is the suitable casting method for making multiple numbers of thin webbed motors?
30. What are the important casting parameters?
31. How do you ensure complete curing of a solid motor?
32. What method can be adopted for curing a composite case solid motor?
33. What are the in-process quality checks to realize defect free motors?

REFERENCES

[1] A method of manufacturing of solid rocketmotors, US patent No. 6101948, Donald Lee Kenaresboro, Forest Ray Goodson, Frank Stephen Inman, Date of Patent August, 15, 2000.

[2] Bhagavan, S.S., Varghese, T.L., *et al.,* "Design and Analysis of Experiments for Selection of Solid Propellant Formulations," *Proc. 3rd International Conference on High Energy Materials,* Trivandrum, 2000.

[3] Bhuvaneswari, C.M., *et al.,* Filled Ethylenepropylene Diene Terpolymer Elastomer as Thermal Insulator for Case-bonded Solid Rocket Motors, *Defence Science Journal,* Vol. 58, No. 1, January, 2008, pp. 94–102.

[4] Crump, Jesse K. and Nilda. Amy, AIAA-91-1851, *27th Joint Propulsion Conference,* pp. 1–9, 1995.

[5] Dombe, Ganesh, Jain, M., Singh, P.P., Radhakrishnan, K.K. and Bhattacharya, B., Pressure Casting of Composite Propellant, *Indian Journal of Chemical Technology,* Vol. 15, July 2008.

[6] Gould, R.F., Propellants Manufacture, Hazards, and Testing, Advances in Chemistry, No. 88, American Chemical Society, Washington, DC, 1969.

[7] Kang, Inpil, *et al.,* Preparation and properties of ethylene propylene diene rubber/multi walled carbon nanotube composites for strain sensitive materials, Composites: Part A 42, 2011, 623–630.

[8] Khan, I.A., Sathiskumar, P.S., Lakshmi, V.M., Sivaramakrishnan, R. and Alwan, S., Analyzing the Occurrence of Voids in Solid Propellant Grains. *HEMCE 2007 Proceedings,* 2007.

[9] Kivity, M., Yoskovich, B.N., Bechar, N. and Schreiber, Y., The Influence of Process Parameters and Composition on Mechanical Properties of Propellants in *42nd AIAA/ASME/SAE/ASEE Joint Propulsion Conference & Exhibit, Sacramento,* California. AIAA 2006–4951.

[10] Knaresboro, Donald L.; Goodson, Forrest R. and Inman, Frank S.; Method of manufacturing solid rocket motors, US patent No. US 6101948 (A), United Technologies Corporation, US (2000).

[11] Krishnamurthy, V.N., "Some Issues in The Development and Production of HTPB. Propellants," *Proc. 2nd International High Energy Materials Conference,* IIT, Madras, 1998.

[12] Krishnan, S., Chakravarthy, S.R. and Athithan, S.K., Propellants and Explosives Technology, ISBN 81-7023-884-6, Allied Publishers Limited, India, 1998.

[13] Kurian, A.J., Varghese, T.L. *et al.,* "Evaluation of Pot-life of HTPB Propellant Using Different Isocyanates," *Proc. 3rd International Conference on High Energy Materials,* Trivandrum, 2000.

[14] Landsem, Eva; Jensen, Tomas, L., *et al.*, Isocyanate—Free and Dual curing of Smokeless Composite Rocket Propellants, *Propellants Explos. Pyrotech Journal*, Vol. 38, 2013, pp. 75–86.

[15] Lembit, Siilats; Collapsible Mandrel, Can. Pat. Appl., CA 2299487 A1 20010824Lembit, siilats, Canada, 2001.

[16] Michel, Royce, *et al.*, "The Advanced Solid Rocket Motor," AIAA Space Program and Technology Conference, Huntsville, March 1992.

[17] Morgan, Richard E., Huntsville and Dye, Charles, B., Methods of and apparatus for casting solid propellant rocket motor, US patent No. 4836961(A), Morton Thiakol, US (1989).

[18] Reghunathan Nair, C.P., Devi Vara Prasad, C.H. and Ninan, K.N., "Effect of process parameters on the viscosity of HTPB/AP/Al based solid propellant slurry," *Journal of Energy and Chemical Engineering*, Vol. 1, 2013, pp. 1–9.

[19] Sau, K.P., *et al.*, Carbon fibre filled Conductive Composites based on Nitrile rubber (NBR), Ethylene Propylene Diene Rubber (EPDM) and their Blend, *Polymer*, Vol. 39 No. 25, pp. 6461–6471, 1998.

[20] Sikder, Arun Kanti and Reddy, Sreekantha, R*eview* on Energetic Thermoplastic Elastomers (ETPEs) for Military Science, *Propellants Explos. Pyrotech*, Vol. 38, 2013, pp. 14–28.

[21] "SNPE-Partener in ARIANE-5 Solid Rocket Motor Development," *News from Prospace*, Vol. 37, May 1995, pp. 45–47.

[22] Subramanian, S., Varghese, T.L., *et al.*, "Hydrogen Augmented Solid Rocket Boosters for Launch Vehicle Applications," *Proc. 3rd International Conference on High Energy Materials*, Trivandrum, 2000.

[23] Sutton, George P., "Book on Rocket Propulsion Element," Volume 7, edition 2001, pp. 27–36, 46–84, 417–453, 474–511.

[24] Varghese, T.L., *et al.*, "Development of a New Generation High Pot-life Hydroxyl Terminated Polybutadiene Propellant," *Proc. 2nd International High Energy Materials Conference*, IIT, Madras, 1998.

[25] Varghese, T.L., *et al.*, Book on Manual of Solid Propellant Chemicals, Special Publications, ISRO-TTG-SP-33-1987.

[26] Varghese, T.L., Chemical Propellants, IITP-2002, Area Specific Intensive Course Module III, General Lectures, Book Published by VSSC, ISRO, August 2002.

[27] Varghese, T.L., *et al.* A Method and Apparatus for Manufacturing Thin Webbed Solid Propellant Grains for Solid Propellant Motors, Patent No. 271/Mas/2001.

[28] Varghese, T.L., *et al.*, Process for Production of HTPB based Composite Solid Propellant with Active Ferric Oxide, Patent No. 1091/Mas/1999.

[29] Varghese, T.L., Krishnamoorthy, V.N., Kurup, M.R., *et al.* "Studies on Composite Extrudable Propellant with Varied Burning Rate Pressure Index," *Defence Science Journal*, Vol. 39, No. 1, Jan. 1989, pp. 1–12.

[30] Verma, Sumit and Ramakrishna, P.A., Activated charcoal, "A Novel Burn Rate Enhancer of Auminized Composite Propellants," *Combustion and Flame*, 157, 2010.

CHAPTER 5

Propellant Characterization and Rheology

1. INTRODUCTION

Propellants are structural component of a rocket motor. Structural failures of the propellant in a rocket motor can cause severe alterations in the performance of the motor. These deviations can result in pressure bursting or burn through of the rocket motor leading to complete motor failure. Hence, propellant should have good physical and mechanical properties to withstand the stresses and strains imposed on it during the various stages of its development—thermal cycling, handling, storage, transportation, ignition, firing and flight. The use of high performance propellants, capable of meeting wider extremes of environmental conditions, stress the importance of maintaining the structural integrity of the rocket motor.

Composite and homogeneous propellants (double base propellants) fall into a class of materials called visco-elastic materials. The dual nature of viscous and elastic properties makes their mechanical response so complex and so very important in maintaining the structural integrity of the motor. Composite solid propellant is a heterogeneous mixture of mainly three basic ingredients, namely, a finely divided inorganic oxidizer (AP), a powdered metallic fuel and an organic polymeric binder. The binder, in addition to its function as fuel, serves as the continuous matrix to hold the discrete particles of oxidizer, metallic fuel and other additives. The binder system consists of pre-polymers, curing agents, plasticizers, catalysts, wetting/bonding agents, stabilizers, etc. and is mainly responsible for the structural properties of solid propellants.

Homogeneous propellants are mainly nitrocellulose plasticized with an explosive plasticizer, NG. They also contain various ingredients like inert plasticizers, stabilizers, ballistic modifiers, darkening agents, high-energy ingredients, combustion instability suppressers, flash and smoke suppressers, etc. Homogeneous propellants are thermoplastics where as composite propellants are generally thermosets. Composite propellants are mainly used for case bonded applications while homogeneous propellants for free standing applications. Hence, their basic requirements of structural properties differ—for example, case bonded propellants require high elongation to withstand the strains during thermal cycling and firing loads and moderate tensile strength to withstand g-loads due to gravity and acceleration while free standing grains require highly rigid propellant with high tensile strength and high modulus to withstand the handling, firing loads and operation loads of the motor under any adverse conditions.

Thus, a solid propellant, being structural material, should meet the structural requirements such as physical, mechanical, thermal and interface properties. It must be able to withstand the thermal stresses during cooling phase, pressurization loads during firing, acceleration and vibration loads during flight, g-loads during storage and must be capable of withstanding different environmental conditions such as temperature, humidity, etc. Also, the propellant must be capable of giving good ageing characteristics and shelf life.

The major factors involved in the characterization of solid propellants are listed below:

1. Mechanical and Physical Characterization
    - Tensile strength
    - Elongation
    - Modulus of elasticity
    - Shore-a-hardness
    - Density

2. Visco-elastic Behaviour
    - Stress-strain curves
    - Stress relaxation and relaxation modulus
    - Creep compliance
    - Shear modulus
    - Failure envelope
    - Dynamic mechanical properties: Master relaxation modulus

3. Dilation in Tension
    - De-wetting strain
    - Poisson's ratio

4. Dilation in Compression
    - Bulk modulus
5. Interface Properties
    - Peel test
    - Tensile bond test
    - Shear bond test
6. Thermal Characterization
    - Specific heat
    - Coefficient of thermal expansion

– Thermal conductivity
– Glass transition
– Calorific value

7. Thermal Analysis
   – Thermo Gravimetric Analysis (TGA)
   – Differential Thermal Analysis (DTA)
   – Differential Scanning Calorimetry (DSC)

8. Rheological Characterization
   – Rheology
   – Viscosity (η)
   – Coefficient of viscosity
   – Methods of determining viscosity
   – Herschel Bulklay Equation
   – Classification of liquids based on rheology:
     (a) Newtonian liquids:
     (b) Non-Newtonian liquids:
        (i) Pseudo plastic
        (ii) Dilatants
        (iii) Thixotropic.

9. Burn Rate Characterization of Solid Propellants
   – Burn rate determination—Acoustic emission technique and Crawford techniques
   – Burn rate laws
   – Temperature sensitivity.

10. Non-Destructive Testing of Propellants
    – Visual Inspection
    – X-ray Radiography (XR)
    – Real Time Radiography (RTR)
    – Computer Aided Tomography (CAT)
    – Neutron Radiography (NR)
    – Ultrasonic Testing (UT)
    – Acoustic Emission Testing (AET)
    – Optical Holography.

2. MECHANICAL AND PHYSICAL CHARACTERIZATION

Mechanical behaviour over the full range of possible environments, including long term storage, is of concern because of the requirement that the structural integrity of the case bonded rocket motors be maintained reliably from the time of manufacture until and during their use as propulsive units. Hence, more attention is usually given to the characterization and improvement of mechanical properties than any other quality, even performance measurements. The basic mechanical property characterization is carried out, assuming the propellant is isotropic. The structural integrity problem with case bonded solid propellant rocket motors is derived from various forces imposed on the propellant grain and on the bonds between grain, insulation, and case. One major propellant requirement is the high modulus to withstand gravitational or inertial forces. The second important requirement of propellant is an elastic response to the high rates of loading necessary to withstand the blast of igniters and the vibration generally encountered during the boost phase of the launch. The case-bonded grain configuration also experiences severe strains like cure shrinkage, difference in coefficient of expansion of propellants and case material, thermal cycling during storage and expansion of case along with propellant in high performance motors at ignition. Case bonding magnifies the strains produced by temperature cycling and rapid pressurization from ignition. A more aggravated condition exists when an internal burning case bonded star port grain is fired at low temperature where strains are concentrated at the star valley, in addition to rapid pressurization strain.

An intensive study of solid propellants as structural materials led to postulate the important parameters of mechanical behavior of propellant binder by Landel. These are: a) glass transition temperature (T_g), b) thermal coefficient of expansion ($\propto$), c) initial molecular weight, d) molecular weight between entanglements, e) average internal viscosity, f) density, g) number of effective chains, h) number of equivalent statistical segments, i) maximum possible volume of filler ($\emptyset_m$), and j) a measure of adhesion between binder and filler (K).

The de-wetting strain (ε_d) is the strain (and corresponding maximum stress) where incipient failure of the interface bonds between small oxidizer crystals and the rubbery binder occur. The de-wetting strain is similar to the yield point in elastomers. The modulus at low strain (E) is used as a quality control parameter. Tensile specimens cut from same cast grain of composite propellant can show 20 to 40% variation in the strength properties between samples of different orientations. The maximum failure stresses of most propellants are relatively low compared to those of plastic materials. The values range from about 0.25 to 8 MPa with average values between 0.3 to 2 MPa and elongation range from 4 to 250% depending on the specific propellant, its temperature and its stress history. Some double-base propellants and binder rich composite propellants can withstand higher stresses. Pressure and strain rate have a major influence on the physical properties. Propellant grains must be strong enough and have elongation capability sufficient to meet the high stress concentrations present during shrinkage at low temperatures and also under the dynamic load conditions of ignition and motor operation.

Major factors associated with maintaining the structural integrity of the propellant in a rocket motor are mechanical properties such as tensile strength, elongation and modulus and physical properties such as shore-A hardness and density. The Tensile Strength (T.S.), elongation and modulus are measured by the uni-axial stress-strain test using universal testing machine with pneumatic clamping grips. Propellant dumbbells conforming to ASTM-D–412–68/C (equivalent to IS 3600) are used for universal tests. Dumbbells are punched out of 5 to 6 mm thick propellant slabs, prepared by casting propellant cartons and guillotine cutting. These dumbbells were conditioned in desiccators over anhydrous $CaCl_2$ for a minimum period of 24 hrs and tested for mechanical properties. The test is commonly done for manufacturing quality control, propellant development and determining failure criteria.

ASTD-D 412 test specimen and Uni-axial Testing Machine (UTM) are shown in Figure 1. A typical stress—strain curve of a propellant dumbbell when tested in a

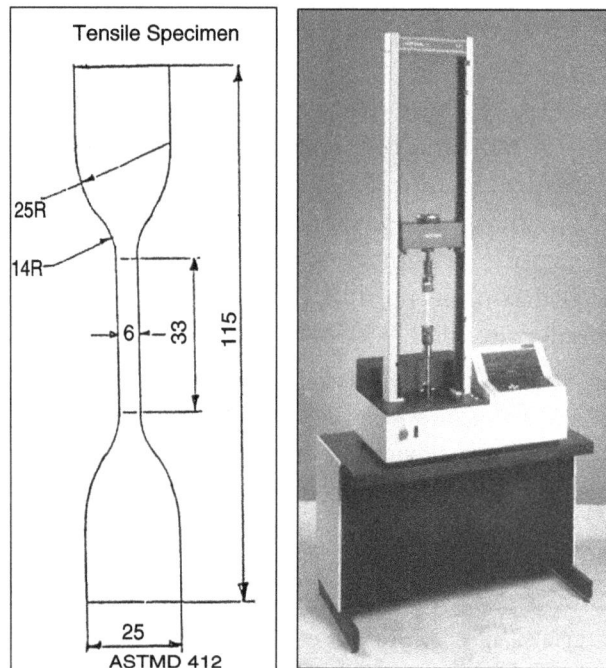

Fig. 1: ASTD-D 412 Test Specimen and Uni-Axial Testing Machine (UTM)

uni-axial tensometer is also shown in Figure 2. Besides standardization of test sample configuration and its manufacturing and handling procedure, the test equipment to be used should be that used for characterization of similar materials where test conditions span over a range of strain rates and over a wide range of temperatures. Displacement rates cover the range of 0.1 mm/min or lower to 500 mm/min. Standard laboratory tests are carried out at 50 mm/min to meet the requirements for quality control, but for characterization, a range of cross heads speeds and test temperatures are used.

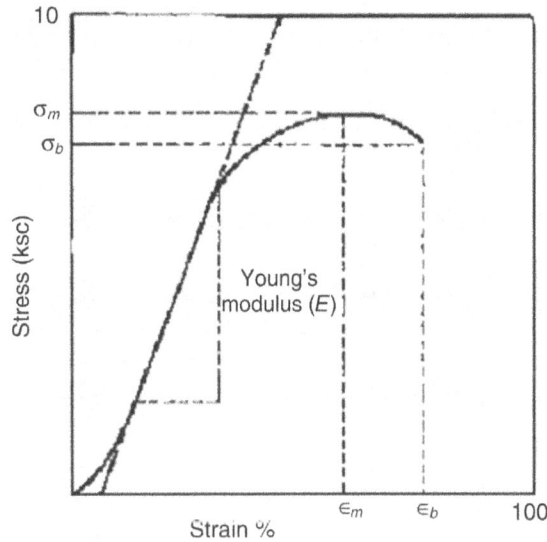

Fig. 2: Typical Stress—Strain Curve

σ_b = stress at break point or T.S. at break, σ_m = Max stress or Max T.S.
ϵ_b = strain at break point or elongation at break point.
ϵ_m = strain at max stress or elongation at max stress.
Modulus E = slope of the initial portions of the curve or Tan–δ.

Tensile Strength (T.S.) is the minimum load required to break a specimen of the material of unit cross section area. It is represented by kg/cm^2 or MPa. T.S. = force required to break the sample/area of cross section. For example., 3.0 kg is the force required to fracture a propellant dumbbell of cross sectional length 0.6 cm and breadth 0.5 cm, then tensile strength = 3.0/(0.6 × 0.5) = 10 kg/cm^2.

Since stress is the force acting per unit area (kg/cm^2), from the stress-strain curve given above, we can see that tensile strength at break point is stress at break point represented by σ_b. Maximum tensile strength is the maximum stress, denoted by σ_m. σ_m is usually taken as the tensile strength of the sample. T.S. of HTPB composite propellant is of the order of 6–10 ksc for case bonding type.

A propellant should have sufficient T.S. to resist deformations or de-shaping and slump characteristics during storage and firing and also to have easy machinability.

Elongation is the percentage extension per unit length of the sample at break point. It is represented by ϵ_b as %. Elongation = Increase in length at break point × 100/Original length of the sample. For example, we can see from the stress-strain curve given above, strain at break point or elongation at break point is the elongation of the sample, ϵ_b. The values for elongation for HTPB composite propellant is of the order of 35–50 % for case bonding type.

A propellant should have good elongation to avoid cracking or de-bonding of the grain during expansion (curing), contraction (cooling) and compression during firing and also to avoid bond separation between propellant and liner while firing or storage at different temperatures due to the difference in thermal coefficient of expansion of the rocket case and the propellant.

Modulus is generally referred as modulus of elasticity which is a measure of the elastic property of the material and is the ratio of applied stress and the resulting strain. Its unit is Kg/cm² or MPa and is represented by E. It is also known by different names such as initial modulus or Young's modulus or tensile modulus. All are one and the same and it is measured as slope of the initial portion of the stress-strain curve or Tan δ. Since stress is the force acting per unit area (kg/cm²) and strain is the % extension per unit length,

Modulus = Slope of the initial portion of the curve
= Tan–δ
= $(Y\,\text{kg/cm}^2)/X\,\%$
= $Y/X\,\text{kg/cm}^2$.

Since modulus is a stress-strain phenomenon, its measurement will give an idea about how much a propellant can withstand the stresses and strains imposed on the propellant during various stages of its development. For example, a high modulus case-bonded propellant (usually having a high T.S. and low elongation) can result in crack during thermal shrinkage. A low modulus propellant (low T.S. and very high elongation) can deform or slump more during storage. The modulus values for HTPB propellant is in the range of 35–45 kg/cm².

Hardness measurement gives the hardness of the propellant, which in turn shows the machinability of the propellant and the extent of curing in a propellant. The instrument used for its measurement is called Durometer. By pressing the durometer on the propellant flat surface, the hardness can be measured from the needle reading.

The values are represented in SAH units (Shore—A Hardness units) and are in the range of 60–80.

Density: Higher density for a propellant is preferred as more propellant can be loaded in a volume limited rocket motor resulting in higher propellant mass fraction.

Usually, solid propellants with high solid loading of 86 to 89% have densities in the range of 1.75 to 1.8 g/cc. The overall performance of a motor is governed by density impulse (I_{sp} × density) rather than I_{sp}. Density measurement at various locations of the propellant grain is also a check on the uniformity of the grain and the internal nature of the propellant like voids, homogeneity, etc.

Density of propellant samples are done by displacement method.

$$\text{Density} = \frac{\text{weight in air} \times \text{density of benzene}}{\text{loss of weight in benzene}}$$

3. VISCOELASTIC BEHAVIOUR

Composite and double base propellants show visco-elastic behaviour. Visco-elastic materials exhibit the deformations of both elastic solids and viscous liquids as seen in Figure 3. Elastic solid has a definite shape which can be deformed by external force to a new equilibrium shape and on removal of the force, it reverts back to the original shape. But a viscous liquid has no definite shape and flows irreversibly under the actions of external force. A propellant sample behaves like a visco-elastic material where stress is proportional to both strain and strain rate.

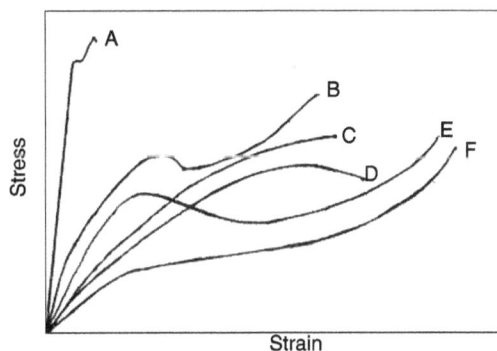

Fig. 3: Stress—Strain Curves of Different Materials

A – Steel (obeys Hook's law)
B – Plastic with yield point (PE), $T_g > RT$
C – Plastic without yield point (HDPE)
D – HTPB propellant, visco-elastic, rate and time dependent
E – PVC propellant (plastic with yield point)
F – Elastomer (rubber), $T_g < RT$.

Stress-Strain Curves of Elastic, Plastic and Visco-Elastic Materials

Steel is a perfectly elastic material up to the yield point, i.e., stress is proportional to strain. Polyethylene is a plastic with yield point and with $T_g > RT$. HDPE is a plastic without yield point. In both cases, stress/strain varies with time. The curve of PVC propellant is similar to that of plastic with yield point. HTPB propellant shows visco-elastic behaviour whereas, elastomer shows a high degree of extension as shown in Figure 3.

The propellant and rubber properties are dependent on strain rate and temperature. Hence, there is a need to correlate this data for obtaining a uniform picture. Since they are activated processes, rate dependence allows to use the generalization,

$$A(T)/Ag = \exp[-Eac/R\{(1/T) - (1/Tr\}]$$

where, Ag is the property at reference temperature Tr, $A(T)$ is the property at temperature T and Eac is the activation energy. The property is dependent on a characteristic temperature of the process as in relaxation.

$Eac/R = T_m.T_g/(T_m - T_r)$ where T_m is melt temperature and T_g glass transition temperature. The area of viscoelastic properties characterization for the effects of time-temperature is carried out on time-temperature superposition principle as a standard practice. This has been extensively used for propellants and rubbers. The reference temperature is chosen arbitrarily as the glass transition temperature of the material. The superposition of the time scale is similar with reduced time reference. The reduced time is being obtained by the relation T/a_T where a_T is the shift factor. The shift factor is obtained from the relationship,

$$\text{Log } a_T + C_1(T - T_g)/[C_2 + (T - T_g)] = 0$$

where C_1 and C_2 are constants. C_1 and C_2 for PBAN propellants are -11 and 135 respectively and -8.6 and 101.6 for SBR rubber.

Stress Relaxation and Relaxation Modulus

If the propellant sample (linear visco-elastic material) is held at constant strain, the stress will decay with time as shown in Figure 4.

Then $\sigma = \epsilon_0 f(t)$ where σ = stress, ϵ_0 = constant strain and t = time.

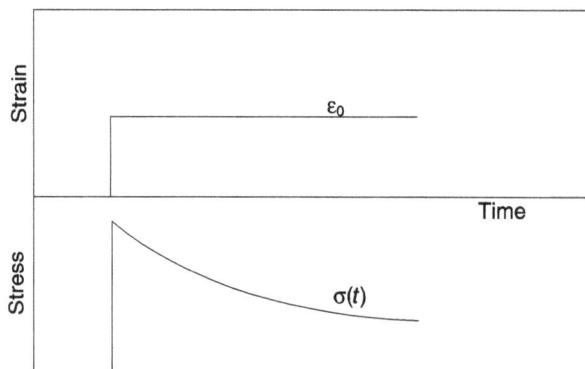

Fig. 4: Stress Relaxation Test

The modulus of material of this type is therefore time dependent. It can be calculated from the time dependent stress at constant strain. The modulus calculated in this way from the stress relaxation test is termed as the relaxation modulus.

Creep Compliance

Similarly, linear visco-elastic materials subjected to a constant stress will exhibit strain which increases with time as seen in Figure 5.

Then $\in = \sigma_o f(t)$

This phenomenon is called creep. Prolongation of creep can result in failure. The constant stress (creep) test provides the creep compliance.

The relaxation modulus and the creep compliance of a propellant give some clue to the behaviour of the propellant towards stresses and strains imposed on the propellant during handling, storage, firing and flight.

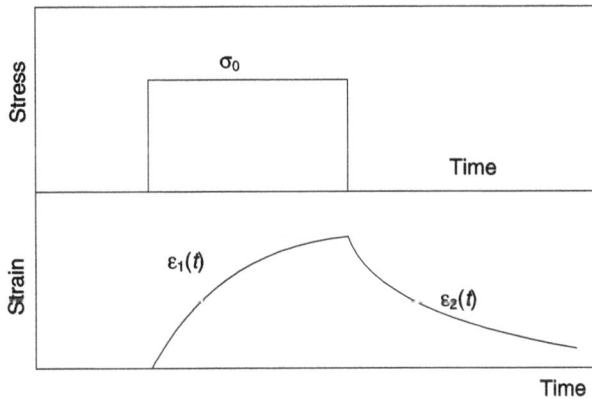

Fig. 5: Creep Test

Shear Modulus (G)

It is the ratio of shear stress to shear strain. It is measured by using UTM at a cross head speed of 50 mm/min. Sample size is $250 \times 50 \times 25$ mm block bonded to rectangular mild steel plates.

G is related to E (Young's modulus) as: $E = 3\,G$

Failure Boundaries

Failure boundaries of solid propellants are generated by plotting failure stress (corrected to reference temp) against failure strain determined over a range of temperatures from $-80°C$ to $+60°C$ and at different strain rates (0.0019, 0.019 and 0.19 sec^{-1}). The failure boundary curves of 2 different propellants are shown in

Figure 6. It may be noted that the failure strain increases counter clock wise around the envelope as the temperature is lowered or strain rate is increased and reaches maximum below room temp and then decreases as the propellant systems approaches their glass transition temp (−50°C) and below also. It could be noticed that at low rates and high temperatures, the failure curve indicates a limiting strain value beyond which failure will result. It can be taken as a long term strain on the grain. A stress-strain combination at any point inside the boundary is considered to be safer while outside the boundary may lead to failure. Out of the 2 curves, the propellant having larger boundary is better with higher margin of safety than the other. Inner area shows safe region and outer region failure.

Failure boundary or failure envelope serves as a useful input for evaluating structural integrity of the propellant under various motor conditions such as ignition pressurization, ageing, etc.

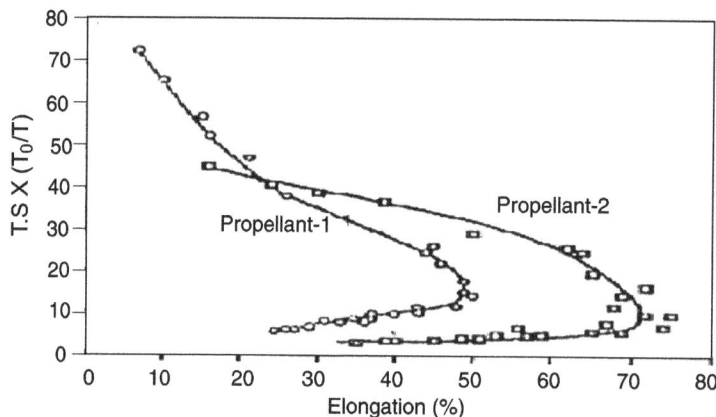

Fig. 6: Failure Boundary Envelope for Propellants 1 and 2

Dynamic Mechanical Properties

Dynamic Mechanical Analysis (DMA) is an important tool to characterize polymers and propellants. The principle is that the sample is mechanically vibrated under which it absorbs vibrational energy in the form of heat resulting in damping. It provides information on storage modulus (E'), loss modulus (E'') and loss tangent (tan δ) as a function of temperature and frequency.

The dynamic mechanical properties are evaluated using a Rheovibron Visco-elastometer at different frequencies (3.5, 11 and 35 HZ) and temperature ranges −80 to +80°C. The dynamic analyzer measures the response of the samples put to sinusoidal strain field. Sample size of 70 × 10 × 4 mm is used at a heating rate of 1°C/minute.

Master Relaxation Modulus

The dynamic modulus determined at different frequencies are converted to master relaxation modulus curve using WLF (William-Lanton-Ferry) equation:

$$\text{Log } a_T = \frac{-C_1 \ (T-T_0)}{C_2 + (T-T_0)}$$

Where a_T is the shift factor at different temps T using $T_0 = 300$ k. The WLF constants used were $C_1 = 6$, $C_2 = 157$ for HTPB propellant.

Log $E \times T_0/T$ plotted against log t/a_T in sec gives the master relaxation modulus curve as seen in Figure 7. This master curve provides modulus data over a wide range of time (from 10^{-12} sec to 10^8 sec) and for finding equilibrium modulus required for structural calculations. The high modulus in the glassy zone is required for the calculations of ignition pressurization. The curve is used as essential input for determining margin of safety calculation for the propellant.

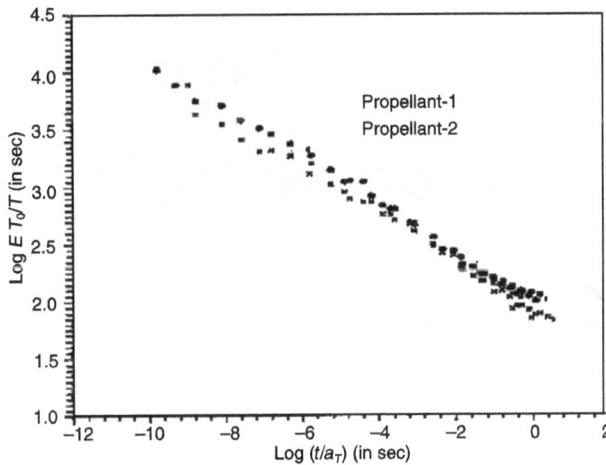

Fig. 7: Master Relaxation Modulus Curves for HTPB Propellants

Storage, Loss and Complex Moduli

Storage modulus E^1, loss modulus E^{11} and complex modulus E^* are calculated using the equation:

$$\text{Complex modulus } E^* = \frac{(L \times 10^{11} \ \text{N/m}^2)}{\partial \times A \times S \times (D-K)}$$

Where L = Length of the sample between the clamps.
    A = Amplitude factor
    D = Dynamic force reading

S = Cross sectional area of the sample
K = Instrumental constant
∂ = Phase difference
Storage modulus $E^{1} = E^{*} \cos \partial$
Loss modulus $E^{11} = E^{*} \sin \partial$.

A computer software can be used to compute E^{*}, E^{1} and E^{11}. Complex moduli is required for vibration analysis of grain structure, ignition shock wave, etc.

Glass Transition

The instrument Dynamic Mechanical Analyser (DMA) directly provides the loss tangent or damping coefficient or loss factor called tan–δ (loss modulus/storage modulus) at different temperatures. From the plot of tan δ verses temperature curve in Figure 8, the temperature of the tan–δ peak shows the glass transition temperature of cross-linked HTPB binder.

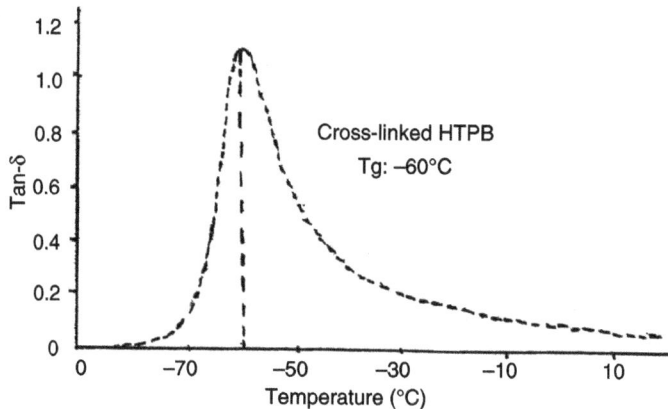

Fig. 8: DMA Variation of tan–δ with Temperature

4. DILATION IN TENSION

De-Wetting Strain (ε_d)

The volume change of propellant specimen under uni-axial tension is measured in the laboratory by attaching a liquid/gas dilatometer to the moving cross head of UTM as shown in Figure 9. Silicone oil is used as the dilatometric liquid. The de-wetting strain is taken as the strain on the $\Delta V/V$ vs. Strain plot where, slope of the curve changes abruptly.

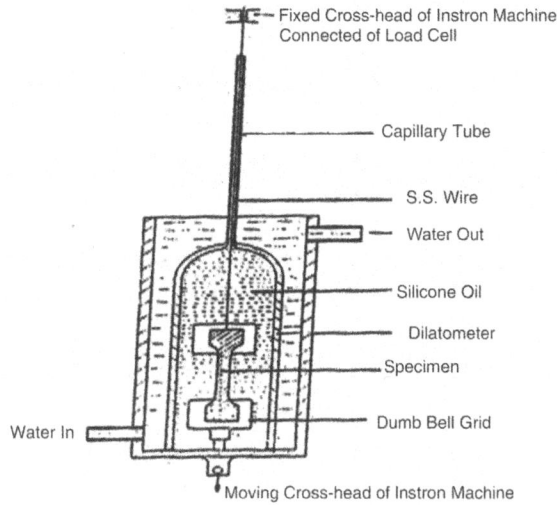

Fig. 9: Liquid Dilatometer

De-wetting strain signifies the threshold above which the propellant is no longer incompressible. A higher value of de-wetting strain (ε_d) shows better bonding between the binder and the filler, thus showing good structural integrity of the propellant. ε_d increases with increase in temperature, increase in pressure and decrease with loading ratio.

The de-wetting strain for the solid propellants is conventionally determined from volumetric change at different strain levels using JANAF dumbbell specimens as seen in Figure 10.

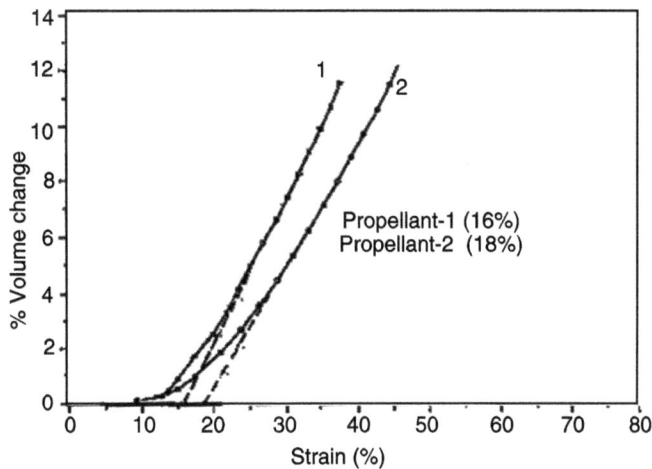

Fig. 10: Dewetting Strains of HTPB Propellant

Poisson's Ratio

Poisson's ratio (υ) is the ratio of lateral contraction to longitudinal elongation when the sample is subjected to tension. It is also indicative of volume changes that can take place in the specimen. From the plot of volume vs. strain in log scale, Poisson's ratio can be calculated using the relation:

$$[d \ln \Delta V/V)/d \ln(\infty)] = 1 - 2\upsilon$$

where, ∞ = extension ratio, υ = Poisson's ratio. For composite solid propellants, $\upsilon = 0.499$.

5. DILATION IN COMPRESSION

Bulk Modulus

Bulk modulus (k) is a measure of incompressibility of a material when subjected to hydrostatic pressure. It is the ratio of stress to volumetric strain.

In this method, volume change of a known volume of non-diffusive impermeable liquid compressed in a piston-cylinder apparatus is measured. A propellant sample is then placed in the same known volume of the same liquid. The volume change of the liquid and specimen is measured by compressing the combination as shown in Figure 11. Then the volume change of the specimen is calculated as:

Bulk modulus (k) = Hydrostatic pressure/($\Delta V/V_0$) Where ΔV is volume change. Bulk modulus is approximately three times that of Young's modulus.

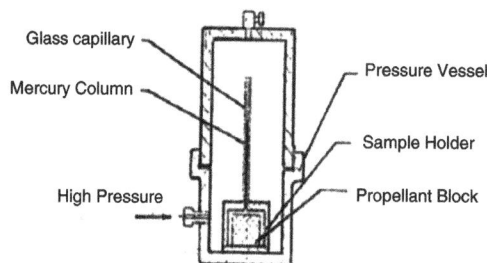

Fig. 11: Bulk Modulus in Compression

6. INTERFACE PROPERTIES

The propellant-insulation interface must be strong enough to encounter different types of loads such as:

1. Stresses caused by thermal contractions of the propellant.
2. Weight of the propellant during storage.

3. Differential movement of the grain and the case during pressurization.

4. Inertial load of the propellant (particularly in upper stage) during acceleration.

Polythene containers (150 mm diameter and 100 mm height) are lined with 2 mm thick Rocasin insulator (NBR based) along the inner and bottom surfaces. The insulator surface is abraded and then liner is applied. Propellant is cast into these cartons on the next day and cured. After curing, top surface of the propellant is trimmed and then inhibited as shown in Figure 12.

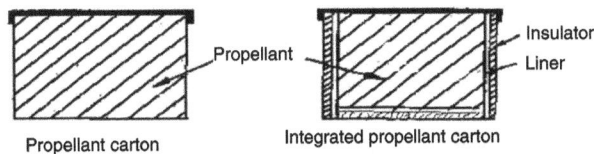

Fig. 12: Types of Propellant Samples

Peel Test

Peel discs of size 20 mm thick, 60 mm ID and 100 mm OD are cut for peel test. Suitable round metallic insert is fabricated to hold the peel discs in the UTM grips. The peel strength values are generally in the range of 0.6 to 2 kg/cm. 90 Degree dynamic peel test assembly is shown in Figure 13.

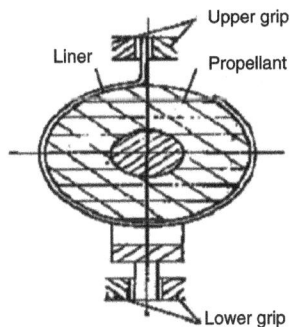

Fig. 13: 90 Degree Dynamic Peel Test Assembly

Tensile Bond Test

Tensile bond specimens are made as cubes of $25 \times 25 \times 25$ mm cut from the integral cartons and later bonded to appropriate metal blocks as shown in Figure 14a. All the specimens are conditioned in desiccators over $CaCl_2$ for a minimum period of 24 hrs and tested. Tensile Bond Strength (TBS) is generally evaluated for the propellant liner-insulator interface and propellant-inhibition interface. TBS values range from 5 to 8 ksc.

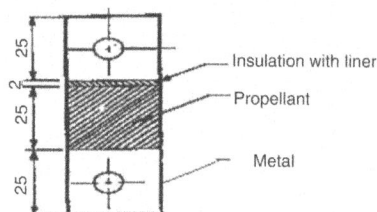

Fig. 14a: Tensile Bond Specimen

Lap Shear Test

The propellant-inhibitor interface is evaluated in terms of lap shear strength, the values of which range from 5 to 8 ksc. Lap shear specimens are bonded to Al metal substrates using epoxy adhesive for testing as shown in Figure 14b.

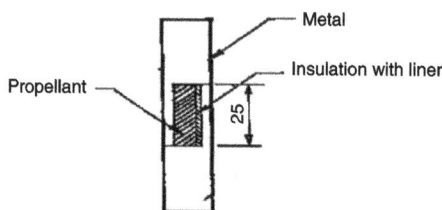

Fig. 14b: Lap Shear Specimen

7. THERMAL PROPERTIES OF PROPELLANT

Specific Heat

Specific heat (C_p) is the quantity of heat required to raise the temperature of unit weight of the substance by 1°C.

$$C_p = Q/\Delta T$$

Where, Q = Quantity of heat, ΔT = Temperature change.

Specific heat is measured using Differential Scanning Calorimeter (DSC). A sample of ≈10 mg is heated at a fixed heating rate (usually 5°C/min) up to 100°C. A blank and a standard material (sapphire) are also heated under same conditions in separate runs. The relative displacement of the DSC curve for the sample and the reference from the blank corrected base line gives the specific heat of the sample at a given temperature. Specific heat verses temperature graph is given in Figure 15.

Specific heat for propellants varies from 0.25 to 0.35 cal/gm/°C depending upon the composition. But most highly loaded aluminized composite solid propellants have C_p values in the range of 0.27 to 0.30 cal/gm/°C.

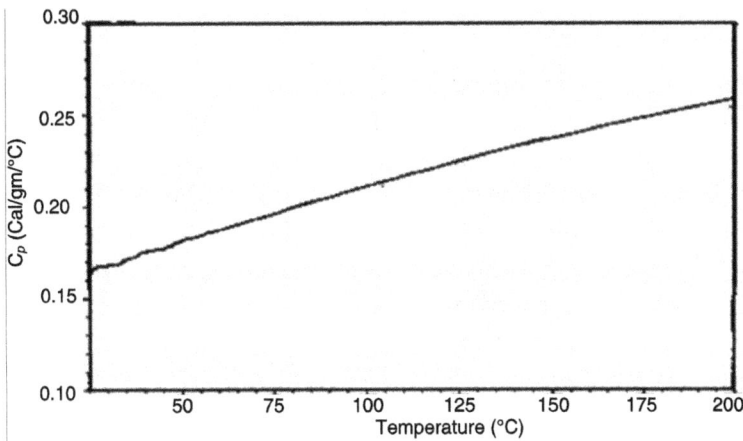

Fig. 15: Specific Heat by DSC

Coefficient of Thermal Expansion (*CTE* or ∝)

Coefficient of Thermal Expansion (*CTE* or ∝) is the ratio of increase in length to the original length for unit rise in temperature °C.

It is measured using a thermo- mechanical analyzer. $5 \times 5 \times 5$ mm sample is heated from ambient to 100°C at a heating rate of 5°C/min in static air atmosphere. The increase in length is measured using a LVDT (Linear Variable Differential Transformer). The linear expansion verses temperature graph is shown in Figure 16.

CTE or $\propto = (\Delta L/L) \times (100/\Delta T)$. Its Unit is per °C.

The value of ∝ for propellants varies from 8×10^{-5} to 1×10^{-4} per °C. Increase in solid loading of the propellant decreases the coefficient of thermal expansion.

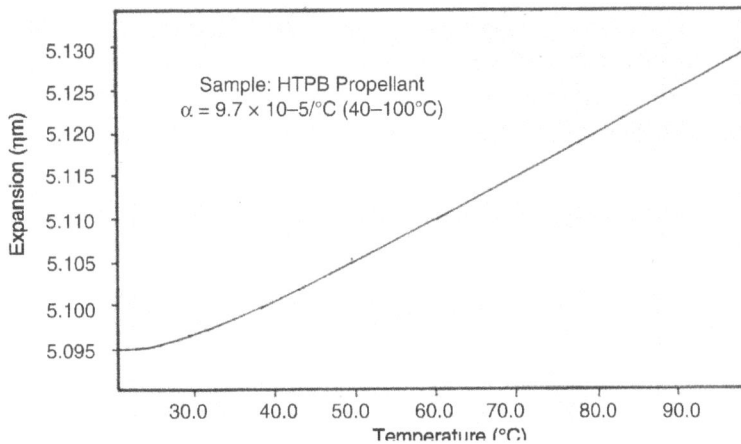

Fig. 16: Coefficient of Thermal Expansion by Perkin Elmer TMA-7

Thermal Conductivity (λ or k)

It is the quantity of heat conducted in unit time per unit area per unit temperature gradient. Different types of equipments are used to measure thermal conductivity. One such apparatus is a C-matic TCHM-DV Thermal conductivity tester. A circular disc of 5 mm diameter with thickness 2 to 10 mm is placed between 2 copper plates maintained at different temps, thus producing a flow of heat from the hotter to the colder plate. The amount of heat is measured with a thin heat flow meter attached to one of the temperature controlled plates and the thermal conductivity is determined using the Fourier heat flow equation,

$$Q = k \times A \times \Delta T / \Delta x \quad \text{or}$$

$$k = \frac{Q \times \Delta x}{A \times \Delta T}$$

Where Q = Heat flow through the sample in W
    A = Area of the disc surface in m^2
    ΔT = Change in temperature in °C
    Δx = Thickness of sample in mm

Thermal conductivity is calculated in $k = W/mK$ or = cal/cm/sec °C

Thermal conductivity values range from 5×10^{-4} to 5×10^{-3} cal/cm/s °C with the higher values for metal wired propellants.

Glass Transition

Glass transition temperature gives valuable information regarding the service temperature, flow characteristics and visco-elastic behaviour of polymers and propellants over a wide temp range.

The instrument used for T_g measurement is DSC, same as that used for measurement of specific heat. 10–20 mg sample is weighed in a standard aluminium pan, then sealed and heated at a rate of 5–10°C/minute. From the heat flow (w/g) vs. temp curve shown in Figure 17, the midpoint of transition is taken as T_g.

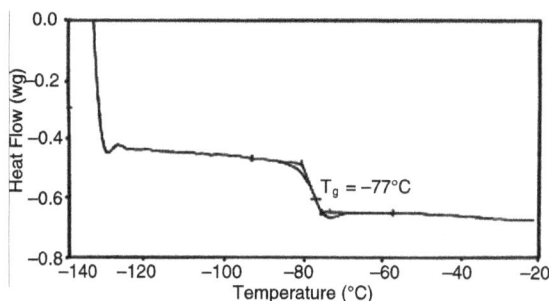

Fig. 17: Glass Transition Temperature of HTPB Propellant by DSC

The Dynamic Mechanical Analyzer (DMA) mentioned earlier also gives directly Tan δ whose peak temp shows T_g.

T_g of polybutadiene class of propellants is in the range –50 to –60°C.

Calorific Value

Calorific value of the propellant is the quantity of heat in calories evolved when 1 gm of the propellant is completely burned. It is expressed in calories per gram (cal/g). Calorific value gives an idea about the energy content of the propellant system.

Increase in the oxidizer and metallic fuel to the optimum level gives the highest calorific value.

Calorific value of solid propellants is determined using Bomb calorimeter in nitrogen atmosphere and is correctly called as calorimetric value. Calorimetric value of a typical composite propellant is 1600 cals/gm. The sample is ignited by passing current through Pt or nichrome wire passing through the propellant specimen. The temperature rise of water is measured.

$$\text{Cal value} = \frac{(w_1 s_1 + w_2 s_2) \times \Delta T}{w}$$

Where　　w_1 = wt. of water
　　　　　w_2 = wt. of bomb
　　　　　s_1 = sp. heat of water
　　　　　s_2 = sp. heat of bomb material
　　　$w_2 s_2$ = water equivalent of calorimeter
　　　　　w = wt. of sample

Heat of combustion or calorific value of a fuel is measured in oxygen atmosphere.

8. THERMAL ANALYSIS

Thermo analytical methods are very widely used to study the physical and chemical changes like thermal decomposition, cross-linking reactions and morphological changes. Thermal decomposition reactions are generally studied using thermo-analytical techniques like Thermo Gravimetric Analysis (TGA), Differential Thermal Analysis (DTA), and Differential Scanning Calorimetry (DSC). The information furnished by TGA, DTA and DSC are to some extent complementary. The rate of heating of the sample and the environment are very important factors to be controlled during thermal studies. Curing reactions of polymeric binders can be studied with the help of DSC.

Thermo Gravimetric Analysis (TGA)

Thermo Gravimetric Analysis (TGA) is the most widely used technique for studying solid state thermal decomposition reactions. TGA has an order of magnitude higher accuracy and precision compared to DSC and DTA, because TGA is based on mass measurement.

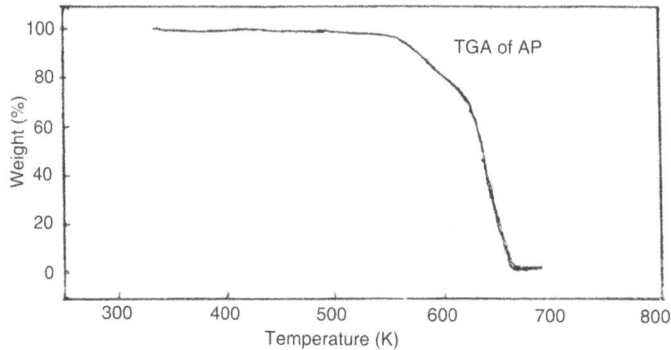

Fig. 18: TGA of Ammonium Perchlorate (AP)

Thermal analysis of the samples are usually done using Du Pont-TA-2000 Thermo gravimetric analyzer at a heating rate of 10°C/minute under N_2 flow rate of 50 ml/minute. 3 to 5 mgs of the sample is heated in a boat shaped platinum crucible using TGA Model 951. Heating rate can be changed depending on the requirement. The weight loss at different temperatures and the major decomposition peak temperatures are evaluated from the thermogram. TGA curves of Ammonium Perchlorate (AP) oxidizer is shown in Figure 18 and HTPB binder in Figure 19, from which the percentage decomposition can be found at different temperature.

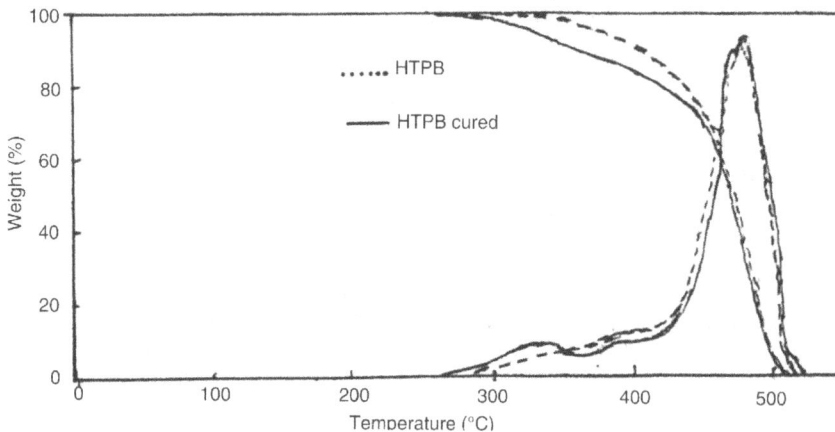

Fig. 19: TGA-DTA of HTPB and HTPB Cured

Differential Thermal Analysis (DTA)

Differential Thermal Analysis (DTA) shows whether a reaction is endothermic or exothermic. Absorption of heat is shown as −ve, i.e., endothermic indicating phase change, melting point, etc. and evolution of heat is shown as + ve, i.e., exothermic indicating decompositions. Simultaneous TG-DTA measurement is also possible. DTA curves of HTPB binder and ammonium perchlorate oxidizer are shown in Figure 19 and Figure 20 respectively. The endothermic peak at 240°C shows phase change of AP, followed by two exothermic peeks which show the low temperature and high temperature decompositions respectively. Thermal decomposition pattern of HTPB shows a two stage process. De-polymerization/cyclization occurs in the first stage corresponding to a mass loss of approximately 10%, followed by decomposition of the product from the first stage. For the cross-linked (diol-triol-TDI) HTPB, the initial stage shows two steps-release of TDI and diol-triol. DTA curve of HTPB propellant in Figure 21 shows endothermic phase change of AP at 240°C, followed by exothermic decomposition of AP and cross-linked HTPB at 285°C, 310°C and 360°C as explained above.

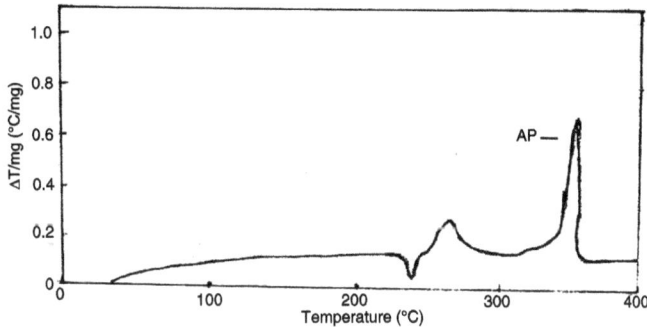

Fig. 20: DTA Curve of Ammonium Perchlorate

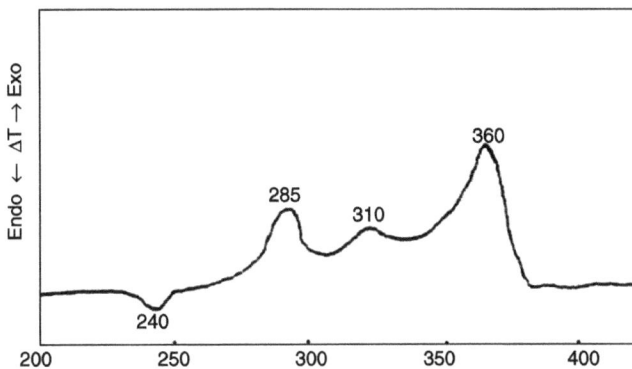

Fig. 21: DTA of HTPB-AP Propellant

Differential Scanning Calorimetry (DSC)

Differential Scanning Calorimetry (DSC) curve shows the quantity of heat flow in unit time against time or temperature. The difference in heat flow in the sample and reference sample is measured as a function of time or temperature when the two samples are heated or cooled at a controlled rate. DSC is used to study cross linking reaction and also for finding T_g of materials. Du Pont-TA-2000 in conjunction with DSC Model 905 using Aluminium cup is used for T_g determination. DSC measurements are more quantitative than DTA and are suitable for reactions involving small energy changes. DSC trace shown in Figure 22 of cross-linked HTPB is found to have T_g –70°C. DSC curve of HTPB propellant is shown in Figure 23 and its pattern is just the reverse of DTA, i.e., endothermic peak pointing upwards and exothermic pointing downwards.

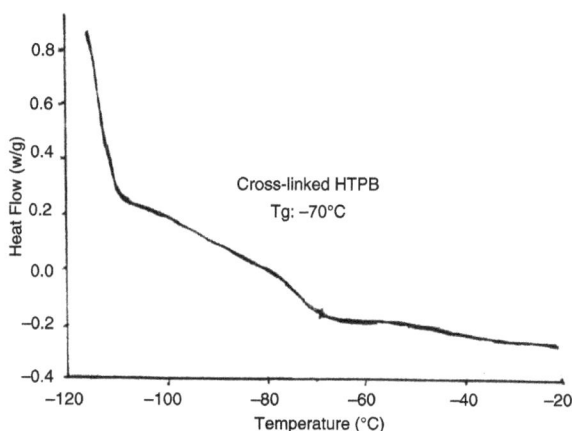

Fig. 22: DSC Trace of Cross-linked HTPB

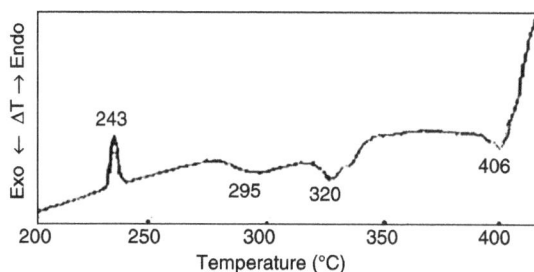

Fig. 23: DSC Trace for HTPB Propellant

The DSC trace of HTPB propellant shows the phase change of AP at 243°C, followed by the three exothermic decompositions of AP and cross-linked HTPB of the propellant.

9. RHEOLOGICAL CHARACTERIZATION

Rheology

Rheology is the study of deformation and flow of materials. Prof. Markers Reiner and Prof. Eugene C. Bingham originated the term 'Rheology'. In Greek 'Rheo' means 'flow'. Rheological characterization of composite propellant slurry is of much importance to propellant casting process as it shows complex rheological characteristics including time dependent non-Newtonian behaviour with yield stress.

Viscosity (η)

Viscosity is the resistance to flow of liquids. It is really a frictional effect due to the passage of one layer of liquid over other.

Units: Pascal seconds (PaS), Poises (Ps), Centi Poises (CPs) or Milli Poises (MPs).

$$0.1 \text{ PaS} = 1 \text{ Ps} = 100 \text{ CPs} = 1000 \text{ Millipoise}$$

Eg: HTPB Resin   : 5000 CPs at 30°C
                             : 1000 CPs at 60°C
      Water   : 1CP at 25°C
HTPB Propellant : 7000 Ps at 40°C

Coefficient of Viscosity

It is the force in dynes/cm^2 required per unit area to maintain a unit difference of viscosity between two parallel layers of liquids one cm apart. Fluidity = 1/viscosity.

Methods of Determining Viscosity

Ostwald's Viscometer

The relative viscosities of liquids w.r.to water is determined by this method. The liquid is sucked by vacuum, then allowed to fall freely from a level A to B.

$\eta_1/\eta_2 = t_1 d_1/t_2 d_2$, where η_2 = viscosity of water, d_1 and d_2 are the densities of the liquid and water and t_1 and t_2 are the time of flow of liquid and water respectively. Knowing the time of flow and density of both liquids and viscosity of one liquid, viscosity of other liquid (η_1) can be found. Time of flow depends on viscosity and density.

Brookfield Viscometer with Helipath Stand

It gives the bulk viscosity and is determined at a particular temp, speed and spindle type. Helipath stand gives helical movement to provide fresh surface of the sample for the spindle to penetrate, there by avoiding cavity effect. The instrument is provided with a temperature bath to maintain specified temperature for viscosity measurement. Disc spindles are used for viscosity measurements of resins, binders, etc. and bar spindles for propellant slurry.

Contrave's Rheometer

The instrument is provided with a concentric cylinder system where the cup is stationary and the bob rotates while the material is sheared between them. The propellant is kept in the cup in a water bath and the bob is rotated. Shear stress vs. shear rate is automatically recorded at specific time intervals. The rheological parameters are derived directly from the equipment.

Herschel Bulklay Equation

$$\tau = KD^n \quad \text{i.e.,} \quad \tau/D = KD^{n-1}$$
$$\tau/D = \eta = \text{Viscosity}$$

i.e., $\quad \eta = KD^{n-1}$

where,

τ = Shear stress, Pascals, D = Shear rate, per sec.

n = Pseudoplasticity index (flow behaviour index)

K = Consistency Index.

Classification of Liquids based on Rheology

Newtonian Liquids

Here viscosity is independent of shear rate and remains constant as shown in the curve. Shear stress is proportional to the shear rate giving a straight line, e.g. water,

$$\eta = KD^{n-1}$$

When $n = 1, \eta = K$,   i.e., Viscosity is constant.

Non-Newtonian Liquids

Here, viscosity is dependent on shear rate. Viscosity can decrease or increase depending on whether the liquid is pseudo plastic or dilatants.

Pseudo Plastic

$$\eta = KD^{n-1}$$
$n = < 1$, Pseudo plastic, e.g., Propellant slurry.

Here viscosity decreases with increase in shear rate as seen in Figure 24. Shear stress verses shear rate curve in Figure 25 shows a decreasing trend. The decrease in viscosity is due to the alignment of random particles offering less resistance to flow of liquid.

$n = 0.6$ for PS-0 propellant

$n = 0.75$ for PS-1 propellant.

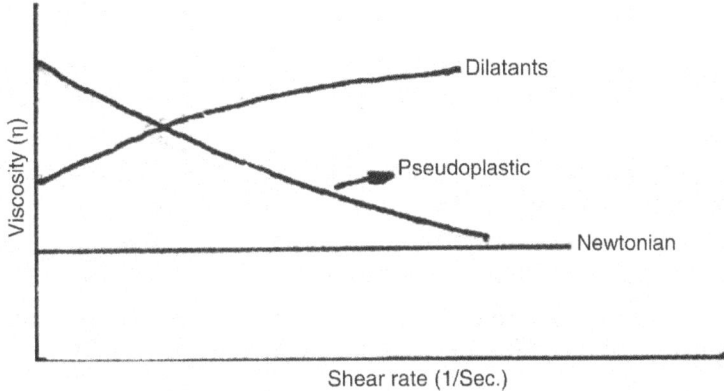

Fig. 24: Viscosity vs. Shear Rate Curve

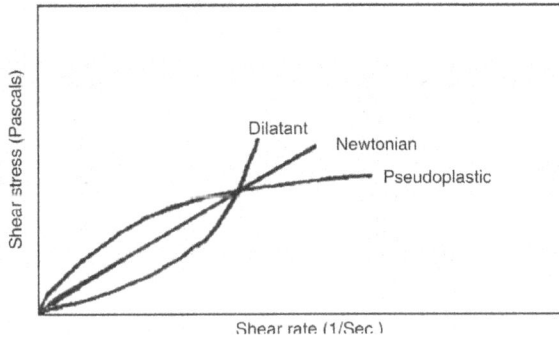

Fig. 25: Shear Stress vs. Shear Rate Curve

Dilatants

$n = > 1$, Dilatant e.g., starch, gelatin.

Here viscosity increases with increase in shear rate as shown in the curve. Shear stress verses shear rate curve shows a increasing trend.

Thixotropic

Certain fluids like uncured propellant slurry when subjected to a constant rate of shear shows a decrease in viscosity as a function of time of shearing, i.e., time dependent. The hysterisis loop in the rheogram of propellant slurry is due to thixotropic nature. This is due to orientation of long molecules in the direction of shear. Thixotropic effect is of not much importance to propellant processing. Composite HTPB propellant slurry is pseudo plastic as well as thixotopic in nature.

10. BURN RATE CHARACTERIZATION OF SOLID PROPELLANTS

Burn Rate Determination—Acoustic Emission Technique and Burn Rate Law

Acoustic emission technique is one of the modern methods for evaluating the burn rate of solid propellants. This method is being widely used to determine the burn rate of solid propellants using cured strands during the development and production stages of solid propellants. Strand burner offers advantage of testing large number of samples at precisely maintained predetermined test pressures to arrive at accurate burn rate and burn rate law. The burn rate at an operating pressure and burn rate law are further confirmed by testing control motors.

Another recent technique to determine burn rate is by Ultrasonic method using a specially designed small solid motor (refer Chapter-6 Internal ballistics of rockets).

The burn rate dependence on pressure of most solid propellants can be expressed by a simple power-law relationship known as St. Robert's Law: $r = ap^n$, where r = burn rate, a = constant known as temperature coefficient, p = pressure and n = pressure index.

The figure on acoustic emission strand burner (Figure 26) illustrates the experimental set up. Strands of size $6 \times 6 \times 80$ mm are ignited electrically by hot wire (fuse wire of 5–10 amperes) and burned under water in a stainless steel bomb. Constant nitrogen pressures ranging from 30 to 70 ksc are maintained in the bomb and surge tank by charging nitrogen from high pressure gas cylinders. An acoustic transducer is mounted on the side of the combustion chamber. The burning time is obtained from the acoustic output.

From the length of the strand and burning time, burn rate at that particular pressure is calculated. A minimum of 5 strands are tested at every pressure to get mean burn rate with standard deviation. In the case of a control motor,

Burn rate = Web thickness/Burn time

Burn rate is a characteristic property of a propellant and is independent of size and geometry of the propellant grain. The factors affecting burn rate are:

1. Pressure—increase in pressure increases burn rate.
2. Initial temperature of the propellant—higher the initial temp, faster the burn rate.
3. Particle size of oxidizer, metallic fuel and ballistic accelerator-decrease in particle size increases burn rate.
4. Concentration of oxidizer and ballistic accelerator—higher concentration increases burn rate.

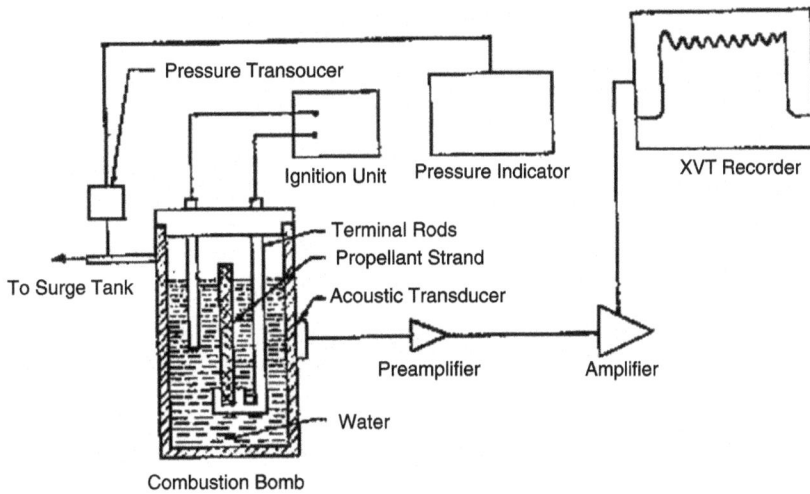

Fig. 26: Acoustic Emission Strand Burner

The plot of log *r* verses log *P* gives a straight line. The slope of the curve gives the value of pressure index '*n*' and from the intercept, a is calculated. The value of '*a*' varies from 0.1 to 0.5 and *n* from 0.2 to 0.5. The value of *n* < 0.5 is preferred in solid propellants. The high value of *n* above 0.5 makes the propellant sensitive to pressure and can lead to explosion.

Crawford Bomb Method

The earlier method of determining burn rate is by using Crawford Bomb in nitrogen pressure with inhibited cylindrical strands of dimension same as given above. The acoustic method is found to be simpler and more accurate than the Crawford Bomb experiment.

Certain propellants show constant burn rate over a range of pressures. Such propellants are called Plateau Burning Propellants and the value of '*n*' is zero. Certain propellants show decrease in burn rate with increase in pressure. Such propellants are mesa burning type and have '*n*' value negative.

Temperature Sensitivity

Rocket motors are generally conditioned for many hours at a particular temperature before firing to ensure uniform propellant grain temperature because burn rate increases with increase in temperature of the grain. This can change the chamber pressure, thrust and burn time of the propellant which in tern affect the performance. Hence, evaluation of temperature sensitivity coefficient σ_p is important for a propellant.

The sensitivity of burn rate to the propellant temperature can be expressed in terms of temperature coefficients.

$$\sigma_p = (1/r)\ (dr/dT)_p \times 100$$

Where σ_p is the temp sensitivity of burn rate, expressed as % change of burn rate per degree change in propellant temperature at a particular value of chamber pressure.

σ_p for a propellant is calculated from the strand burning rates determined over a temperature range of 0°C to 55°C at a particular pressure. σ_p values for composite solid propellants are generally less than 0.4%/°C and range from 0.2 to 0.3%/°C and 'n' value 0.30 to 0.45.

Temperature sensitivity of pressure π_k is expressed as % change of chamber pressure per degree change at a particular value of area ratio k between the burning surface area and the throat area. It is obtained from motor test data.

11. NON-DESTRUCTIVE TESTING (NDT)

Non-Destructive Testing (NDT) is the testing of materials to find out the defects in it without destroying it. Various NDT methods are used in rocketry for the inspection and quality control of all kinds of solid propellant rocket motors and related materials such as nozzles, igniter, pyrotechnic devices, fasteners, etc. The common defects are cavities, de-bonds, cracks, voids, agglomeration, porosity, foreign material inclusions, etc.

The main NDT methods are:
1. Visual Inspection
2. X-ray Radiography (XR)
3. Real Time Radiography (RTR)
4. Computer Aided Tomography (CAT)
5. Neutron Radiography (NR)
6. Ultrasonic Testing (UT)
7. Acoustic Emission Testing (AET)
8. Optical Holography.

Visual Inspection

Visual inspection of motor case, propellant grain, rocket motor, or any article is the preliminary step in NDT and is done using borescope and fibrescope. All finished products are inspected for accurate dimensions using precision instruments. Any visible defects such as cracks, de-bonds, blowholes, etc. with respect to location are well recorded. Detailed inspection and quality control of rocket components are done using various NDT techniques detailed below.

X-ray Radiography (XR)

X-ray radiography is the major NDT method used for the inspection and evaluation of rocket motors. When X-rays (0.001 A° to 1 A°) penetrate the component, they undergo absorption, scattering, and transmission. Scattering and absorption causes attenuation, i.e., reduction in the intensity of the X-ray beam. The transmitted beam falls on the film kept very close to the object. Higher density objects such as lead or steel give more attenuation (less transmission) than lower density materials such as propellant, insulation, and inhibition systems. The film gets lesser radiation and lighter shade for high density materials compared to lower density materials.

The exposed film is processed to get a radiograph which gives a permanent image of the object. The degree of blackness of the shade gives optical density of the radiograph which is in the range of 2.0 to 3.0. Depending on the thickness of the component, exposure to correct amount of radiation is required to get a clear image on the film. Conventional low energy X-ray tubes up to 450 kv are used for inspection of smaller to medium size items such as small propellant grains, igniters, small nozzles, etc. High energy X-ray machines such as linear accelerators (linacs) up to 15 MeV (Million Electron Volts) are used for big solid propellant boosters.

Two types of defects can occur in a rocket motor:
1. Propellant defects such as cracks, tears, porosity, voids, foreign materials, inhomogenity and slumping.
2. Interface defects such as debonds, delamination and defects in loose flap filling.

The detection of these defects is very essential as these may lead to shoot up of chamber pressure leading to mission failure. Among the defects, propellant crack is the most critical one as the flame can penetrate through it resulting in the detonation of the motor.

Defects in the propellant is detected by propellant grain coverage technique in which the X-rays are directed normal to the propellant grain geometry. De-bond between steel/fibre glass-insulation-liner is detected by interfacial coverage technique in which the X-rays are directed tangential to the interface. Defects in big size rocket motors are detected by triangulation technique. In this method, the imaging plane is kept constant and the X-ray beam is directed through the motor at three different angles and generating the corresponding radiographs.

Factors affecting radiographic inspection:

1. X-ray energy and output
2. Source to film distance
3. Quality of film
4. Exposure duration
5. Type of object
6. Thickness of the object
7. Optical density.

Medium speed X-ray film D7 as well as fine grain film D4 are used for both propellant grain and interfacial radiography of motors.

Real Time Radiography (RTR)

Real time radiography offer fast and nearly 100% coverage for the various defects in the motor. RTR avoids the use of expensive X-ray films and film processing.

The principle involved is that X-ray images are electronic in nature and can be digitized and displayed. X-rays transmitted by a component falls on a fluorescent screen and its light output is coupled to a CCD (charge coupled device) camera. The corresponding electronic signals are then digitized and the radiographic image is displayed on a high resolution monitor of a computer.

Both interfacial and propellant grain coverage can be done by rotating the SRM on a rotary table or raising/lowering of X-ray machine. X-ray images are instantly displayed on the monitor and 100% coverage of SRM possible at all angles. However, NDT man has to face the problem of interpretation of hundreds of images to locate the various defects like voids, cracks, de-bond, etc.

Computer Aided Tomography (CAT)

Computer Aided Tomography (CAT) is the most modern development in radiography and it provides a three dimensional image of the internal structure of the component. CAT can give contrast sensitiveness as low as 0.1% to 0.2%. This makes CAT an ideal choice for detection of cracks.

The principle involved is that CAT gives an image of a cross section of thin slice of a component under inspection. Tomographic system exposes only thin cross sectional slice of the component at a time. The component is rotated 180° in small increments and the X-ray energy transmitted by the component are taken at each angle and converted to the cross sectional image. CAT is commonly known as scanning in medical field and it is being introduced now in industrial radiography.

Neutron Radiography (NR)

X-rays are highly absorbed or attenuated by high density materials while neutrons are attenuated severely by low density materials like explosives used in pyrotechnic devices. Thus, neutrons readily image low density explosive materials encased in high density materials used in pyrotechnic devices. Thermal neutrons in the energy range of 0.01 to 0.5 eV are used in neutron radiography. Since neutrons are neutral particles, they are to be converted to radiations or charged particles to make them sensitive to photographic action on a film to form image.

Neutron radiography is more suited than X-rays for inspection of: i) light materials encased in dense materials, ii) thin samples of light materials and iii) highly radioactive materials. Neutron radiography is the only suitable method for the inspection of pyrodevices such as detonators, explosive bolts, cable cutter, etc.

Ultrasonic Testing (UT)

Ultrasonic Testing (UT) is a versatile NDT method for testing the bonding between the rocket chamber (metallic) and insulation, metal to ablative in nozzles and even non-metal to other materials. UT is very sensitive to planar defects like cracks. UT can also detect defects like crack on the case and defects in the weld region.

Ultra sound is the vibration produced by the oscillation of a piezo-electric crystal by the application of electric pulse. The ultrasonic wave from the piezo-electric probe is coupled to the motor case. The reflected energy reach back to the crystal to produce electric pulse which is monitored by means of an oscilloscope. A portion of the energy is being transmitted to the bonded insulation also. This difference gives an idea about the bond condition. Signal processing technique in UT can detect the size and shape of the defect.

Since UT gives high attenuation, it is not suitable for testing propellants. Radiography is ideal for NDT of propellants.

Acoustic Emission Testing (AET)

Acoustic Emission Testing (AET) method is used to detect any defect that gives acoustic emission under stress. AET method is found to be a very effective technique to assess the structural integrity of pressure vessel like rocket motor and detect flaws during proof pressure tests. The rapid release of energy prior to failure prevents any catastrophic failure. During pressure testing, piezo-electric transducers pick up the AE signals, amplified and recorded by the analyzer. The pressurization is done in steps of 10 kgf/cm^2 and up to the maximum operating pressure of the motor. AET along with other NDT techniques helps in improved flaw detection.

Optical Holography

Holographic NDT is used mainly for smaller rocket grains and other components. Holography is basically an optical interferometric technique which is very sensitive to inner cracks, de-bonds, de-lamination, etc. Laser light is split into an object beam and reference beam. The object beam strikes the object surface and gets scattered. The scattered beam interferes with the reference beam to give interference pattern which is recorded on a photographic plate as hologram. The presence of voids, cracks, de-bonds, etc. produce abrupt changes in the interference pattern.

Typical Questions

1. Draw a stress-strain curve of a solid propellant and explain how T.S., elongation and young's modulus are found?
2. What is the importance of mechanical properties for solid propellants?
3. What are dynamic mechanical properties? How is T_g measured?
4. What is Visco-elasticity? Explain with Suitable stress-strain curve.
5. Explain the following: a) Bulk modulus, b) De-wetting strain, c) Herschel Bulklay Equation.
6. What are the interface properties required for a solid motor? How are they determined?
7. What is the density of HTPB propellant? How is it measured?
8. What are the thermal properties required for a solid propellant? Give their values for HTPB propellant.
9. Explain the following:
   (a) Calorific Value (b) Glass Transition (c) Rheology, and (d) Fluidity.
10. Explain acoustic emission technique for determining burn rate of solid propellants?
11. Explain 2 methods of determining Viscosity of fluids?
    What is the viscosity of HTPB resin and HTPB propellant?
12. What are Newtonian and Non-Newtonian liquids? Explain the following:
    (a) Pseudoplasticity (b) Thixotropy
13. How is burn rate law of a solid propellant determined?
14. How is temperature sensitivity of solid propellants determined?
15. Explain the following: a) Neutron Radiography, b) Optical Holography, c) Computer Aided Tomography, d) Ultrasonic Testing.
16. What is X-ray Radiography and Real Time Radiography? What are the factors affecting it?
17. What are the NDT methods for detecting cracks and de-bonds in a solid propellant motor?

18. What is the ideal method for detecting defects during proof pressure testing of a steel motor case?
19. What is meant by the term Thermal Analysis? Explain the various thermo-analytical methods used in thermal analysis?
20. Explain the TGA-DTA-DSC curves of ammonium per chlorate and a composite propellant containing it?

REFERENCES

[1] AGARD: AR350: Structural Assessment of solid propellant grains: Dec. 1997.
[2] Alderman, N., Applied Rheology Newsletter, An update on yield stress, Warren Spring Laboratory Dec. 1991.
[3] Bohn, M.A. and Elsner, P., *Prop, Explos, Pyrotech*, 24, 199, 1999.
[4] Brown, M.E., Introduction to Thermal Analysis: Techniques and Application, New York, Kluwer Academic Publishers, 2001.
[5] Cauty, F., "Ultrasonic Method Applied to Full-Scale Solid Rocket Motors," *Journal of Propulsion and Power*, Vol. 16, No. 3, 2000, pp. 523–528.
[6] Charsley, E.L., Thermal Analysis—Techniques and applications, Warrington, S.B. (ed.), Royal society of Chemistry, London, 1987.
[7] Dealy, J.M., Wissbrun, K.F., Melt rheology and its role in plastics processing—Theory and application. Chapman and Hall, 1995, pp. 340–341.
[8] Drake, G., Hawkins, T., Brand, A., Hall, L., McKay, M., Vij, A. and Ismail, I. *Propellants, Explos., Pyrotech.*, 28, 174, 2003.
[9] Ferry, J.D., "Viscoelastic Properties of Polymers," Wiley Publications, 1980, pp. 130–154.
[10] Gould, R.F., Propellants Manufacture, Hazards, and Testing, Advances in Chemistry, No. 88, American Chemical Society, Washington, DC, 1969.
[11] Guo, C.J. and Uhlherr, P.H.T., Yielding Behaviour of Viscoplastic Materials: Review, *J. Ind. Eng. Chem.*, Vol. 12, No. 5, 2006, 653–662.
[12] Gupta, B.L., Varma, M. and Goel, S.B., "Rheological characterisation of fuming nitric acid gel," *J. Prop. Explos. Pyrotech,* Vol. 11, 1986, pp. 85–90.
[13] Hatakeyama, T. and Liu, Z., (Eds), Hand book of thermal analysis, John Wiley and Sons, New York, 1998.
[14] J.-P., "Solid Propellant Burning Rate Measurement Methods Used within the NATO Propulsion Community," AIAA Paper 2001–3948, July 2001.
[15] KA, Tejasvi and Panigrahia, S.K., *et al.*, Estimation of Pressure index from the Single Ballistic Evaluation Motor, *Proc. 9th International High Energy Materials Conference,* Feb. 13–14, Trivandrum, 2014.
[16] Krishnan, S., Chakravarthy, S.R. and Athithan, S.K., Propellants and explosives technology, ISBN 81-7023-884-6, Allied Publishers Limited, India, 1998.
[17] Kubota, Naminosuke/Wiley-VCH, Propellant and explosives: Thermochemical aspects of combustion, 2002, ISBN 3527-302107.
[18] Lakshmi, R. and Athithan, S.K., "An empirical model for the viscosity build-up of HTPB based solid propellant slurry," *Polymer composite*, Vol. 20, 1999, p. 346.
[19] Lohrmann, M., Hubner, "Influence of the Marrix—Filler Interaction on the Mechanical Properties of Filled Elastomers," Energetic Materials, ICT, 1996.

[20] Manelis, G.B., Nasin, G.M. and Rubtsov, Yu I., Eds Taylor and Francis, Thermal decomposition and combustion of explosives and propellants, 2003.

[21] Mathammal, B. Rani *et al.*, Methodology for Calculating Modulus for Composite Solid Propellants, *Proc. 9ᵗʰ International High Energy Materials Conference*, Fb. 13–14, Trivandrum, 2014.

[22] Mathew, S., Manu, S.K. and Varghese, T.L., *Propellants, Explos., Pyrotech.*, 33, 146, 2008.

[23] Mohan, Y.M., Raju, M.V. and Raju, K.M., *J. Appl. Polym. Sci.*, 93, 2004.

[24] Muthiah, R.M., Krishnamurthy, V.N. and Gupta, B.R., Rheology of HTPB propellant. 1. Effect of solid loading, oxidizer particle size, and aluminum content. *J. Appl. Poly. Sci.*, 44 (1992), 2043–2052.

[25] Muthiah, R.M., Manjari, R., Krishnamurthy, V.N. and Gupta, B.R., Rheology of HTPB Propellant: Development of Generalized Correlation and Evaluation of Pot Life, Propellants, Explosives, Pyrotechnics 21, 1996, 186–192.

[26] Nawaze, Q. and Nizam, F., "Viscoelastic responses of HTPB based solid fuel to horizontal and vertical storage slumping conditions," *Key engineering materials*, Vol. 510–511, pp. 22–31, 2012.

[27] Pinumallaa, Kiran and Jeenu, R., *et al.*, usage of ultrasonic burning rate for the performance prediction of large solid booster motors, *Proc. 9ᵗʰ International High Energy Materials Conference*, Feb. 13–14, Trivandrum, 2014.

[28] Rajan, Manjari *et al.*, Structure property relationship of HTPB based propellants: Effect of hydroxyl value of HTPB resin, *Journal of Applied Polymer Sci.*, Vol. 48, 1993, 271–278.

[29] Reshmi, S., Varghese, T.L. and Ninan, K.N., *et al.*, *34ᵗʰ International Annual Conference of ICT.* Germany, 2003.

[30] Reshmi, S., Varghese, T.L. and Ninan, K.N., *et al.*, *35ᵗʰ International Annual Conference of ICT.* Germany, 2004.

[31] Sekkar, V., Krishnamurthy, V.N. and Jain, S.K., *J. Appl. Polym. Sci.*, 66, 1795, 1997.

[32] Sekkar, V., Venketachalam, S. and Ninan, K.N., *Eur. Polym. J.*, 38, 169, 2002.

[33] Selim, K., Ozkar, S. and Yilmaz, L., *J. Appl. Polym. Sci.*, Vol. 77, 538, 2000.

[34] Shull, Peter J. Non-Destructive Evaluation, Theory Techniques and Applications, New York: Marcel Dekker, 2002, pp. 503–506.

[35] Sreelatha, S.P. and Ninan, K.N., *Thermochim Acta*, 290, 191, 1997.

[36] Sutton, Rocket Propulsion Elements, Sixth Edition, John Wiley, New York, 1992.

[37] Toncu, Gheorghita; Stanciu, Virgil and Toncu, Dana-Cristina, "Solid Propellant Rocket Motor Thrust Correlation with Initial Temperature of Propellant Charge," *U.P.B. Sci. Bull.*, Series D, Vol. 73, Iss. 1, 2011.

[38] Turi, E.A., Ed, Thermal Characterization of Polymeric Materials, Vol. 1 and 2, 2ⁿᵈ Edition, Academic Press, San deigo, CA, 1997.

[39] Varghese, T.L, Ninan, K.N. and Krishnamurthy, V.N., *et al.*, Performance Evaluation and Experimental Studies on Metallized Gel Propellants, *Defence Sci. J*, Vol. 49, No. 1, 77–78, 1999.

[40] Varghese, T.L. *et al.*, "Development of a New Generation High Pot-life Hydroxyl Terminated Polybutadiene Propellant," *Proc. 2ⁿᵈ International High Energy Materials Conference*, IIT, Madras, 1998.

[41] Veit, P.W., Landuk, L.G., Simpson Jr., J.W. and Svob, G.J., Evolution of an Ageing Programme— Minuteman Stage II Solid Rocket Motor, AIAA-88-3328, July 1988.

[42] Wunderlich, B., Thermal Analysis, Academic Press, San Diego, 1990.

[43] Yang, Vigour, Solid Propellant Chemistry, Combustion and motor internal ballistics, AIAA-2000, ISBN 1 56347-442-5.

CHAPTER 6

Internal Ballistics of Rockets

1. INTRODUCTION

Ballistics is the science of mechanics that deals with the motion, behavior and effects of projectiles, especially bullets, rockets or the like. In other words, it gives the dynamical aspects of propelling, free flight and reaction at or near the terminal point within the atmosphere and the instrumentation associated with various types of measurements. Ballistics is often broken down into the following four categories: i) internal ballistics, ii) transition or intermediate ballistics, iii) external ballistics, and iv) terminal ballistics. The study involves the processes originally accelerating the projectile, for example, the passage of a bullet through the barrel of a gun or mortar or a rocket; right from ignition of propellant upto the point of burnout of propellant for rockets or upto the point of exit from gun/mortar barrel.

A solid propellant rocket motor consists of a combustion chamber containing the propellant, an igniter usually assembled at the rear end and a nozzle with throat insert at the other end. On igniting the motor, the combustion gases are accelerated at the convergent portion and exit through the divergent portion of the nozzle. The propulsive force is obtained by ejecting propellant gases through convergent/divergent nozzles at high velocities. To deliver a specific thrust for a time duration i.e., the total impulse to meet the mission requirement, the burn rate and mass discharge rate of propellant are very important parameters. The propellant grain can be free standing type (without bonding to the case) or case bonding type (with liner bonding to the case) with suitable internal geometry of the grain. The motor burning pattern can be progressive, regressive or neutral depending on the type and shape of the propellant grain. The burning surface of a propellant grain recedes in a direction generally perpendicular to the surface. The rate of regression, expressed in cm/sec or mm/sec or in/sec, is the burning rate.

2. SOLID PROPELLANT BURN RATE

Burn rate of a solid propellant is an important factor affecting the performance of the motor and the mission. Burn rate is the characteristic property of a propellant and is independent of size and geometry of the propellant grain. For composite propellants, it can be increased by changing the propellant characteristics like catalyst percentage, oxidizer percentage and particle size, increasing the heat of combustion of binder/plasticizer, chamber pressure increase, increasing the initial temperature of propellant and acceleration and spin. Burning rates of solid propellants are obtained by three ways viz., a) strand burner method, b) small scale ballistic evalution motors, and c) full scale motors. Burn rate of the propellant sample is evaluated using 6 mm square strands by acoustic emission technique or using 6 mm round inhibited samples by Crawford Bomb experiment under nitrogen pressure. Strand burner burning rate is usually lower than that obtained from motor firing by about 10% as it does not simulate the hot chamber environment. The ballistic evaluation motors also give lower burning rate values compared to full scale larger motors because of scale up factors. Strand burning data are useful in screening propellant formulations and in quality control operations.

For all solid propellants, burning rate is a function of chamber pressure, at least over a limited pressure range. For most propellants, the empirical equation relating chamber pressure (P_c) with burning rate (r) is as follows:

$$r = a\, P_c^{\,n}$$

Burning rate r is expressed in cm/sec or in/sec and chamber pressure P in MPa or psia, a is a constant dependent on initial grain temperature, and n is called burning rate exponent or combustion index which is dependent on initial grain temperature and chamber pressure. For stable operation, n has values greater than 0 and less than 1. High values of n give rapid increase of burning rate with pressure. As n approaches 1, burning rate and chamber pressure become very sensitive to one another. When n is low and approach closer to zero, burning becomes unstable and may extinguish. The methods of determining burn rate, burn rate laws and temperature sensitivity of burn rate are explained under Chapter-V.

3. MOTOR BURNING PATTERN

Depending upon the internal geometry of the grain and the type such as case bonded or free-standing, there are three different patterns of motor burning. They are progressive, regressive and neutral burning and are represented by Figure 1.

Progressive Burning Motor

In the case of a case bonded radial burning motor, the burning area increases with time during burning. Hence, the thrust or pressure increases with time of burning and hence, progressive as shown in the first curve in Figure 1.

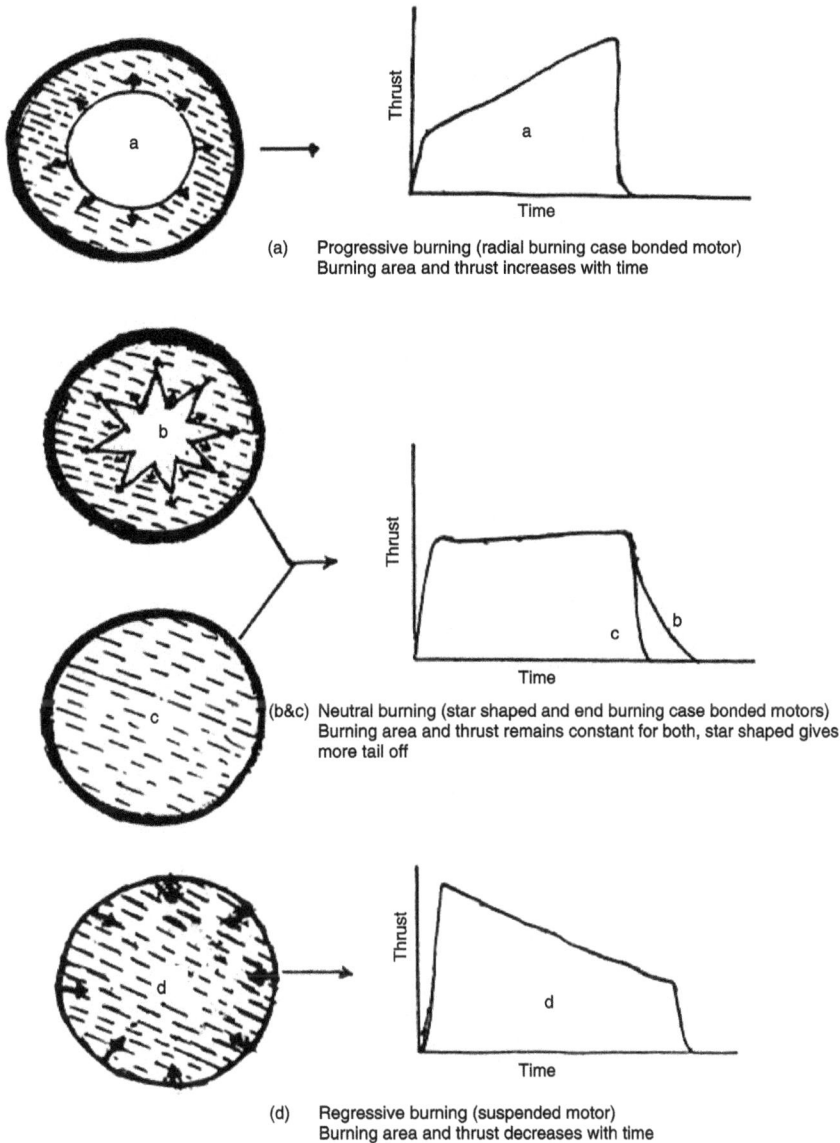

(a)   Progressive burning (radial burning case bonded motor)
       Burning area and thrust increases with time

(b&c) Neutral burning (star shaped and end burning case bonded motors)
       Burning area and thrust remains constant for both, star shaped gives
       more tail off

(d)   Regressive burning (suspended motor)
       Burning area and thrust decreases with time

Fig. 1: Progressive, Neutral and Regressive Motor Burning Patterns

Neutral Burning Motor

In the case of a motor with star shaped internal geometry as well as end burning case bonded motor, the burning area remains almost constant. Hence, the thrust or pressure remains constant with time of burning and hence, neutral burning as shown in the second curve in Figure 1. Star shaped motor gives higher tail off as shown in curve b due to the slivers left behind.

Regressive Burning Motor

In the case of a free-standing or suspended motor, the burning area decreases with time of burning. Hence, the thrust or pressure decreases with time of burning and hence, regressive burning as shown in the third curve in Figure 1.

4. PLATEAU AND MESA BURNING PROPELLANTS

When progressive, regressive and neutral burning pattern of motors are due to internal geometry of the motor, plateau and mesa burning are due to specific additives in the propellant formulations. In the case of a normal propellant like HTPB-AP-Al propellant, burn rate increases with increase in pressure and hence, progressive burning. In this case, burn rate law $r = aP_c^n$ where, $n =$ around 0.4. Addition of ballistic modifier like copper chromite further enhances the value of n.

Certain propellants show constant burn rate over a range of pressures, i.e., propellant is insensitive to pressure. Such propellants have n value zero and are called plateau burning propellants. For example, use of lead stearate in double base propellants gives plateau burning.

Certain propellants show decrease in burn rate with increase in pressure. Such propellants have negative n value and are called mesa burning propellants. For example, use of ammonium sulphate fine particles in HTPB propellant gives mesa burning.

The progressive, plateau and mesa burning behaviour is depicted in the Figure 2.

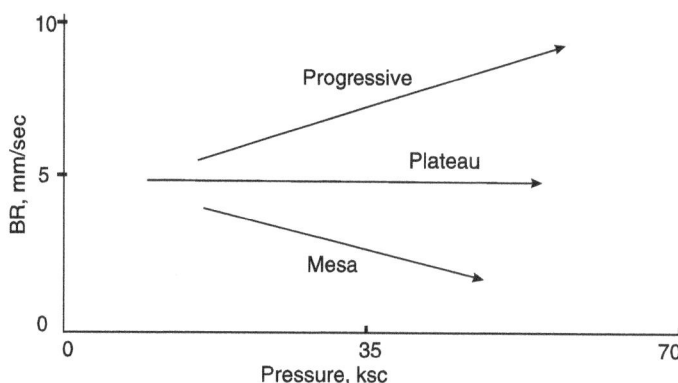

Fig. 2: Types of Burn Rates

5. INTERNAL BALLISTICS

A solid propellant motor burns by what is called a deflagration process. In this process, the propellant gets ignited over its exposed surface and then decomposes at and above the burning surface. During the release of hot gases, a part of the heat energy is conducted back into the freshly exposed propellant surface. This helps in further decomposition in a self sustained manner until all the propellant is burned out. The burning takes place in parallel layers, i.e., layer by layer.

The combustion of solid propellant in a rocket is slow compared to the rate of explosion. Hence, propellant is classified as low explosive and thus, distinguished from a high explosive like nitroglycerine. Hence, combustion of solid propellant is referred to as a burning or deflagration process while that of a high explosive is referred to as explosion or detonation.

The pressure in a rocket is due to the delicate balance between the two main actions:

1. The motor burning tending to increase the pressure.
2. The combustion gas escaping through the nozzle tending to decrease the pressure.

Hence, the "study of the internal ballistics of rockets is an accurate study of the rate of burning of propellant grain and the rate of discharge of combustion gases through the nozzle."

The two different laws for the dependence of burn rate 'r' on pressure 'P' in use are:

1. $r = a + b\,P$ Linear law
2. $r = a\,P^n$ Power law

The constants a and b in $r = a + b\,P$ depend of course on the composition of the grain and also strongly on the grain initial temp. The 'a' in $r = a\,P^n$ also strongly depends on the grain initial temperature.

In the case of a motor with cylindrical port in Figure 3,

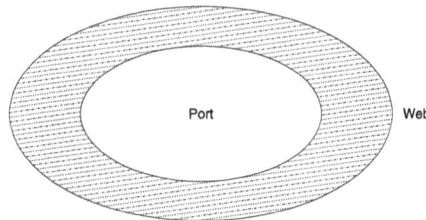

Fig. 3: Cylindrical Port

Motor av. burn rate ($\bar{r}$) = web thickness/web burning time

                = w/t_b for single port, case bonded

                = $(w/2)/t_b$ for single port, free standing.

Where t_b is burn time and web thickness $w = (OD-ID)/2$. OD and ID are grain outer diameter and inner diameter respectively.

As the burning occurs in parallel layers, the regularity of burning can be assessed by the formula:

$$M_b = S\rho r$$

where M_b = rate at which the mass is burned, kg/sec
         S = burning surface area, cm^2
         ρ = propellant density, g/cm^3
         r = burn rate, cm/sec.

For a free-standing radial burning motor, S is almost constant because the decrease in the outer surface area is balanced by an increase in the internal surface area. However, there is a small decrease in the surface area due to the burning of ends. Still it gives near constancy of surface area of burning.

In the case of case bonded radial burning motor, surface area increases and hence, thrust or pressure increases. In the case of star shaped motors and end burning motors, the burning area remains more or less constant and hence, thrust or pressure will be constant as referred in the progressive-regressive-neutral burning motors detailed above.

Ultrasonic Burn Rate Determination

Ultrasonic method is a recently developed technique for the determination of burn rate of solid propellants using a specially designed small motor. Cylindrical propellant grain of size of about 4 cm long and 3.5 cm diameter with epoxy inhibition on the cylindrical portion is used for the test. One end of the grain is bonded to a tapered 7 cm long PMMA block. PMMA acts as a coupling agent between the propellant and the ultrasonic transducer. It also protects the transducer from the high temperature and pressure of the motor. Ultrasonic transducer is used for the pressure-time measurements. This assembly is bonded to one end of the combustion chamber. Nitrogen gas is used for the initial test pressure. The propellant open end is ignited using nichrome wire. This is a closed chamber test and as the propellant burns, chamber pressure increases. The burn rate of the propellant is thus continuously monitored using ultrasonic method. The burn rate at the required pressure is computed from it. This method gives a better prediction of burn rate for bigger motors. A case study in ISRO on ultrasonic burn rate shows an augmentation factor of 1–2% for cylindrical grains and about 0.5% for star grains of bigger size motors.

Pressure or Thrust Curves

Typical pressure or thrust time curves of a solid propellant rocket motor is given in Figure 4.

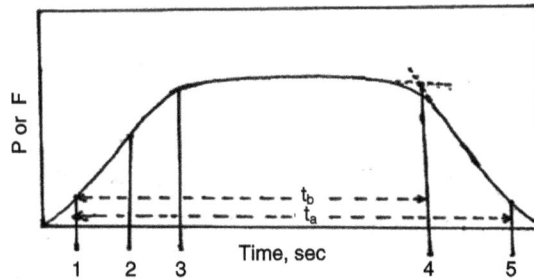

Fig. 4: Typical Pressure-Time Curve of a Solid Motor

Pressure-time curves are similar to thrust-time curves since,

Thrust, $F = C_F \times P_c \times A_t$

where C_F is thrust coefficient, P_c = chamber pressure and A_t = throat area.

Let us understand the various terms associated with the p-t curve of a motor test. They are ignition delay, ignition rise time, pressure/thrust build up, action time, web burn time, specific impulse, total impulse, thrust coefficient, characteristic velocity, discharge coefficient, mass flow rate or gas discharge rate, equilibrium pressure, erosive burning, chuffing and resonance burning.

Ignition Delay

0–1 marked in the p-t curve is the ignition delay usually measured in milli-seconds. Zero time refers the time at which ignition voltage is applied and 1 represents the time at which 10% of P_{max} is achieved.

Ignition Rise Time

1–2 shown in the curve represents ignition rise time. It is the time after ignition delay i.e., from 10% of P_{max} to 75% of P_{max}.

Pressure-Thrust Build Up

2–3 given in the curve shows the pressure-thrust build up time. It is the time from 75% of P_{max} to the time of attaining initial equilibrium pressure or thrust. Equilibrium pressure is the pressure at which the rate of gas production exactly balances the rate of gas discharge.

Web Burn Time

1–4 refers to the web burn time. It is the time from 10% of P_{max} to the intersection point of pressure-time curve marked at point 4. Web burn time is denoted by t_b and is used for calculating motor burn rate.

Action Time

1–5 shown in the curve gives the action time. It is the time between the two points of 10% of P_{max} marked as 1 and 5 in the p-t curve. Action time is denoted by t_a and is usually used for calculation of specific impulse.

Thrust (*F*) and Total Impulse (*I*)

Thrust is nothing but the force developed by a rocket as a reaction force when the hot gases from the combustion chamber get expanded to atmospheric pressure at the divergent section of the nozzle.

Total Thrust (*F*) or Total Impulse (*I*)

The area under the f-t curve gives the total thrust F or total impulse (I). It increases with the size of the motor.

$$F = \bar{F} \times t_a \quad \text{and} \quad \bar{F} = F/t_a$$

where $\bar{F} = Av.$ Thrust

Also $I = \bar{F} \times t_a \quad \text{and} \quad \bar{F} = I/t_a$

Specific Impulse (*I*$_{sp}$)

Specific Impulse is an index of energy. It is the thrust produced per unit weight of propellant burned. Total thrust or total impulse divided by weight of the propellant gives the specific impulse.

$I_{sp} = F/W$ or I/W = Total Impulse (I) or Total thrust (F)/Total wt of propellant (W)

Units of I_{sp} = kg. force. sec/kg mass, for simplicity sec. is used or Newton. sec/kg.

Exhaust Velocity (*V*$_e$)

Specific impulse multiplied by acceleration due to gravity gives the exhaust velocity of the motor.

$$V_e = I_{sp} \times g$$

where g = acceleration due to gravity, 9.81 m/sec.[2]

Thrust Coefficient (*C*$_F$)

Thrust coefficient (C_F) shows the efficiency of the nozzle. Total thrust divided by (average chamber pressure × throat area) gives the thrust coefficient. The values of C_F varies from 1.2 to 1.8. Total thrust is related to thrust coefficient by the formula,

$$F = C_F \times P_c \times A_t$$

i.e., $C_F = F/(P_c \times A_t)$

where P_c = Chamber pressure and A_t = Throat area.

Characteristic Velocity (C^*)

Characteristic velocity represents the combustion efficiency of the motor and is related to chamber pressure, nozzle throat area and propellant mass flow rate by the following equation,

$$C^* = (P_c \times A_t)/M_b$$

where $M_b = \rho_p \times r \times S$

    M_b = Mass burning rate or mass flow rate

    ρ_p = Propellant density

    r = Propellant burn rate

    S = Propellant burning area.

Characteristic velocity of rockets using HTPB propellant is = 1600 m/sec.

Mass Flow Rate or Mass Burning Rate (M_b)

Mass flow rate or mass burning rate (M_b) is related to propellant density, propellant burn rate and propellant burning area by the relation:

$$M_b = \rho_p \times r \times S$$

where, M_b = Mass burning rate or mass flow rate

    ρ_p = Propellant density

    r = Propellant burn rate

    S = Propellant burning area.

Mass Discharge Rate (M_D)

The fundamental formula for Mass discharge rate or gas discharge rate (M_D) is given by the equation:

$$M_D = C_D \times A_t \times P_c$$

where C_D = Discharge coefficient

    A_t = Throat area

    P_c = Chamber pressure

    M_D = Mass discharge rate or Gas discharge rate.

Discharge Coefficient (C_D)

Discharge coefficient (C_D) represents both combustion efficiency and nozzle efficiency. C_D is related to mass discharge rate (M_D) by the equation,

$$C_D = M_D/(A_t \times P_c)$$

If P_c is measured in Kgs/cm^2, A_t in cm^2 and M_D in Kgs/sec., then C_D of rockets using HTPB propellant is 6.15×10^{-3}/sec.

C_D is also related to C^* by the equation:

$$C_D = g/C^* \qquad \text{or} \qquad C^* = g/C_D$$

Characteristic velocity of rockets using HTPB propellant is $= 1600$ m/sec.

Equilibrium Pressure, Idealised p-t Curve, Importance of K and Factors Affecting Equilibrium Pressure

Equilibrium Pressure (P_{eq})

Equilibrium pressure (P_{eq}) is the pressure at which the gas generation rate exactly balances the gas discharge rate. It is also called steady state pressure. Since the rate of gas production is not a constant and it varies with pressure, achieving exact steady state or equilibrium pressure is difficult. However, very close to a steady state can be achieved.

If we draw mass burning M_b against pressure and mass discharge M_D against pressure together, both curves meet at a point which can be taken as the equilibrium pressure or steady state pressure P_{eq} denoted by P_e.

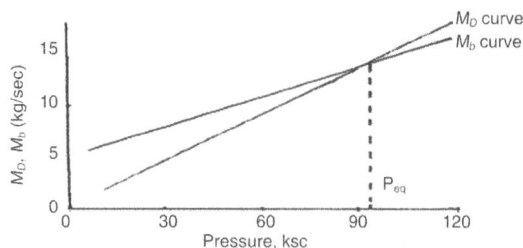

Fig. 5: M_D-M_b Interception Curve and P_{eq}

Mass burning and mass discharge are same at equilibrium pressure. That is, at P_{eq}, $M_b = M_D$, i.e., pressure is constant. If the pressure in the rocket is less than P_{eq}, $M_b > M_D$, i.e., the gas is produced faster than being discharged, hence pressure will rise. If the pressure in the rocket is above P_{eq}, $M_D > M_b$, i.e., pressure will fall.

Thus, during burning, the pressure will rise to P_e and then level off and remains at this value until all the propellant is burned and then drop rapidly to 1 atm.

Idealised p-t Curve

In idealized p-t curve, pressure remains constant throughout burning. That is, pressure is at equilibrium or steady state and under this condition, mass burning rate and mass discharge rate are same.

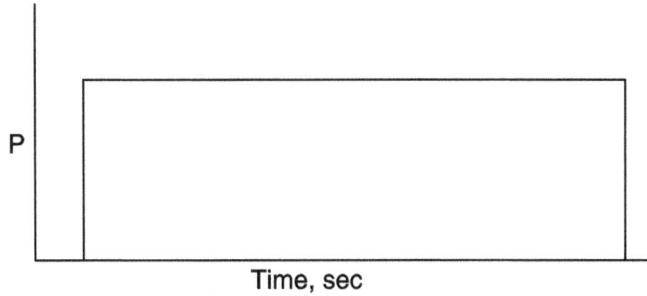

Fig. 6: Idealized p-t Curve

i.e., at P_e, $M_b = M_D$

i.e., $S \times \rho \times r = C_D \times A_t \times P$

Introducing $r = a\,P^n$, $S \times \rho \times a\,P^n = C_D \times A_t \times P$.

At P_{eq}, $S \times \rho \times a \times P_e^n = C_D \times A_t \times P_e$.

$$P_e/P_e^n = (S/A_t) \times (\rho \times a)/C_D$$

i.e., $P_{eq} = [(S/A_t) \times (\rho \times a)/C_D)]^{1/1-n}$

since, burning surface area/throat area, i.e., $(S/A_t) = K$,

$$P_{eq} = [K \times \rho \times a/C_D]^{1/1-n}$$

Equilibrium pressure (P_{eq}) depends on propellant constants K, ρ, a, n and C_D.

Importance of K or (S/A_t)

From the equation for equilibrium pressure, $P_{eq} = [(S/A_t) \times (\rho \times a)/C_D)]^{1/1-n}$

or $\quad P_{eq} = [K \times \rho \times a/C_D]^{1/1-n}$

K = burning surface area/throat area, i.e. (S/A_t) and it is of fundamental importance in rocket ballistics. The value of K alone gives a fair indication of the pressure to be expected in any rocket. Thus, K determines the size or structure of the rocket itself and the propellant quantity.

A 10% increase in K or 'a' increases P_{eq} by 40% in the case of double base propellants. Though the volume available to gases increases as the grain burns, the effect of volume on P_{eq} is very small since the density of gas is very small (0.03 g/cc) compared to 1.63 g/cc for propellant.

Factors Affecting Equilibrium Pressure

A number of factors affect equilibrium pressure. K alone gives a good indication of the expected pressure and size of the rocket. Increase in the value of K, increases equilibrium pressure significantly. Since $K = (S/A_t)$, an increase in burning surface area or a decrease in nozzle throat area enhances the pressure. Propellant density 'ρ', burn rate constant 'a', pressure index 'n' and 'C_D' depend only on the composition of the propellant.

Erosive burning, Factors Affecting Erosive burning and How to Validate Erosive Burning

Erosive Burning

Erosive burning is the increase in burn rate caused by a high velocity gas stream across the burning surface.

The amount of the increase in burn rate depends on:
  1. the velocity of the gas stream
  2. composition of the propellant.

Why and Where?

Erosive burning happens due to increased heat transfer to the propellant burning area. It occurs mainly in the radial burning motors.

Factors Effecting Erosive Burning

We have seen eartlier that mass flow rate,

$$M_b = \rho_p \times r \times S$$

This can also be written as,

$$M_b = \rho_g \times V_g \times A_p$$

where ρ_g is gas density, V_g is gas velocity and A_p is port area.

Also we have seen earlier that mass discharge rate,

$$M_D = C_D \times A_t \times P_c.$$

Equating mass flow rate to mass discharge rate,

$$\rho_g \times V_g \times A_p = C_D \times A_t \times P_c$$

i.e., $V_g = (C_D \times A_t \times P_c)/(\rho_g \times A_p)$

Since P_c/ρ_g is nearly constant and C_D is constant for a propellant, so V_g is nearly proportional to A_t/A_p.

Thus, the amount of erosion depends on A_t/A_p and it increases with increase in throat area and decreases with increase in port area (smaller web). Erosion also lowers the tail end of the p-t curve. Burn rate is also a factor affecting erosion. A decrease in velocity decreases the burn rate. Hence, an increase in burn rate of the propellant also enhances erosion.

Since erosive burning causes the propellant near nozzle end of the rocket to burn faster than the propellant at the head end, erosion causes the grain to become tapered during the burning. The tapering effect of erosion is partially compensated by an exactly opposite tapering due to pressure drop between head end and nozzle end.

How to Validate Erosive Burning?

A case bonded radial burning motor is made with port machined and web measured to correct dimensions. It is fitted with a device which blows one end of the chamber at some stage of burning. The motor is then ignited. At certain time of burning, one end of motor is blown out using the device. The drop in pressure extinguishes burning.

The partially burned motor is cooled, disassembled and the web is measured at both ends. The difference between the burn rates at the head end and nozzle end is found out, which gives the erosive burning. Thus, total burn rate can be expressed by the relation

$$r = r_0 \left(1 + k \left(A_t/A_p\right)\right)$$

where, r = total burn rate,

        r_0 = erosion free burning (burn rate in an atmosphere at rest)

        k = erosion constant

A_t/A_p = throat area/port area.

The erosion constant K is numerically about 0.5 for most American rockets.

Chuffing

When the pressure in a rocket motor decreases below a certain critical value, the chamber pressure may fall suddenly to atmospheric pressure and the charge apparently ceases to fire. After a delay, the charge reignites. This repetition gives a series of explosions or chuffs.

In chuffing, the motor propellant clearly gets extinguished spontaneously, followed by re-ignition. The re-ignition of motor takes place after an appreciable time interval. A typical pressure-time graph showing chuffing is shown Figure 7.

Chuffing is a particularly unpleasant phenomenon in flight as it causes the rocket to fly off in an unpredictable direction.

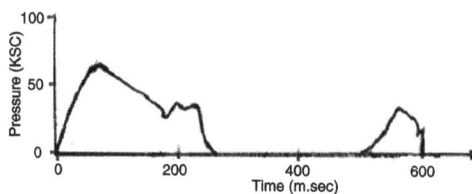

Fig. 7: p-t Curve Showing Chuffing

Factors Affecting Chuffing

1. Small ratio of burning surface area to the nozzle throat area, ie smaller port area/throat area ratio.
2. Low propellant temperature.
3. Due to regressive burning causing a fall in the temperature of the burning surface.

Resonance Burning

Resonance burning causes secondary peaks which are irregular peaks and appear in the p-t curve of only perforated grains. The resonance or pressure waves in the perforation are responsible for the irregular peaks. A typical p-t graph showing resonance burning is shown in Figure 8.

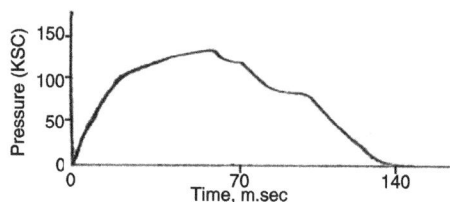

Fig. 8: p-t Curve Showing Resonance Burning

This can be eliminated by drilling radial holes through the web of the grain into the perforation. A typical example is a CTPB based composite propellant free-standing type grain with epoxy-cotton tape insulation and radial port with slotted configuration and drilled holes, developed in ISRO for special applications.

6. COMBUSTION INSTABILITY

Combustion instability, also called oscillatory combustion, occurs when a solid propellant rocket motor experiences unstable combustion as the pressure in the interior gaseous cavities oscillates by at least 5% (sometimes by about 30%) of the chamber pressure. When instability occurs, the heat transfer to burning surfaces, the nozzle and the insulated case walls is greatly increased, thereby increasing the burning rate, chamber pressure and thrust and reducing the burning duration. The change in thrust-time profile causes significant changes in the flight path, and at times can lead to failure of mission. If prolonged and if the vibration energy level is high, the instability can cause damage to hardware such as overheating the case and causing a nozzle or case failure.

The instabilities of solid propellants is different from that of liquid propellants. In the liquid propellant, there is a fixed chamber geometry while in solid propellant motors, the geometry of oscillating cavity increases in size as burning proceeds and with stronger damping factors like solid particles and energy absorbing viscoelastic materials. Hence, instabilities are classified by the type of mode which becomes unstable. There are: i) longitudinal modes of instability, and ii) traverse modes of instability. These acoustic modes are given by the solutions in the acoustic wave equation for the combustion cavity. For example, in a motor with cylindrical port geometry, longitudinal, tangential or radial modes of oscillations and their small combinations can occur possibly in the motor. The frequencies of oscillations vary widely depending on mode of oscillation. In solid motors, the transverse mode frequencies are much higher than longitudinal mode frequencies.

Typical Questions

1. Draw a typical p-t curve of a solid motor and explain the following:
   a) Ignition delay, b) Action time, c) Web burning time.
2. Explain the following with suitable formula:
   a) Specific Impulse, b) Total Impulse, c) Thrust coefficient,
   d) Discharge coefficient, e) Characteristic velocity, f) Mass flow rate,
   g) Mass discharge rate.
3. What is meant by internal ballistics?
4. What are the burning rate laws?
5. How is motor burn rate determined from p-t curve?
6. Explain acoustic emission technique for determining burn rate?
7. What are the factors affecting burn rate?
8. What are plateau and mesa burning?
9. What is temp. sensitivity of burn rate?
10. What is erosive burning? What are the factors affecting erosive burning?
11. How is erosive burning validated?
12. Explain progressive, regressive and neutral burning behaviour with suitable examples.
13. What is idealized p-t curve?

14. Explain chuffing with suitable p-t curve. What are the factors affecting chuffing?
15. Explain resonance burning? What are the remedial measures?
16. What is meant by Equilibrium pressure? What are the factors affecting it?
17. What is the importance of K on equilibrium pressure?
18. Explain the ultrasonic method for the determination of burn rate of solid propellants?
19. Explain combustion instabilities in solid and liquid motors?

REFERENCES

[1] Amareswara Sainadh Ch, Srinivasan, V., *et al.*, "Variation of Scale Factors in Solid Rocket Motors," *9th International High Energy Materials Conference,* Feb. 13–14, Trivandrum, 2014.

[2] Cauty, F., "Ultrasonic Method Applied to Full-Scale Solid Rocket Motors," *Journal of Propulsion and Power,* Vol. 16, No. 3, 2000, pp. 523–528.

[3] Charles E. Roger, Erosive Burning Design Criteria for Solid Rocket Motors, Nevada Aerospace Science Associates, Part 6, 2005.

[4] Fry, R.S., "Solid Propellant Subscale Burning Rate Analysis Methods for U.S. and Selected NATO Facilities," The Johns Hopkins Univ., CP TR 75, Chemical Propulsion Information Agency, Whiting School of Engineering, Columbia, MD, Jan 2002.

[5] Gany, A., Israel Institute of Technology, Israel, Article on Solid Rocket Performance, Propellants and explosives technology, ISBN 81-7023-884-6, by Allied Publishers Limited, India, 1998.

[6] J.-P., "Solid Propellant Burning Rate Measurement Methods Used Within the NATO Propulsion Community," AIAA Paper, 2001–3948, July 2001.

[7] Jeenu, R., Pinumalla, Kiran and Deepak, Desh, "Industrial adaptation of ultrasonic technique of propelant burning rate measurement using specimens," *Journal of Propulsion and power,* Vol. 29, No. 1, 2013, pp. 216–226.

[8] Kiran, Perumalla, Srinivasan, V., *et al.*, "Usage of Ultrasonic Burn Rate for the Performance Prediction of Lsarge Solid Boosters, *9th International High Energy Materials Conference,* Feb. 13–14, Trivandrum, 2014.

[9] Maag, H.J. and Klingenberg, G, Gun Propulsion Concepts-Part 11: Solid and Liquid Propellants, *Prop, Explos, Pyrotech,* Vol. 21, 1996.

[10] Michel, Royce, *et al.*, "The Advanced Solid Rocket Motor," AIAA Space Program and Technology Conference, Huntsville, March 1992.

[11] Patil, Prajakta R.; Krishnamurthy, V.N. Joshi and Satyawati, S., Effect of nano-copper oxide and copper chromite on the thermal decomposition of ammonium perchlorate Propellants, *Explosives, Pyrotechnics,* 33(4), (2008).

[12] Pein, DLR, Research Centre, Germany, Article on Introduction to Analysis of Rockets, Propellants and explosives technology, ISBN 81-7023-884-6, by Allied Publishers Limited, India, 1998.

[13] Pinumallaa, Kiran, Jeenua, R., *et al.*, usage of ultrasonic burning rate for the performance prediction of large solid booster motors, *Proc. 9th International High Energy Materials Conference,* Feb. 13–14, Trivandrum, 2014.

[14] Roger, Charles E., *Solid Propellant Grain Design and Internal Ballistics*, Vol. 33, No. 5–6, October/November 2002.

[15] Sutton, Rocket Propulsion Elements, Sixth Edition, John Wiley, New York, 1992.

[16] Tejasvi, K.A. and Panigrahia, S.K., *et al.*, Estimation of Pressure index from the Single Ballistic Evaluation Motor, *Proc. 9th International High Energy Materials Conference,* Feb. 13–14, Trivandrum, 2014.

[17] Toncu, Gheorghita; Stanciu, Virgil and Toncu, Dana-Cristina, "Solid Propellant Rocket Motor Thrust Correlation with Initial Temperature of Propellant Charge," U.P.B. Sci. Bull., Series D, Vol. 73, Iss. 1, 2011.

[18] Warren, Francis, A., Solid Propellant Technology-chapter-3 internal ballistics of Rockets, by R.B. Kershner, 1970.

[19] Yang, Vigour, Solid Propellant Chemistry, Combustion and motor internal ballistics, AIAA-2000, ISBN 1 56347-442-5.

CHAPTER 7

Advanced Solid Propellants

1. INTRODUCTION

The design of future launch vehicles and weapon systems require the use of propellants and explosives having enhanced performance (energy output) and reduced vulnerability during storage and transportation. The important design considerations for such formulations include improved mechanical properties, decreased signature, extended service life and reduced environmental impact in the manufacture, use and disposal. In particular, extensive programmes have evolved worldwide for the development and introduction of Insensitive Munitions (IM), ordinance that fulfills performance expectations but in which the response to unplanned hazardous stimuli is reduced.

During the last four decades, composite propellants based on AP/Al/HTPB have gained wide acceptance to power launch vehicles and missile systems. There have been significant progress in the synthesis and development of new energetic compounds and its use in explosives and propellants in recent years. These include nitramines—RDX, HMX, CL-20, TNAZ, advanced oxidizers—ADN, HNF, energetic binders—GAP, BAMO, Poly NIMMO, PGN, and energetic plasticizers having groups like $-NO_2$, $-N_3$, $-ONO_2$, $-NNO_2$, $-NF_2$, $-F$, etc. These materials not only increase the specific impulse but also the density, in addition to providing eco-friendliness and better thermal stability compared to double base compositions.

Calculations tell that the use of an improved oxidizer (like ADN) in a propellant system can give up to a 50% increase in range of ground-to-air missile over a conventional AP/Al/HTPB system. Similar calculations tell the use of ADN in Inertial Upper Stage (IUS) orbit transfer from Low Earth Orbit (LEO) to Geosynchronous Orbit (GSO) would provide 8.9% increase in payload. Using ADN in the booster and IUS system would give a payload increase of 17.4%. Introduction of more advanced system using AlH_3 in place of Al and use of ADN in the IUS would provide a 12.4% increase in LEO to GSO transfer step for Titan IV. The dramatic payload gains can be traded off or a smaller launch vehicle, thus decreasing the size of the system and its cost. Also, ADN is an oxidizer and is chlorine free.

The use of energetic binder can have a major impact on the solids loading of the propellant. The reduction in solids loading is likely to vastly improve the safety of the overall system, possibly taking it from a sensitive 1.1 category to an insensitive 1.3 category systems. For example, using BAMO/AMMO binder to replace conventional binder with AP and also other ingredients gives a reduction in solids from 90% in the conventional system to 80% in the advanced system while having same energy density. A GAP binder system may give similar results.

New propellants are required not only to increase the specific impulse (I_{sp}) but also meet environmental and toxicity constraints and improved safety. New propulsion materials will significantly reduce overall weight and hence, the cost of propulsion systems. They also permit innovative manufacturing techniques which will yield revolutionary rocket designs. Finally, a better understanding of the chemistry and material properties for propulsion systems will lead to solutions to problems that continue to plague the propulsion industry today.

2. ADVANCED OXIDIZERS

The present day work-horse oxidizer is AP. However, composite propellants based on it produces large quantities of Hydrogen Chloride (HCl) as the major exhaust product, which affects the global environment. Each of the solid booster of space shuttle having 503 tonnes of propellant liberates about 100 tonnes of HCl and other chlorine products during its burning, thereby polluting the atmosphere and hence, cause ozone depletion in the atmosphere. The large amount of HCl emission can result in acid rain. Also, military rockets need to have a low signature of smoke. Both space exploration and defence needs dictate the development and use of new oxidizers in place of AP that are both environmental friendly as well as more energetic to meet future space missions.

One possible successor is Ammonium Nitrate (AN). However, AN is less energetic, highly hygroscopic and has a crystal phase transformation around 32.1°C (ambient temperature in many South Asian countries) with a large volume change. Hence, only Phase Stabilized Ammonium Nitrate (PSAN) with its phase transition at 32.1°C either suppressed or shifted to higher temperatures is used. This stabilization is achieved by either co-crystallizing or doping of AN with metal salts of nickel, potassium, etc. To boost the energetics, various other oxidizers (AP, HMX, RDX) have been included in the formulations. The other energetic oxidizers that can be thought of are included in Table 1.

HMX has high heat of formation but has a negative oxygen balance. AN, AP, HP_1, and HP_2 have negative heat of formation with positive oxygen balance. These oxidizers exhibited lower heats of formation compared to AP, produced energetic propellants that were detonatable, in addition to being friction and impact sensitive. The main characteristics in demand for new oxidizers are high density, high oxygen balance and also no chlorine content.

Table 1: Characteristics of Various Oxidizers

Oxidizer	Oxygen Balance (%)	Density (g/cc)	ΔH_f (kJ/mol)
AN	20.0	1.73	−365.04
AP	34.0	1.95	−296.00
HP_2	41.0	2.20	−293.30
HP_1	24.0	1.94	−177.80
RDX	−21.6	1.82	70.63
HMX	−21.6	1.96	74.88
ADN	25.80	1.82	−150.60
HNF	13.10	1.87	−72.00
CL-20	−10.95	2.04	381.20
TNAZ	−16.66	1.84	33.64
ONC	0	2.10	413.80

(HP_1 = Hydrazine perchlorate, HP_2 = Hydrazine diperchlorate, ONC = Octanitrocubane, TNAZ = Trinitroazetidine)

Ammonium Dinitramide (ADN) $NH_4^{+-}N(NO_2)_2$

ADN is emerging as one of the most promising high energy oxidizer for new generation propellants since 1970 in Russia. ADN based propellants are reported to be used in the Topol-M (second and third stages) intercontinental missile, SS-24 (second generation, first and third stages) and SS-20-N. ADN when used as oxidizer in space shuttle booster in place of AP, it is expected to increase the lift-off capacity of space shuttle by 8%. It is chlorine free oxidizer. It has positive oxygen balance and an enthalpy of formation superior to AP and AN. It is stable to moderate shock, pressure and heat and does not detonate in small quantities. Only extremes of temperature or pressure can cause its explosion. The calculated specific impulse of propellants with an energetic binder and Al shows ADN is lower than that of HNF but better than that of AP. In the case of ADN based formulations, a maximum I_{sp} of 275 s was obtained for a 76% oxidizer content and show a gradual decrease in I_{sp} as the oxidizer content is further increased. ADN is classified as 'class C' or minimum hazard material. Pyrolysis and spectroscopic studies indicate ADN decomposition yields mostly ammonia, water, N_2O and NO_2. One of the major decomposition products of ADN is Ammonium nitrate. Hence, ADN is envisaged as a suitable successor to AP.

ADN is made from aliphatic monoisocyanate using stoichiometric quantities of nitronium tetrafluoborate and nitric acid in acetonitrile as the nitrating mixture, followed by ammonia treatment.

Synthesis of ADN

1. $R\text{-}NCO \xrightarrow[\text{HNO}_3/\text{CH}_3\text{CN}]{\text{O}_2\text{NBF}_4} R\text{-}N(NO_2)_2 \xrightarrow{\text{Base (NH}_3)} NH_4^{+-}N(NO_2)_2$

ADN is also made by nitrating the ammonium sulphamate with concentrated nitric acid and sulphuric acid (nitrating mixture) at minus 35 to minus 45°C with stirring. In the laboratory, a white precipitate was formed with increase in viscosity of the medium. After 45 minutes of stirring the solution was poured into crushed ice. The diluted acidic solution was neutralized with cold liquor ammonia maintaining the temperature below 0°C. The pH was checked and kept around 7.5 to 8. The neutralized solution was evaporated in a rotary evaporator under vacuum to remove water. The solid was further dried in vacuum, followed by extraction with isopropanol. The isopropanol solution obtained was filtered and evaporated under vacuum to yield pale yellow crystals of pure ADN.

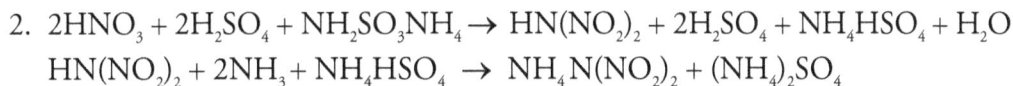

2. $2HNO_3 + 2H_2SO_4 + NH_2SO_3NH_4 \rightarrow HN(NO_2)_2 + 2H_2SO_4 + NH_4HSO_4 + H_2O$

$HN(NO_2)_2 + 2NH_3 + NH_4HSO_4 \rightarrow NH_4N(NO_2)_2 + (NH_4)_2SO_4$

In another method, ADN was selectively adsorbed on activated charcoal and eluted with hot water to get the ADN.

The Table 2 given below gives the performance characteristics of few propellant combinations using conventional and new energetic binders and oxidizers.

Table 2: Performance Characteristics of Advanced Propellants

Composition (%)	$I_{sp}(s)$	I_{sp} *Density (s.g/cc)*	*HCl (%)*
HTPB14/AP 68/Al 18	264.5	465.2	21
PNIMMO/AP 58/Al 20	263.5	484.1	17.6
GAP/AP 58/Al 20	266.0	491.9	17.5
GAP 22/AP 56/Al/Mg (70:30)	260.3	471.3	10.7
GAP/ADN 55/Al 20	276.5	479.3	nil
GAP/CL-20 55/Al 20	267.1	505.8	nil
GAP/HNF 60/Al 15	277.2	486.4	nil

(Pressure 7 MPa, $A_E/A_T = 10$, Equilibrium flow and sea level conditions)

For conventional propellant based on HTPB, maximum specific impulse obtained is 264 seconds while for a new oxidizer-binder system such as GAP/ADN or GAP/HNF, the specific impulse obtained is around 276 to 277 seconds. This clearly indicates the improved performance for a new binder/oxidizer system. Compared to AP based formulations, there is no HCl emission with the new oxidizer/binder system. Major problems encountered in ADN application are sub-zero reaction condition, undesirable hygroscopicity and low melting (90°C) and decomposition (123°C) temperatures.

Hydrazinium Nitro Formate (HNF) $N_2H_5^+\{C(NO_2)_3^-\}$

HNF is another potential oxidizer. Its production was initiated in late 80's in the Netherlands which has a pilot plant of 3 to 6 kg per batch. The Ariane 5 boosters presently use a propellant based on AP/HTPB with a specific impulse of 266s. Using HNF, an increase in I_{sp} of more than 7% may be achieved, raising the I_{sp} to 284.6s. The payload mass increases by 10% by using HNF based propellants in Ariane 5 boosters. HNF has higher heat of formation than ADN. HNF, salt of nitroform, is reported to be discovered in 1951. It is one of the most powerful solid oxidizer and a number of patents appeared till 1970s when the research on HNF was stopped suddenly. The major reasons are: i) the hazardous nature of synthesis method of nitroform. Several fire and explosions accidents have been reported. ii) The incompatibility between HNF and well known binders like HTPB due to attack on binder double bonds by HNF, leading to propellant swelling and deterioration of mechanical properties. The use of saturated binders circumvents this problem, but leads to formulations that give similar performance to that of AP/HTPB based propellants.

Currently, a safe synthesis method for HNF has been established by TNO, starting from nitroform and hydrazine. Nitroform is manufactured safely by Rockwell method. The process has been scaled up. HNF has higher melting point and less hygroscopic compared to ADN but impact, friction and shock sensitivities are similar for both oxidizers. The particle size of HNF produced is between 5 to 20 μm. HNF has been recrystallized to get 200 to 300 μm in order to use bimodal distribution in propellant formulations. The stability of HNF mainly depends on purity of HNF which is a function of hydrazine content. However, the major concern is its high friction sensitivity. Phlegmatization of HNF with special gels is being attempted to overcome the sensitivity problem.

The burning rate and burning rate exponent reported are high and formulations based on HNF/GAP/Al have been reported. Though the new propellant looks promising, a number of critical issues still have to be addressed especially the compatibility of curing agents with HNF, the possibility to make spherical or cubic crystals in place of needle shaped ones, possibility to manufacture castable propellants with high solid fraction, the overall chemical properties of propellant and improvement of safety and hazard properties of the propellant and its ingredients.

HNF has been synthesized by the action of hydrazine with trinitromethane in ethylene dichloride. Trinitromethane itself is made from acetic anhydride and nitric acid followed by alkaline hydrolysis or by nitration of acetylene in presence of mercury catalyst or from isopropyl alcohol and nitric acid. The purity of the HNF produced plays a major role both in its stability and the stability of propellants. The synthesis of HNF is given below.

Synthesis of HNF

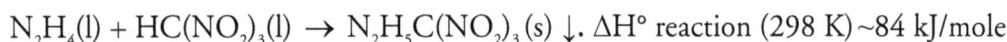

$N_2H_4(l) + HC(NO_2)_3(l) \rightarrow N_2H_5C(NO_2)_3(s) \downarrow$. $\Delta H°$ reaction (298 K) ~84 kJ/mole

Because the reaction is exothermic, the process vessel has to be cooled well and must be kept below 5°C in order to exceed the flash point of the solvent used in the reaction. The HNF crystals are dried in vacuum at elevated temperatures. Washing the crystals free from nitroform or hydrazine adsorbed to the crystals is very essential before drying. This avoids any reaction after production of HNF and stabilizes the final product. The crystals are subjected to recrystallization process to improve the purity of raw HNF and to obtain different sizes of HNF crystals.

Hexanitro Hexaazaiso Wurtzitane (HNIW or CL-20)

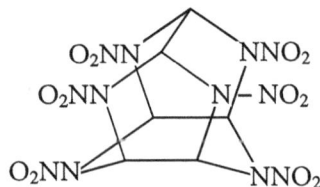

Hexanitro hexaazaisowurtzitane (HNIW or CL-20) was considered a laboratory curiosity a decade ago and is currently produced at Thiokol, USA, SNPE, France and Premier Explosives, India. CL-20 with its cage like structure can release energy at a much higher level than current explosives like HMX or RDX. High heat of formation (+454 kJ/mole) is due to strain introduced by the cage like structure and the presence of six $N-NO_2$ groups, in comparison with +63 and +76 kJ/mole for RDX and HMX respectively. The cage like structure of the molecule also results in close packing of the constituent atoms, leading to high density (>2.04 g/cc). CL-20 is the most powerful explosive today and also acts as a clean, combustion efficient oxidizer for future propellants. In addition, it has a good oxygen balance (–11%) and useful in propellants and explosives.

It exists in different polymorphs and its impact sensitivity is similar to PETN. CL-20 with particle size distribution in the range of 3 to 5 µm is found to be least sensitive to impact. It is chemically stable upto a temperature of 120°C for at least 32 hours. However, if a reaction occurred with CL-20, it is always violent leading to detonation and is independent of temperature. This is different from what is seen with HMX and RDX. Several promising plastic bonded explosives containing 85 to 95% CL-20 have been prepared and studied to improve the war head performance. CL-20 formulations showed higher burning rate and increased energy compared to HMX based ones at comparable hazard properties.

HNIW or CL-20 is synthesized from glyoxal and benzylamine to give benzyl derivative of hexaazaisowurtzatane, which on nitration gave CL-20.

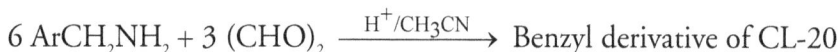

$$6\ ArCH_2NH_2 + 3\ (CHO)_2 \xrightarrow{\ H^+/CH_3CN\ } \text{Benzyl derivative of CL-20}$$

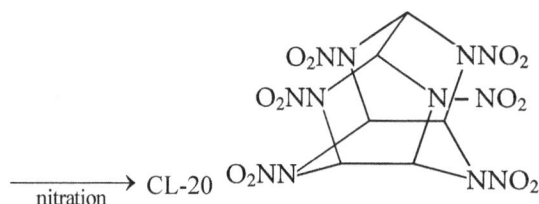

$$\xrightarrow[\text{nitration}]{} CL\text{-}20$$

A series of nitramines arranged about a cyclobutane ring as Cis and trans forms. Cis configurations has been synthesized and used to boost the energy of propellants. These nitramines have comparable energetics as that of HMX or PETN for use as explosives but with higher density compared to HMX and PETN. Because of its better hydrolytic and thermal stability compared to PETN, it can be used as exploding bridge wire detonators.

1,3,3-Trinitro-azetidine (TNAZ) $(CH_2)_2N\ (NO_2)C\ (NO_2)_2$

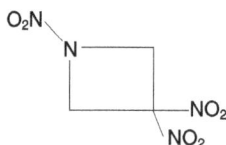

The other nitramine that has higher oxygen balance and density compared to HMX with half the impact sensitivity of HMX, is 1,3,3-trinitro-azetidine (TNAZ). TNAZ has been synthesized starting from t-butyl amine and epichlorohydrin followed by nitration with $NaNO_2$, $C\ (NO_2)_4$ and HNO_3/AC_2O. It is a powerful and thermally stable energetic compound, finds use in melt cast explosives in place of TNT.

Synthesis: t-Butyl amine + epichloro hydrin

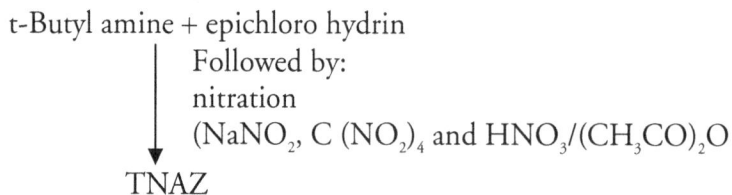

Followed by:
nitration
$(NaNO_2,\ C\ (NO_2)_4$ and $HNO_3/(CH_3CO)_2O)$

TNAZ

Nitrocubanes

Highly nitrated cubanes are the latest entrants to the class of high energy materials. The most powerful compound of the series, Octanitrocubane (ONC) has superior performance even over CL-20. It has a density greater than 2 g/cc and a heat of formation exceeding +400 kJ/mole, which gives a detonation velocity of around 11,000 m/sec. It has been synthesized starting from expensive materials involving several synthetic steps under stringent reaction conditions. Its application is limited because of the very high cost.

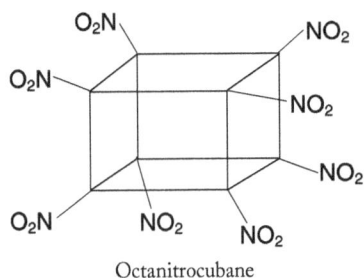

Octanitrocubane

High nitrogen content materials are a unique class of novel energetic materials, deriving from their very high positive heat of formation rather than oxidation of the carbon backbone, as in case of oxidizers. These materials increase the density and lead to generation of a large volume of gas per gram. The list includes furazans, furoxanes, tetrazoles, triazines and tetraazines, etc.

3. ADVANCED BINDERS

Currently, hydroxyl terminated polybutadiene cross linked with isocyanates and containing a plasticizer such as Dioctyl Adipate (DOA) is used in solid propellants. Other polymers used include carboxyl terminated polybutadiene, ter polymer of butadiene, acrylic acid and acrylonitrile (PBAN) and hydroxyl-terminated polyethers. These binders have excellent physical properties and reduced vulnerability of explosive charges, but they are inert, the binder dilutes the explosive, reducing overall energy output and performance of composition. HTPB, for example, requires a theoretical solid loading of 92% by weight of ammonium perchlorate for complete combustion, but 15 to 20% of binder is required for getting a processible formulation. One way to overcome this problem of high solid loading is to change the manufacturing method from cast cure to extrusion or pressing, thereby reducing the quantity of inert binder required. The other approach is to use polymers and/or plasticizers which contribute to the overall energy of the formulation. This has been successful in the development of advanced rocket propellants and high performance explosives.

A radical approach is inclusion of energetic functional groups, such as the azido group ($-N_3$), nitro group ($C-NO_2$), nitrato group ($O-NO_2$), nitramino group ($N-NO_2$) or less common difluoroamino group ($-NF_2$), along the polymer backbone and in the plasticizer. Incorporation of these groups increases the internal energy of formulation, in addition to improving the overall oxygen balance. New energetic polymeric binders include glycidyl azide polymer (GAP), polybis azido methyl oxetane (poly-BAMO), poly nitrato methyl methyl oxetane (poly-NIMMO), poly azidomethyl methyl oxetane (poly-AMMO) and poly glycidyl nitrate (poly GLYN). Energetic plasticizers include the oligomers of the above mentioned polymers as well as a wide range of nitrate esters, nitroaromatics and azido plasticizers. The following Table 3 gives a comparison of the physico chemical properties of some most promising advanced energetic binders in comparison with the workhorse binder (HTPB).

Table 3: Comparison of Physico-Chemical Properties of Advanced Energetic Binders

Energetic Binder	ΔH_f kJ/mole	OB %	ρ g/cc	T_g °C
GAP	+117	−121	1.30	−50
Poly BAMO	+413	−124	1.30	−39
Poly AMMO	+18	−170	1.06	−35
Poly NIMMO	−335	−114	1.26	−25
PGN	−285	−61	1.39	−35
HTPB	−62	−324	0.92	−65

Prominent among the energetic groups is the azido group, which contributes a heat release of around 355 kJ/N_3 unit.

Glycidyl Azide Polymer (GAP)

The first polymer developed in this category is Glycidyl Azide Polymer (GAP), first synthesized in 1972. GAP is synthesized by a two step route as follows:

Epichlorohydrin is polymerized with Lewis acid catalyst like BF_3 eherate or $SnCl_4$ and an initiator like ethyleneglycol or butanediol. The molecular weight ($\overline{M}n$) of the product (PECH) ranges from 2000 to 5000. In the absence of catalyst, low molecular weight around 500 to 700 is formed. The azidation is carried out in polar solvent like DMF or DMSO. When DMSO is used as solvent, the complete conversion of PECH to GAP occurs in 18 hrs at 90°C. In water, polymerization can be effected using phase transfer catalyst in 7 days at 90°C. The conversion of PECH to GAP was confirmed by the appearance of a sharp band in IR at 2100 cm^{-1} and total disappearance of band at 747 cm^{-1} due to CH_2-Cl.

Gap Synthesis

GAP is synthesized by cationic polymerization of epichlorohydrin, followed by condensation with sodium azide at high temperature.

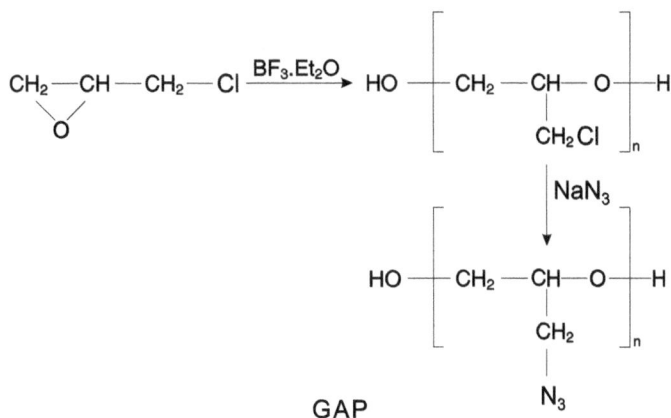

$$CH_2-CH-CH_2-Cl \xrightarrow{BF_3.Et_2O} HO-\left[CH_2-CH-O\right]_n-H$$

with O (epoxide), CH_2Cl substituent

$$\downarrow NaN_3$$

$$HO-\left[CH_2-CH-O\right]_n-H$$

$$CH_2$$

$$N_3$$

GAP

The functionality of linear GAP is nearly two. To achieve the desired level of crosslinking to produce a tough and elatomeric rubber, the functionality has to be raised by addition of triols or use of triisocyanate crosslinkers. GAP with increased functionality has also been reported. If instead of a diol, glycerol is used as initiator, GAP triol is formed. GAP may then be crosslinked by reaction with isocyanates to give an extended polymeric matrix. Branched GAP with long GAP backbone chain to which shorter GAP chains have been grafted, is synthesized by simultaneous degradation and azidation of high molecular weight solid polyepichlorohydrin with sodium azide in the presence of a base cleaving agent and polyol in a polar solvent. Adjusting catalyst/polymer weight ratio controls the molecular weight. Degradation catalysts include lithium methanolate and sodium hydroxide.

GAP is hard and brittle at low temperatures as a consequence of the rigid, conjugated $-N_3$ groups limiting the flexibility of the polymer backbone. Gas evolution on curing the liquid GAP with isocyanates is also a serious problem. Isocyanates react with moisture to generate carbon dioxide, which remains trapped in the voids of the cross-linked binder, resulting in decreased mechanical properties, performance and safety of the propellant composition. The energetic properties of GAP are not due to its oxidation products, but due to chain scission of the azide group, which gives nitrogen gas with a heat of reaction of +685 kJ/gmole at 5 MPa. GAP also contains a relatively large concentration of carbon atoms, and hence, has a large combustion potential, burning smoothly at elevated temperatures and pressure without explosion.

Several copolymers of GAP based thermoplastic elatomers have been reported. The GAP-ehylene oxide copolymer is one such polymer. It is prepared by initially copolymerizing Epichlorohydrin (ECH) with Ethylene Oxide (EO) to poly epichlorhydrin-ehthylene oxide copolymer followed by azidation to yield GAP-EO copolymers. The GAP-EO copolymers cured with isophorone diisocynate showed higher elongation at break and lower T_g compared to the one made from GAP homopolymer. Similarly, GAP-THF copolymers have been prepared exhibiting better mechanical properties, better thermal stability and lower hazard sensitivity compared to the GAP polymer.

Polyoxetanes

Energetic polyoxetanes were made from monomers such as nitratomethyl-3-methyl oxetane (NIMMO), 3,3-bis-(azidomethyl) Oxetane (BAMO) and 3-azidomethyl-3-methyl oxetane (AMMO). NIMMO is prepared by the acyl nitrate nitration of 3-hydroxy-methyl-3-methyl oxetane. Nitration can also be carried out using dinitrogen pentoxide (N_2O_5) in a flow nitration system. BAMO is synthesized by treating 3,3-bis (chlormethyl)oxetane with sodium azide in dimethyl formamide at 85°C for 24 hrs. NIMMO is a low T_g monomer suitable for munition applications, BAMO is a solid ideal for use as the hard block in TPE manufacture, while AMMO is an unsymmetrical monomer used to provide amorphous character. These energetic monomers are readily polymerized to liquid curable prepolymers by use of boron trifluoride ehterate/1,4-butanediol initiator. Polymerisation occurs by donation of proton from initiator to the oxetane, which then undergoes propagation with more oxetane monomers to generate the polymer chain. The polymer chain is then terminated either with water or alcohol to give the hydroxyl-terminated polymer.

The poly-NIMMO prepared as above, is difunctional. The molecular weight can be adjusted by changing monomer feed rates and the ratio of diol to Lewis acid co-initiator. Generally, the polymeric chains terminate at both ends with primary hydroxyl groups which is more reactive towards isocyanate groups compared to secondary or tertiary hydroxyl groups during cure. However, such primary hydroxyl reactivity can have adverse effects in regards to final cure stability, ageing and decomposition. For good elongation, at least one isocynate equivalent has to be used. For crosslinking purposes, an isocyanate of higher functionality or a separate crosslinking agent must be used.

Bis-Azidomethyl Oxetane (BAMO)

Synthesis

BAMO, bis azidomethyl oxetane is prepared by cationic polymerization of 3,3-bis chloromethyl oxetane, followed by condensation with sodium azide.

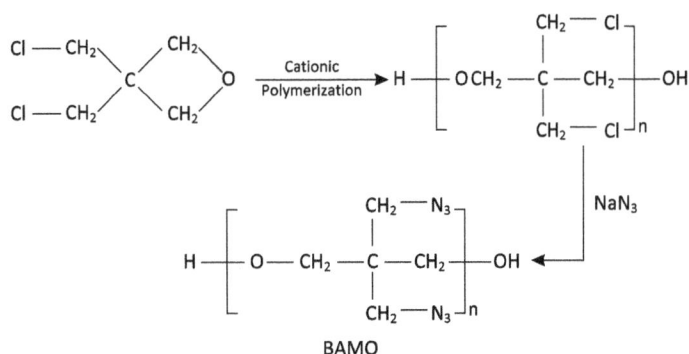

BAMO

Poly-Nitrato Methyl Methyl Oxetane (*Poly* NIMMO) or PLN

Synthesis

PLN, poly Nimmo, i.e., poly nitrato methyl methyl oxetane, is synthesized by first nitrating 3-hydroxy methyl-3-methyl oxetane with N_2O_5 to get 3-Nitrato methyl 3-methyl oxetane which is cationically polymerized to get poly Nimmo. PLN is available from ICI, Scotland. The nitration can also be done using acyl nitrate.

The classical cationic polymerization of oxetanes suffers from several short-comings like lack of molecular weight control and product reproducibility, poor initiator incorporation and the failure of adequate form of copolymers of desired structure. Use of activated monomer method changes the mechanism from a cationic polymerization process to the attack of a hydroxyl-terminated polymer on an activated monomer. The salient feature of the activated polymerization process is attack of an alcohol on an activated oxetane monomer, thereby ensuring rapid and complete initiator incorporation into the polymer chain. The activated monomer polymerization yields copolymers with good molecular weight control and reproducibility, requires only catalytic quantities of a stable acid catalyst and is carried out at ambient temperatures.

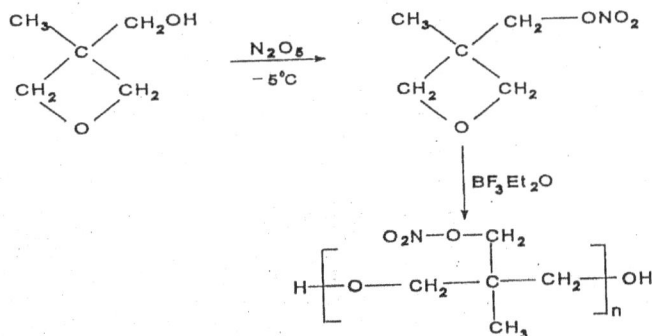

PLN (Poly Nimmo)

Oxetane polymers, such as poly-NIMMO have low glass transition temperatures and missible with similar plasticizers and cured with conventional isocynates. Poly-NIMMO has an intrinsic energy content of 818 kJ/kg and is classified as a non-explosive in UK. Aging and degradation studies of poly-NIMMO showed gassing, although it can be reduced by addition of stabilsers like diphenylamine and 2-nitrophenylamine (1% wt/wt).

BAMO-THF Copolymer

BAMO, unlike GAP, could not give a liquid polyol, hence, its copolymerization products are being developed. BAMO-THF copolymers are reported by various researchers. Copolymerization of BAMO with THF is carried out using BF_3 etherate and 1,4-butanediol. BAMO: THF (50:50) copolymer is a liquid polyol with a molecular weight of around 7000 and a functionality of 1.99. The polyol has good thermal stability and formulations based on it have been processed with good mechanical properties. However, their heat of formation is lower than that of poly BAMO.

These copolymers of BAMO paved way for their application as energetic thermoplastic elastomers. These thermoplastic elastomers have hard and soft segments in their backbone. When heated, the hard segment melts and allows the thermoplastic elastomer to be blended with other propellant ingredients. When cooled the hard segment crystallize and produce physical crosslinks, which can be reversed by heating or salvation. Thus, they are recyclable. This will reduce the amount of waste propellant generated by casting–curing route and hazards involved in the disposal of waste propellants. A BAMO-AMMO thermoplastic elastomer has been developed. Similarly oxygen rich nitrato group containing polymers has led to the development of Polyglycidyl Nitrate (PGN) and poly nitratomethyl methyl oxetane (poly-NIMMO).

PGN (Poly Glycidyl Nitrate)

Oxirane polymers have similar physical and chemical properties to the oxetane polymers, the major difference being fewer methylene groups in the repeating unit. PGN is poly glycidyl nitrate. It is prepared by first nitrating glycidol with N_2O_5 to get Glycidyl Nitrate (GLYN) which is then cationically polymerized to hydroxyl terminated Poly GLYN called PGN.

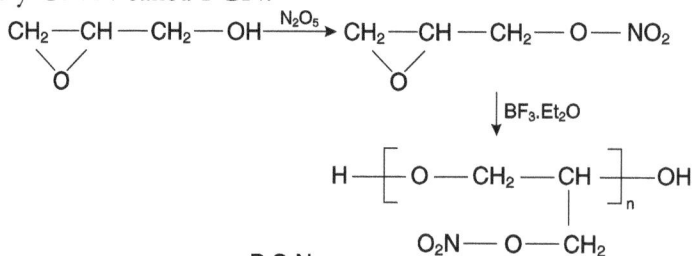

$$CH_2-CH-CH_2-OH \xrightarrow{N_2O_5} CH_2-CH-CH_2-O-NO_2$$

$$\downarrow BF_3.Et_2O$$

$$H-[O-CH_2-CH]_n-OH$$
$$O_2N-O-CH_2$$

P G N

GLYN, like NIMMO, is prepared using N_2O_5 in dichloromethane solvent, in a flow reactor to high purity product in high yields, requiring no purification before polymerization. Polymerization employs tetrafluroboric acid etherate initiator, not boron fluoride etherate as for poly NIMMO, combined with a difunctional alcohol to give difunctional polymer. Slow addition of the monomer solution generates an activated monomer species, which combines with alcohol in a ring opening process, generating a proton. The proton activates further monomer unit to add on to the polymer chain. Reaction is terminated by quenching in excess water, followed by neutralization.

PGN is a clear, yellow liquid with high density and low T_g. Its sensitiveness is low and has a calculated energy of 2661 kJ/kg compared to 2500 and 818 kJ/kg of GAP and poly NIMMO respectively. Though poly GLYN prepolymer shows good chemical stability, isocynate cured polyurethane rubbers show poor stability on aging, not shown by poly NIMMO rubbers. This degradation is not prevented by the addition of stabilizers or by exclusion of oxygen. This degradation of poly GLYN rubber is associated with chain scission at the urethane linkage and has little to do with normal nitrate ester degradation process. This kind of chain scission is not possible with poly NIMMO which has no labile hydrogen atom attached to carbon atom carrying the nitrato group for transfer. This instability appears an inherent property of poly GLYN pre-polymer, and is not dependent on the isocyanate used.

Miscellaneous nitrogen containing polymers for use as energetic binders include polyvinyl nitrate, polynitro phenylene, nitramine polyethers, N, N'-bonded epoxy binders, poly (furfuryl glycidyl ether) and nitrated HTPB (NHTPB) binders. For example, NHTPB has low viscosity, low T_g and is cured with isocyanate.

Fluorinated Polymers

Fluorinated polymers give advantages over inert hydrocarbons such as HTPB. Fluorinated polymers have higher densities which result in formulations with higher densities and inherently higher performance. Secondly, replacement of hydrogen (a fuel) on the backbone by fluorine (an oxidant) increases the overall oxygen balance of the composition, thereby increasing the performance. One of the main type of fluorinated polymer is polyformals, synthesized from dihydric alcohols containing fluorine with formaldehyde to yield hydroxyl-terminated polymers,

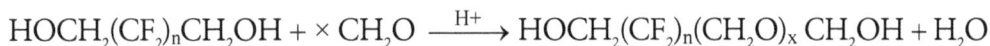

$$HOCH_2(CF_2)_nCH_2OH + \times CH_2O \xrightarrow{H+} HOCH_2(CF_2)_n(CH_2O)_x CH_2OH + H_2O$$

In addition to fluorine itself, several functional groups such as fluorodinitroethyl, difluoroamine groups etc. have been considered. Molecules with a single difluroamino group tend to be sensitive and unstable, and has been attributed to lability of the α-hydrogens. Geminal bis (difluroamines) are less sensitive and more stable. The difluoroamino group is also a potent oxidizing group, although the univalent fluorine is half as effective as divalent oxygen in providing a stoichiometric combustion balance. Further, difluoroamine groups are relatively sensitive to impact.

Copolymers of fluorinated polymers have found uses as inert binders with high densities for energetic compositions. Copolymers are typically vinylidine fluoride and hexafluoro propylene, known as viton A. Such copolymers have high fluorine content and high thermal stability. These polymers are used also in plastic bonded explosives, allowing for processing methods such as ram extrusion and compression molding, without loss of performance. A disadvantage of using these polymers with energetic ingredients is the use of flammable solvents, as it is an environmental and safety issue to remove such inflammable solvents from energetic processing to get solvent free energetic systems.

4. ENERGETIC PLASTICIZERS

The major role of plasticizers in energetic material formulations is to modify the mechanical properties of the propellant/explosive to improve the safety characteristics. This is made possible by softening the polymer matrix and making it more flexible. In addition to improving the tensile strength, elongation, toughness and T_g, the plasticizer can reduce the propellant mix viscosity to ease processing, modification of oxygen balance and energy content and burning rate modification to tailor ballistics. Plasticizers by nature are oligomers that have a molecular weight of 200 to 2000. Plasticizers with molecular weights more than 2000 tend to be viscous and their properties more akin to the polymer matrix. Plasticizers with less than 200 molecular weight are highly volatile and migrate out of formulation readily, though they reduce the T_g.

Energetic plasticizers do this job in propellants and explosives formulations. Molecules having nitro/nitrato substituted esters are preferred plasticizers because of superior oxygen balance and energy content. Structural similarity with the energetic polymer should facilitate the incorporation, but one of the common problems is exudation-migration of low molecular weight plasticizers to the surface of the formulation. A promising approach is to increase the structural similarity, and hence miscibility, by using low molecular weight oligomers of the polymer matrix as the plasticizer. Low molecular weight azido polymers (GAP oligomer and a cyclic tetramer of NIMMO) are used as effective plasticizers.

The first energetic plasticizer known is nitroglycerine or glycerol trinitrate. It is unstable at temperatures of over 70°C and explodes when heated above 200°C. In spite of the plasticizer exhibiting several significant physiological effects, it is used as an effective plasticizer for many applications. The nitrate esters have proved to be a source of energetic plasticizers and the major nitrate esters used today include

Trimethylol Ethane Trinitrate (TMETN), Triethylene Glycol Dinitrate (TEGDN), Butanetriol Trinitrate (BTTN) and Ethylene Glycol Dinitrate (EGDN). TMETN is chemically stable, insoluble in water and has low volatility. It is less impact sensitive than NG. EGDN is a better plasticizer than NG for nitrocellulose, and has more energy but less sensitivity compared to NG. However, it is more volatile and has less density than NG. BTTN is often used in place of NG in propellants in view of its better stability compared to NG. Most of the nitrate ester plasticizers are 1.1 explosives that have low critical diameters, high volatility and high sensitivity, making them difficult to handle and transport.

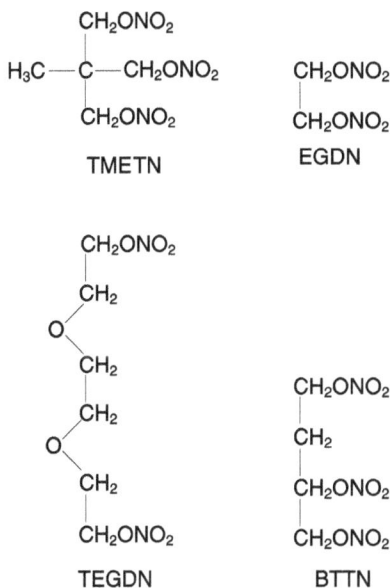

Bis (2,2-dinitropropyl) acetal (BDNPA) and bis (2,2-dinitropropyl) formal (BDNPF) have found application in energetic formulations as in US Army M900 program, Low Vulnerable Ammunition (LOVA) gun propellants and HMX based insensitive explosives. BDNPA/F plasticizers are 50:50 mixtures and are used to form an eutectic to lower the melting point, making the plasticizer useable at lower temperatures. BDNPA/F was made by oxidative nitration of nitroethane to 2,2-dinitropropanol using sodium nitrate, persulphate salts and catalytic amount of potassium ferricyanide. BDNPA/F exhibits poor plasticizing effects on uncured explosive formulations. The plasticizer is also unstable under severe conditions like elevated temperature (> 74°C) and high shock loading.

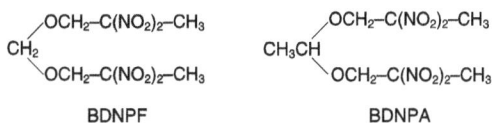

Synthesis of BDNPF and BDNPA

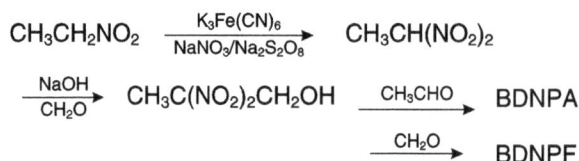

The poor mechanical properties of GAP like polymers can be improved by incorporation of energetic azido functional plasticizers. These low molecular weight GAP plasticizers are synthesized in a single step involving azide replacement of chlorine from epichlorohydrin monomer, followed by polymerization without catalyst. These GAP plasticizers are compatible with GAP polymers. The plasticizing effect of the plasticizer will be lost on terminal hydroxyl groups reacting with isocyanate crosslinking agent. For this purpose, a plasticizer called GAPA (azide terminated glycidyl azide plasticizer) has been prepared with no hydroxyl groups.

GAPA is a low moleculat weight plasticizer, with low T_g and good stability.

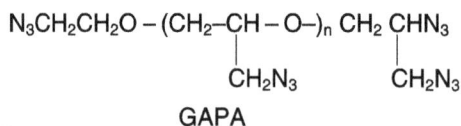

GAPA

A new class of azido-acetate ester energetic plasticizers giving polymers low T_g, good thermal stability and compatibility, have been reported. They are ethylene glycol bis(azidoacetate) (EGBAA), diethyleneglycol bis(azidoacetate) (DEGBAA), trimethylol nitromethane tris(azidoacetate) (TMNTA) and pentaerythritol tetrakis (azidoacetate) (PETKAA).

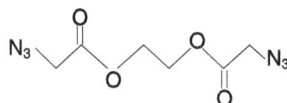

EGBAA

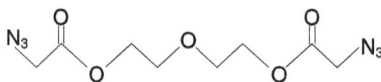

DEGBAA

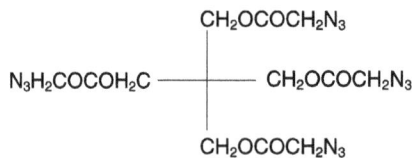

PETKAA

TMNTA

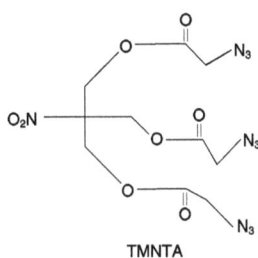

K10 or Rowanite 8000 is a nitroaromatic plasticizer consisting of a mixture of 2,4-dinitroethyl benzene and 2,4,6-trinitroethyl benzene (65/35). K10 is a clear, yellow/orange coloured liquid used in plastic bonded explosives in UK. The nitroethyl nitramines (NENAs) are effective plasticizers with nitrocellulose systems. They contain both nitrate ester and nitramine functionalities in their structure.

$$RN(NO_2)CH_2CH_2ONO_2$$

where R = methyl, ethyl, propyl, iso-propyl and pentyl.

NENAs are made by nitration of commercially available alkyl ethanolamine.

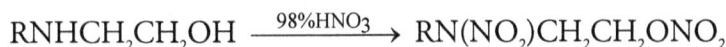

$$RNHCH_2CH_2OH \xrightarrow{98\%HNO_3} RN(NO_2)CH_2CH_2ONO_2$$

The use of NENAs as plasticizers in rocket and gun propellants give excellent properties like high burning rates, reduction in flame temperature and low product gas molecular weight and higher specific impulse. NENAs also possess good thermal stability, readily plasticizes the polymers and give good impact sensitivity. One of the major disadvantages of NENAs as plasticizers is migration from compositions on long standing or aging. They also volatilize on standing because of their low molecular weight. Thus, the propellant formulations using NENAs have difficulty in achieving a 10 year service life. Use of BuNENA with poly NIMMO and poly GLYN binders significantly reduced the T_g and showed no plasticizer migration. Another disadvantage of NENAs is the non-compatibility with ammonium perchlorate. Use of alkyl urea has been suggested as stabilizer for NENAs when used with ammonium perchlorate.

Linear NIMMO oligomers (polymers consisting of between 1 to 10 monomer units) and cyclic tertramers were prepared for use as plasticizers in poly NIMMO binder systems. Cyclic tertramer has high T_g and was found not effective as plasticizer. To remove the terminal hydroxyl groups in the linear oligomers, they were further nitrated with N_2O_5 at –10°C to get nitrato terminated oligomers without any chain degradation. Nitration also enhances the oxygen balance and enhances the energy content of the binder system. Nitration with excess N_2O_5 also converts residual NIMMO monomer into TMETN.

A dimer of glycidyl nitrate (GLYN) has been prepared for plasticization of poly NIMMO and poly GLYN. A linear GLYN dimer is prepared by end capping a

1,4-butanediol spacer unit with GLYN, and then nitrating the terminal hydroxyl groups. GLYN dimer is mixture of oligomers and has low T_g (–64.9°C) and impact sensitivity compared to nitrate esters such as BTTN and TMETN. Studies of poly GLYN with GLYN dimer plasticizer have given well cured binders with good mechanical properties. It is expected that GLYN dimmer will be less susceptible to migration than the more conventional plasticizers.

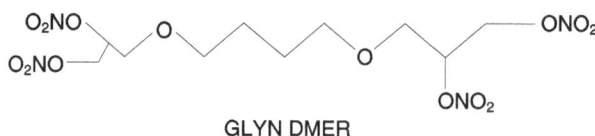

GLYN DMER

5. ADVANCED SOLID PROPELLANTS

A combination of energetic polymers with modern oxidizers like ADN, HNF or CL-20, can lead to tremendous technology upgradation in the field of solid propellants in the coming years, despite the fact that all energetic polymer systems are processable with solid loading only up to 80%, unlike HTPB which offers solid loading exceeding 86%. This is due to superior oxygen balance and heat of formation of energetic binders. ADN and HNF with GAP, poly-BAMO, poly-NIMMO and PGN show theoretical specific impulses of more than 300 sec as seen from Table 4 given below.

Table 4: Specific Impulse of Advanced Propellants

Oxidizer	Binders			
	GAP	*Poly-BAMO*	*Poly-NIMMO*	*PGN*
ADN	310	312	309	306
HNF	314	317	313	310

(Pressure 5 *MPa*, A_e/A_T = 60, Equilibrium flow and vacuum conditions)
(HTPB/AP/Al in 11/71/18% Wt ratio under the same conditions give 295 s)

Realization of such propellants is pursued fervently and major problems associated with processing as well as compatibility are being tackled. A CL-20 formulation with GAP/TMETN/BTTN offers a 7% higher specific impulse compared to RDX based formulations. A few formulations of CL-20 with GAP/BDNPF/A have shown sensitivity comparable to that of HMX formulations. Suitable ballistic modifiers capable of enhancing burning rates below 70 kg/cm^2 pressure, leading to reduction in pressure exponent "*n*" have also been identified.

Ballistically modified AP-BAMO/NIMMO propellant with a specific impulse of ~265 sec has been reported. Formulations of CL-20 (76%) with BAMO/AMMO has resulted in a gun propellant giving higher density (1.77 g/cm^3) and impetus (F = 1297 J/g) compared to RDX based formulations (density = 1.64 g/cm^3 and

$F = 1182$ J/g). Poly-NIMMO based composite LOVA propellant has established the possibility of high energy gun propellant ($F = 1250$ J/g) with improved insensitivity to the accidental impact of ballistic projectiles, bullets and metal fragments. An improved artillery gun propellant containing 58% RDX, 17% NQ and 25% BAMO/AMMO (35:65) has been reported. BDNPF/A based double base propellants with an I_{sp} of 246 sec compared to that of conventional triacetin based compositions ($I_{sp} = 232$ sec) have been reported.

Due to the implementation of Environmental Protection Agency (EPA) restrictions, the more recent trend is to develop eco-friendly HEMs using green nitrating agent like dinitrigen-pentoxide. Another way to achieve higher energy and density is to tailoring of the molecular architecture either by introducing exothermically decomposing groups or by introducing strain/cage like geometry. Prominent energetic group, such as azide, contributes a heat release close to 355 kJ/N_3 unit.

High nitrogen compounds have few or no nitro groups and rely on their efficient gas production and also on their high positive heats of formation for energy release since elemental nitrogen, which has a zero heat of formation, is the major product of decomposition. The limited amount of carbon and hydrogen in these materials mainly form methane upon decomposition especially when the compounds contain only the elements of carbon, hydrogen and nitrogen. High nitrogen compounds also produce more gas per gram than most high explosives and since nitrogen is the major decomposition gas, the products are inherently cooler. As a consequence of this, high nitrogen materials offer distinct advantages over conventional energetic materials and may be used as ingredients in pyrotechnics, gas generators and low signature propellants.

Cyclic Polynitrogen Compounds

Advances in computational materials science and theoretical condensed matter physics, coupled with the power and speed of modern super computers, have led to design and study of novel materials and molecules. In spite of numerous theoretical studies predicting that certain catenated-nitrogen compounds might be stable, only two homoleptic polynitrogen species are known that can be prepared on macro scale viz., dinitrogen (N_2) and the azide anion (N_3^-). In view of extensive theoretical investigation indicating that molecules such as N_4, N_8, $N(N_3)_2^-$, $N(N_3)_3$ and $N(N_3)_4^+$ are vibrationally stable, the lack of single successful synthesis on a macroscopic scale may be due to greater experimental difficulties resulting from their high endothermicities.

The high energy content of polynitrogen candidates stems from the N–N single and double bonds they possess. The average bond energies of 160 and 418 kJ/mole respectively are much less than one-third or two-thirds of nitrogen triple bond energy of 954 kJ/mole. Therefore, any transformation of a polynitrogen compound to N_2 molecules is accompanied by a very large energy release and any new metastable polynitrogen compound will be isolable and manageable only if it possesses a sufficiently large energy barrier to decomposition. Consequently, storing the maximum energy in a polynitrogen molecule would mean having largest number of single bonds. The azido group is highly energetic ligand, which adds about 87 kcal/mole of endothermicity to a hydrocarbon compound. The triazocarbenium cation $C(N_3)_3^+$ was first prepared in 1966 as its $SbCl_6^-$ salt and was characterized in 1975.

One of the most important cyclic polynitrogen compound is octaazacubane. The explosive properties of it are given in Table 5.

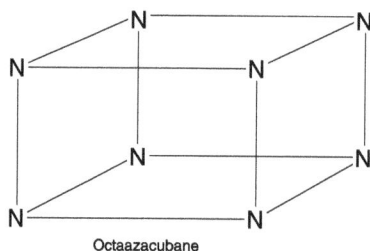

Octaazacubane

Table 5: The Explosive Properties of Octaazacubane and HMX

	Properties	*Octaazacubane*	*HMX*
1.	Density (g/cm^3)	2.65	1.90
2.	VOD (km/s)	15	9.1
3.	PCj (k bar)	1370	390
4.	ΔHf (kcal/mole)	530	18

Fullerenes are large carbon molecules with from 32 to as many as 600 atoms. A nitrogen fullerene, according to Riad Manaa might be possible by joining six bicyclic 10-atom nitrogen molecules into a soccer-ball shaped molecule, similar to carbon based buckminster fullerene. With their high energy density, large nitrogen molecules would be prime candidates for new high explosives or perhaps for novel propellants. Calculations show that N_{60} is purely single bonded with N–N = 1.43 A. and has three active IR bonds at 68, 701, 1153 cm^{-1}. The pentagonal and 6-membered ring bond lengths switched, with r_1 = 1.43 and r_2 = 1.44A where r is bond length.

Most of our solid propellant systems were developed in the late 1950's with some development continuing into the early 1970's. But no significant new energetic material has been introduced into the propellant area since then. However, many new materials and technologies are now available that we need to employ. New materials are inherently expensive to buy until they go into a system. There is no production capability to allow economics of scale to operate. The early high price inhibits timely evaluation and development. The problem will be even greater in the future because we have so few systems coming along. There is a need to maintain a research effort that will be adequate to provide a strong technology base. This requires developing materials and ingredients without necessarily having an immediate use for the materials, but rather knowledge that having qualified materials on the shelf will result in the next system being developed using today's technology, not yesterdays.

Typical Questions

1. Why do we need advanced oxidizers and fuels?
2. What are the desirable characteristics of advanced oxidizers and fuels?
3. Explain the synthesis of the following advanced oxidizers?
   i) ADN, ii) HNF, iii) CL-20.
4. Explain the synthesis of GAP? How does it improve the energetic of the propellant?
5. Explain the synthesis of BAMO, AMMO and NIMMO? Compare and contrast their chemical properties?
6. What are the safety precautions to be adopted while synthesizing advanced oxidizers and fuels?
7. How do we preserve these advanced oxidizers and fuels?
8. How do we dispose off the advanced oxidizers and fuels which are not qualified for use?
9. How is PGN prepared? What are its chemical properties?
10. Compare the specific impulse of GAP/AND/Al and GAP/HNF/Al with HTPB/AP/Al composite propellants?
11. Name few energetic plasticizers with nitro groups and azide groups? Explain how do they improve the energetic of solid propellants?
12. Give the chemical structures of the following and explain their energetic features?
    i) Octanitro cubane, ii) Octaaza cubane.
13. What are the important design considerations for development of advanced propellant formulations?
14. List 5 energetic groups which enhance the performance of fuels and oxidizers?
15. How does the use of ADN contribute towards upper stage motors, pay load capability and eco-friendliness?

REFERENCES

[1]   Agrawal, J.P. and Hodgson, R.D., Organic chemistry of explosives, John Willey and Sons Ltd., The Atrium, Chichester, England, 2007.

[2]   Badgujar, D.M. and Mahulikar, P.P., Advances in science and technology of modern energetic materials: An Overview, *J. Haz. Mat.*, 151, 2008, 289.

[3]   Borman, S., "Advanced Energetic Materials Emerge for Military and Space Applications," C&EN, January 1994, pp. 18–22.

[4]   Bottaro, J.C., Recent Advances in Explosives and Solid Propellants, *Chem. Ind.* 10, 1996, 249.

[5]   Bottaro, J.C., Schmitt, R.J., Penwell, P. and Ross, S., U.S. Patent 5254324, 1993, SRI International, USA.

[6]   Chen, M., Sui, Y.K. and Yang, Z.G., *Initiators & Pyrotechnics,* No. 5 (2007), pp. 5–8.

[7]   Cumming, Adam, "New Directions in energetic Materials," *J. of Defence Science,* I, 319–327, 1997.

[8]   David, Ryder, GAP cured with acrylate, US Patent No. 6143103, 7 November 2000.

[9]   DeLuca, L.T., New Energetic Ingredients for Solid and Hybrid Rocket Propulsion, *Proc. 9th International High Energy Materials Conference,* Feb. 13–14, Trivandrum, 2014.

[10] Eroglu, M.S. and Guven, O., "Thermal decomposition of Poly (Glicidyl Azide) as study by high temperature FTIR and Thermogravimetry," *Journal of Applied Polymer Science,* 61, 201–206, 1996.

[11] Fischer, N., Klapötke, T.M., Matecic Musanic, S., Stierstorfer, J. and Suceska, M. TKX-50, New Trends in Research of Energetic Materials, Prat II, Czech Republic, 2013, 574–585.

[12] Frankel, M.B., Grant, L.R. and Flanagan, J.E., "Historical Development of Glycidyl Azide Polymer," *Journal of Propulsion and Power,* 8 (3), 560–563, 1992.

[13] Gadiot, G.M.H.J.L., Mul, J.M., Muellenbrugge, J.J., Korting, P.A.O.G., Shoyer, H.F.R. and Schnorhk, A.J., "New Solid Propellants based on Energetic Binders and HNF," *Acta Astronautica,* Vol. 29, No. 10/11, 1993, pp. 771–779.

[14] Gaur, Bharati; Lochab, Bimlesh; Choudhary, V. and Verma, I.K., "Azido Polymers-Energetic Binder for Solid Rocket Propellants," *Journal of Macromolecular Science Part-C,* 43, 4: 505–45, 2003.

[15] George, P. Sutton, "Book on Rocket Propulsion Element," Vol. 7, edition 2001, pp. 27–36, 46–84, 417–453, 474–511.

[16] Ghee, Ang How and Santhosh, G., Advances in Energetic Dinitramides—An Merging Class of Inorganic Oxidizers. Singapore: World Scientific, 2008.

[17] Gordon, S., Mc Bride, B.J. and Zeleznik, F.J., "Computer Program for Calculation of Complex Chemical Equilibrium compositions and Applications, supplement 10 Transport Properties," NASA TM 86855, Washington, DC, Oct., 1984.

[18] Gould, R.F., Propellants Manufacture, Hazards, and Testing, Advances in Chemistry, No. 88, American Chemical Society, Washington, DC, 1969.

[19] Gupta, Manoj, "High Performance Propellants based on Energetic Binders," *Proc. 9th International High Energy Materials Conference,* Feb. 13–14, Trivandrum, 2014.

[20] Jeffrey, C. Bottaro, "Recent Advances in Explosives and Propellants," Chemistry and Industry, April 1996, pp. 249–252.

[21] Johannessen, B., "Low Polydispersity Glycidyl Azide Polymer," US Patent No. 5741997, 1998.

[22] Krishnamoorthy, V.N., Energetic Materials for the New Millennium, *Proc. 3rd International High Energy Materials Conference*, Dec. 6–8, Trivandrum, 2000.

[23] Landcrs, L.C. and Sanley C.B., Propellant development for the Advanced solid rocket motor, in 27th Joint Propulsion Conference June 24–26,/Sacramento, CA. AIAA/SAE/ASME, 1991, 9 1–2074.

[24] Lewars, E., Computational Chemistry Introduction to the Theory and Applications of Molecular and Quantum Mechanics, Dordrecht, Kluwer Academic Pub, 2003.

[25] Lo, R.E. and Thierschmann, "Propellant Trends in Advanced Rocket propulsion," ESA-SP-293, 1989.

[26] Manu, S.K., Sekkar, V., Scariah, K.J. Varghese, T.L. and Mathew, S., *Journal of Applied Polymer Science*, 110, 2008, 908–914.

[27] Manzara, A.P. and Johannessen, B., "Primary Hydroxyl Terminated Polyglycidyl Azide," US Patent No. 5, 164, 521, Nov. 1992.

[28] Miller, R.S., "Research on New Energetic Materials," *Material Research Society Symposium Proceedings*, Vol. 418, 1996, pp. 3–14.

[29] Muthiah, R.M., Varghese, T.L., Rao, S.S., Ninan, K.K. and Krishnamurthy, V.N., Realization of an eco-friendly solid propellant based on HTPB-HMX-AP system for launch vechile application, *Propellants, Explosives, Pyrotechnics*, Vol. 23, 1998, pp. 90–93.

[30] Nair, U.R., Sivabalan, R.R., Gore, G.M., Geetha, M., Asthana, S.N. and Singh, H., Hexanitro hexaaza isowurtzitane (CL-20) and CL-20-Based Formulations (Review), Combustion, Explosion, and Shock Waves, 41, 2013, pp. 121–132.

[31] Nazare, A.N., Asthana, S.N. and Singh, H., "Glycidyl Azide Polymer (GAP)—An Energetic Component of Advanced Solid Rocket Propellants—A Review," *Journal of Energetic Materials*, Vol. 10, 43–63, 1992.

[32] Oommen, C. and Jain, S.R., "Ammonium Nitrate: A Promising Rocket Propellant Oxidizer," *J. Hazardous Materials*, A67, 253–281, 1999.

[33] Reshmi, S., Varghese, T.L., Ninan, K.N., *et al.*, *34th International Annual Conference of ICT*. Germany, 2003.

[34] Sadavarte, V.S., Singh, R.V. *et al.*, Advanced High energy Rocket Propellants Containing Hexanitrohexaaza isowurtzitane (CL-20), *Proc. 9th International High Energy Materials Conference*, Feb. 13–14, Trivandrum, 2014.

[35] Schoyer, H.F.R., Ashnorhk, A.J., Kortig, P.A.O.G., Van lit, P.J., Mul, J.M., Gadiot, G.M.H.J.L. and Meaulenbrugge, J.J., "High Performance propellants based on Hyrazinizum Nitroformate," *AIAA Journal*, 1995, pp. 855–868.

[36] Schoyer, H.F.R., Schnorhk, S.J., Korting, P.A.O.G., Vam Lit, P.J., Mul, J.M., Gadiot, G.M.H.J.L. and Mulenbrugge, J.J., "Advanced Solid Propellants Based on HNF—A Status Report," ESA/ESTEC, 1995, pp. 1–8.

[37] Sekkar, V., Krishnamurthy, V.N. and Jain, S.K., *J. Appl. Polym. Sci*, 66, 1795, 1997.

[38] Sema, Keskin and Saira, Ozkar, Kinetics of Polurethane formation between GAP and trisocyanate, *Journal of applied Polymer Science*, Vol. 81, 918–23, 25 July 2001.

[39] Silva, G., Rufino, S.C. and Iha, K., Green Propellants: Oxidizers, *J. Aerosp. Technol. Manag.*, 5(2), 2013, 139–144.

[40] Singh, Haridwar, "High Explosives—Past, Present and Future," In S. Krishnan, S.R. Chakravarthy and S.K. Athithan (eds)," Propellants and Explosives Technology, ISBN 81-7023-884-6, Allied Publishers Limited, Chennai. 1998, pp. 245–270.

[41] Soman R.R., Athar, J. *et al.*, Curing Studies of Gap diol with Different Isocyanates for Propellant Application, *Proceedings of the 9th international high energy materials conference*, 2014, Trivandrum.

[42] Varghese, T.L. and Ninan, K.N., "Gelled Propellants and New Energetic Materials as Propellant/Explosive Ingredients," India, 1998, pp. 419–448, in S. Krishnan S.R. Chakravarthy and S.K. Athithan (eds.), book on Propellants and Explosives Technology, ISBN 81-7023-884-6, Allied Publishers Limited.

[43] Varghese, T.L., Chemical Propellants, IITP-2002, Area Specific Intensive Course Module III, General Lectures, Book Published by VSSC, ISRO, August 2002.

[44] Varghese, T.L., Manu, S.K., Mathew, S. and Ninan. K.N., *J Applied Polymer Science*, 114, 2009, 3360–3368.

[45] Varghese, T.L., Prabhakaran, N., Rao, S.S. and Ninan, K.N., "Development of a New generation high pot life hydroxyl terminated polybutadiene propellant." *Proc. Second international HEMCE and Exhibit, IIT Madras,* 1998.

[46] Varghese. T.L., Ninan, K.N., Krishnamurthy, V.N., *et al.*, Performance Evaluation and Experimental Studies on Metallized Gel Propellants, *Defence Sci. J.*, Vol. 49, No. 1, 77–78, 1999.

[47] Venkatachalam, S., Santhosh, G. and Ninan, K.N., *Prop. Expl. Pyro.* 29 (3), 2004, 178–187.

[48] Y. and Mace, H. New energetic molecules and their applications in the energeticmaterials. *In Proceedings of 29th International Annual Conference of ICT, Karlrushe,* Germany, 1998, pp. 3/1–3/17.

CHAPTER 8

Safety, Quality and Reliability
in Solid Propellants

1. SAFETY

Solid propellants have been extensively used in launch vehicles and missiles. The process of manufacturing large boosters for launch vehicles involve various steps such as handling of tonnes of raw materials, mixing of raw materials in horizontal or vertical mixers, propellant casting, curing, storage, transportation and flight. Most of the high energy solid propellants for launch vehicles use high solid loading of oxidizer like ammonium perchlorate and fine particles of aluminum powder in the range of 86–88% along with polymeric binder like HTPB, diisocyanate curing agents, plasticizer, ballistic modifier, etc. Some of the components as well as processing operations are hazardous and the hazardous nature increases with use of more energetic formulations containing advanced oxidizers like Ammonium Dinitramide (ADN), Hydrazinium Nitroformate (HNF), Hexa Nitro Hexaza Isowurtzitane (HNIW), etc. and more energetic binders like Glycidyl Azide Polymer (GAP), Poly Glycidyl Nitrate (PGN), Poly Nitrato Methyl Methyl Oxetane (poly NIMMO), etc. and explosive materials like RDX (cyclo trimethylene trinitramine), HMX (cyclo tetra methylene tetra nitramine), etc.

Propellant industries are faced with not only normal occupational hazards of chemical industry but also with the necessity of working with materials which are combustible and explosive in nature, toxic (iscyanates, MAPO) and corrosive (RFNA). Some materials are not only hazardous to buildings and equipments but also to the workers. Hence, safety is a must for the survival, growth and progress of this industry.

Safety is antonym to danger. Higher safety level ensures lower danger level. Hazard is defined as the inherent potential threat of causing damage or loss. The concept of safety is linked with accidents which are unintended events which cause heavy damage to materials and life. For energetic materials like propellants, even though absolute safety is not possible, still they have to be safe, limiting the danger level to an acceptable level. Explosive safety is one of the most difficult and demanding tasks due to requirements of striking a balance between operational and explosives safety considerations. The advances in explosives-missiles technology must go hand in hand with safety engineering and research and testing. Hence, there is a need to understand the ability and necessity to identify hazards and design to control them or avoid them. Explosive safety has been understood by accidents that have occurred in the past. All these experiences have been framed in to rules and laws describing what is correct way to do and what is not.

It is fact that high explosives that are detonable cause en masse explosion whereas propellants which are non-detonable have mass fire risk under normal conditions. During the production of rocket propellants and explosives, the assembled rocket motor can get ignited and start combustion when it is not expected to do so leading to production of hot gases, local fires or ignition of adjacent rocket motors. The pressure build-up can be high enough to cause explosion. If the motor is not constrained, its thrust will increase suddenly and the erratically flying motor can cause damage. The exhaust will be toxic and corrosive. It is necessary to understand fully the hazards and the reasons that caused and the methods for preventing hazardous situations from arising. The damage caused by the blast during casting of composite propellants based on PBAN at Morton-Thiokol, USA is worth recalling here. Two shuttle segments and two casting pits were lost. Blast and incendiary effects, unexpected from propellant were observed several thousands of meters away from the site.

Inadvertent ignition of the rocket motor could be the reason for the incident. The inadvertent ignition can be caused by, i) stray or induced currents activating the igniter, ii) electrostatic discharge, iii) fires causing the motor temperature to go beyond auto ignition point, iv) impact(bullet hitting or dropping of the motor on hard surface) and v) energy absorption on continuous mechanical vibration. To prevent stray current from activating the igniter, "safe and arm system" is provided. Electrostatic Discharge caused by lightening or friction of insulating materials is prevented by ensuring sufficient electric conductivity and grounding or earthing the motor or the container having the explosive. Propellant being a viscoelastic material can absorb vibration energy and become locally hot when vibrated for long periods at particular frequencies. This can happen in a motor where a segment of grain is not well supported and vibrates at natural frequencies.

Explosions have taken place even with black powder and Alfred Nobel witnessed all possible failure during his research with NG and NC. From literature, it can be seen that maximum accidents have taken place in pyrotechnic and pyro devices, followed by high explosives. Most important reason for the accidents is friction.

Safety is represented by green triangle with environment, enforcement and operation as the three arms. Any deficiency on any one of the three arms of safety triangle will break, thereby leading to calculated or unimaginable consequences.

Safety is everybody's business, not only for explosive industries but also in every walk of life. Everyone have to accept that accidents are totally avoidable and absolute safety is within the reach, provided a firm determination exists. Studies have indicated that human factor is the cause of 90% of the incidents. Some important factors for the accidents are: i) lethargy and haste, ii) sense of complacency or over dependence on auto controls, iii) sense of excellence with half knowledge, and iv) environment. Training including fire drill and safety awareness programmes is an important step in safety policy. To reiterate the importance of safety, National safety week and fire service day are observed in all explosive industries or organizations to rededicate the working staff towards safe working practices.

2. HAZARDS CLASSIFICATION

Hazardous materials have nine hazard divisions as per United Nations (UN) classification. They are explosives, gases, inflammable liquids, inflammable solids, poisonous substances, radioactive substances, corrosive substances and miscellaneous materials. Propellant comes under UN classification 1. This is further divided into six subdivisions:

1. Propellants that is prone to a transition from deflagration to detonation. The major hazards of this class are blast, high velocity projection and flame. Structural damage is directly proportional to quantity and type of the material involved. Propellant mixing, high explosives and mines come under this category.

2. Materials under this category have projection hazard with minor explosion hazard but not mass explosion hazard. The projected pieces can explode or detonate on impact. Rockets, grenades and assembled munitions come under this category.

3. Under this class, we have finished propellants where the case may burst if the chamber pressure is too high, but the propellant will not detonate. It may burn with great violence and intense heat resulting in dangerous fragment formation or thermal radiation hazard.

4. This class includes substances with primarily moderate fire hazard as with small arms ammunition, caps, etc. The hazard is mostly confined to the packages.

5. This group is for very insensitive substances which have mass explosion hazard. There is very little chance of initiation or transition from burning to detonation under normal conditions.

6. This class contains substances which have chances of explosion limited to a single article. No mass explosion hazard, negligible accidental initiation or propagation.

Each of subdivisions from 1.1 to 1.4 has typical symbols containing numbers in different red colored geometries inside a square. This classification determines the method of labeling and the cost of shipping rocket propellants, loaded missiles, explosives or ammunition. It also determines the required limits on the amount of that propellant stored or manufactured in any one site or building and the minimum separation distance of that site or building to the next building or site.

Fire is one of the hazards during processing, storage, transportation and end use of propellant. Fire requires oxidizer, fuel and initiation and propagation. Fire is classified under four categories—A, B, C and D. Class A fire includes solid materials of organic nature like wood, paper, textiles, rubber, etc. Class B fire involves liquids and liquefiable solids like alcohol, oil, petroleum, paints, etc. Class C fire exists in gases and liquefied gases in the form of gas leak or liquid spillage as in acetylene, propane and hydrogen. Class D includes metals and their alloys like potassium, sodium, aluminum, zirconium, etc.

3. SAFETY TESTS

Safety during Propellant Processing

All operations in propellant processing must be supervised independently by safety personnel. The testing of energetic propellants is important with respect to three main aspects:

1. Response to a given stimulus
2. Type and level of response and
3. The environment.

The main hazards involved in the operations of solid propellants are:

1. Impact
2. Friction
3. Auto-ignition and
4. Shock or Detonation.

Among these, impact, friction and auto-ignition predominate in the processing of composite solid propellants. Shock or detonation is related to propellants containing explosives like RDX, HMX, NG (Nitroglycerine), NC (Nitrocellulose), etc.

Impact Test

Impact test is done using an impact tester similar to the one used in the Naval Ordnance Laboratory (NOL). Here, cured propellant samples of size 5 mm diameter and 1–2 mm thick are used. The sample encapsulated with Al foil or open is kept between two hard anvils and a 3 kg weight is dropped freely from various heights.

The impact initiation energy test gives the % decomposition of the sample at various energy levels (heights). This helps to determine the energy level at which maximum fraction of the sample decomposes. Percentage ignition is plotted against the height of fall and the 50% ignition height in cm is evaluated. This method is known as Bruceton method. The drop weight × distance at which 50% of the sample get fired is taken as the impact energy in kg × cm. 10 samples are tested for a propellant formulation. Audio-visual techniques are used to detect ignition of the samples. The 50% ignition height (cm) in each case is evaluated.

Impact energy at which 50% of HTPB/AP/Al 86% solid propellant get fired is found to be in the range of 55–60 kg × cm. Some of the studies done on HTPB propellants show that impact sensitivity increases with decreasing particle size of AP i.e., fine particles are found to be more sensitive compared to coarse particles. This could be attributed to large surface area available for the finer particles. Also, impact sensitivity increases with increase in the loading of AP.

Friction Test

The friction test is done using a friction tester similar to the one used by NOL. The test sample size is similar to the one used for impact testing. The test sample is placed between a fixed rough stainless steel plate and a movable stainless steel plate. The movable plate is pulled at varying lever loads to create the required friction.

As in the case of impact tester, here also ten samples are tested for each propellant formulation to detect ignition of the samples using audio-visual techniques. Percentage ignition is plotted against the friction force in kg × cm and the 50% ignition friction force in kg × cm is evaluated. This method is known as Bruceton method. The 50% ignition weight in each case is evaluated. The distance covered by the movable plate multiplied by the weight at which 50% ignition of the sample gets fired is taken as the friction energy in kg × cm.

Friction energy at which 50% of HTPB/AP/Al 86% solid propellant get fired is found to be above 175 kg × cm. Some of the studies done on HTPB propellants show that friction sensitivity increases with increasing particle size of AP i.e., coarser particles are found to be more sensitive compared to fine particles. This could be attributed to friability of larger AP particle having more crystal defects than fine particles. Also, friction sensitivity increases with increase in the loading of AP.

Auto-Ignition Test

Air heating type auto-ignition tester is used to determine the auto-ignition temperature of the sample. Samples are of 5 mm diameter, 2–3 mm thick, weighing 80–120 mg is used. The sample is kept in an Aluminum sample holder which is suspended in a hot air current under heating rates of 10–15°C/minute. The temperature at which the sample gets ignited is taken as the auto-ignition temperature. Two samples are tested at each heating rate.

Due to temperature constraints in the auto-ignition test, DSC (Differential Scanning Calorimetry) can be used for samples having auto-ignition temperature above 300°C. In this case, 3–5 mg of the propellant sample is evaluated using Du Pont-TA-2000 in conjunction with DSC model 905 or equivalent using platinum crucible and the temperature at which the sample explodes during DSC test is taken as the auto-ignition temperature.

Auto-ignition temperature of HTPB/AP/Al 86% solid propellant determined by the DSC method is found to be 325°C. Auto-ignition temperature of HTPB propellant is found to decrease with increase in the particle size of AP due to the higher friability of coarser particles. Auto-ignition temperature decreases with increase in the loading of AP.

Shock or Detonation Test

Shock or Detonation Test is done using card gap tester similar to the one used by NOL. In this test, cellophane cards of 0.3 mm thick are used for the test. Samples of size 40 mm diameters and 140 mm length are used for the test. This test shows whether a propellant sample is susceptible for detonation. It also gives the critical diameter of the sample below which it will not detonate. The number of cards required to ignite 50% of the sample is evaluated using the same Bruceton method. Shock sensitivity tests show that HTPB composite propellants do not detonate.

4. SAFETY APPROACH

In view of high hazards inherent in the manufacture, assembly and testing of large size rocket motors, mandatory safety criteria have to be adopted in process equipments, process buildings, electrical aspects and general safety features. Inbuilt safety features are to be incorporated in equipments or processes. The A, B, C of safety first approach demands safety attitude (A), safety behavior (B) and safety culture (C) which are some of the safety tools to be developed. Other safety tools are safe operating procedures, inspection in the work place, engagement of safety trained staff in the work area, etc. Some of the personnel safety measures to be adopted in the work spot are use of cotton apron, hand gloves, safety goggles, first aid box, etc. The toxic level of toxic gases/vapors is to be monitored as and when required in the work spot.

Safety During Propellant Processing

All operations in propellant processing must be supervised by independent safety personnel to ensure safety as per check list. The concern of safety hazard peaks up during oxidizer grinding and propellant mixing. Perchlorate is sensitive to impact and friction. It is also capable of detonation under favorable conditions. Ultrafine particles are particularly sensitive to electrostatic discharge, mass explosion or even detonation. Solid propellant manufacture involves the association of different ingredients during the mixing stage to form final propellant slurry. Unusual hazards may exist in unusual combination of two or more ingredients. These hazards may have direct impact on processing and a corresponding effect on costs. Knowledge of chemical reaction between ingredients is established in the laboratory and based on the information, the addition sequence of ingredients and mix conditions are determined and followed. Example of potential combination hazards are MAPO-AP and Al-AP powders. These are suitably avoided by selecting the proper mixing sequence.

Propellant mixing comes under class I or II and has major explosion hazard. The major process equipments for composite solid propellants are horizontal sigma and vertical change can mixers. The bowls are made of stainless steel 304, and the blades and shafts are made of phosphor-bronze, all with fine smooth finish, free from any surface defects. Boiler quality steel is used for jackets and non-sparking gun metal is used for bushes. The horizontal mixers are provided with lids with counter weights. The gap between the blade and the bowl and between the blades is very critical in mixing operations. Flame proof limit switches are provided for opening and closing the lids of horizontal mixers. Worm reduction gear and speed variators are used to control the blade speed and their ratios. Mixing in vertical mixer is safer than in horizontal mixers as the slurry does not come in contact with glands. Also, variable quantity can be mixed in vertical mixers and the cleaning operations are easy. Several kinds of hazards triggering stimuli are simultaneously present in the propellant mixers. Friction, impact, hotspots due to mechanical work and chemical interaction are the usual types. A larger mixer, being capable of higher shear work, has higher potential for hotspots.

Mixing is carried out in a traversed building with remotely monitored feeding and operation controls. The whole mixing operation is remote controlled from a centralized control panel, with CCTV surveillance system. Sigma blade mixers are being phased out for safety reasons. Complete sweep inside the vessel needs very low gaps between mixer blades and vessel wall. Any accidental fall of foreign material can result in fire hazard. The ingredients are, in general, added to the mixer through a sieve to avoid occurrences of such events. In addition, the mixer blade is designed to rotate in only one direction and reverse sweep is generally avoided. Water deluge, bowl drop mechanism in case of fire is the other safety features. The closed remote ingredient feeding system prevents exposure to fine dust of solid ingredients and toxic vapors of liquid ingredients.

Propellant curing, machining and trimming are done by proper remote operated machines with several safety precautions. Fool proof limit switches are provided in view of man-machine interaction along with emergency shut-down switches. Hand trimming of the grains is done by personnel wearing safety uniforms and using non-sparking tools.

Avoidance of static electricity hazards is possible by efficient earthing of working personnel and machines used in processing of propellants. Use of non-conducting materials in the working place must be avoided when working with propellants and explosives. Synthetic clothes, non-conducting flooring and non-conducting shoes must be avoided as much as possible when working or handling propellants. Conducting flooring are preferred for buildings where high energy materials with ignition energy less than 45 millijoules are handled. Dry conditions favor accumulation of static charge.

Since propellant does not need external oxygen for sustaining combustion, water is the best and only fire fighting medium. Automatic deluge/sprinkler systems are provided for all propellant processing operations. Hydrant systems supplement deluge system along the external boundaries of the concerned facilities. As the fire protection has to be ready throughout the process operations, redundancy is provided for pumps, pipe lines and water storage. For processes like mixing, casting, curing and trimming, appropriate fire detectors are installed for initiating immediate fire control action automatically. Readiness of fire protection systems is a pre-requisite for facility readiness for propellant processing.

The motor case will break or explode if the chamber pressure exceeds the case pressure. The release of high pressure gas energy can cause explosion and motor pieces could be thrown out into the adjacent area. The sudden depressurization from chamber pressure to ambient would extinguish the propellant burning. Large pieces of unburnt propellant pieces can be seen after the burst. This type of failure can be due to grain over-aging, porosity, severe cracks, or large amount of nonbonded areas. The other reasons could be due to motor damages, obstruction of the nozzle due to insulation pieces, or degradation of propellant strain capabilities due to moisture absorption. Motors are sealed to prevent humid air access.

An integrated safety plan is followed for assembly and testing of rocket motors. It covers the safety aspects of operating staff, test motor, test facility and test data. The major operations involved are handling of motor segments, dimensional check up, interface preparation, segment alignment, assembly of segments, sub-systems and overall assembly of the motor on test stand. This is followed by calibration of pressure and thrust pick ups, instrumentation, static test and data collection.

The hazard potential of modern rocket propellants has increased because of the use of RDX/HMX and NG/NC systems to boost the energy content or performance.

Hence, the ranking of high energy propellants and explosives are narrowing down. Past incidences of high explosives are relevant to the present and future incidents of advanced solid propellants.

During non-destructive testing of finished propellants, X-ray sources are used to detect flaws, cracks, porosities and voids. This needs X-ray source and the building for the activity and it must be planned as per the Radiation Protection rules. The rules ensure safe working conditions for all radiation workers and operational dosage limits monitoring.

A number of propellant ingredients and a few resins used in the manufacture of FRP motor cases and nozzles can be dermatological or respiratory toxins. Some of them are carcinogens or suspected carcinogens. These chemicals and the uncured propellant slurry have to be handled carefully to prevent the exposure of operators to these fumes /vapors. This means use of gloves, face shields, etc. and good ventilation in working rooms. The exhaust plume gases can be very toxic because of chlorine, hydrochloric acid, hydrofluoric acid and some fluorine compounds. For large motors, the hydrochloric acid in the exhaust can be many tons and can cause acid rains. Test facilities and launch facilities with toxic plumes require special precautions and special decontamination processes.

The best way to control hazards and prevent accidents is: a) to train the workers in the hazards of each propellant, teach them how to avoid unsafe conditions, prevent accidents and manage accidents and recover from the accidents, b) to design the facilities, equipments and motors to be safe, and c) to enforce rigid safety rules during design, manufacture and operation, for example, earthing of propellant loaded motor cases, wearing of cotton uniforms, and use of non-sparking tools, no smoking in propellant storage or handling areas, etc.

In any time bound development programme, to increase the assurance that delays are not caused by accidents, specific safety requirements will be applied to operations. Therefore, knowledge of safety requirements of the developmental task is essential. As in many other cases, every research effort should give due consideration to the hazardous properties of ingredients in use and handling and processing the ingredients during various operations.

Cardinal Principles

Certain cardinal principles should be used in establishing hazardous operations involved in solid propellant handling which has proven to be hazardous both from the point of fire and explosion. These are: a) where operationally feasible, separate each handling operation from all others in a manner that if an accident occurs in one, it will not propagate by fire, blast over pressure or fragments to another operation.

This can be achieved by isolation or shielding, b) do not deploy more personnel in the operation than is absolutely required for its accomplishment, and c) the propellant or other hazardous material at the point of operation must be maintained at the minimum necessary quantity for an efficient operation.

Processing of large motors involve handling of large quantities of propellant ingredients, propellants and large motor case, all undergoing various process steps resulting in the final solid propellant rocket motor ready for use. It is essential that all the process steps and description of process along with specification of raw materials, intermediate and final products are well laid out and documented. This is called integrated process document, contains details of non-explosive operations carried out on the motor case prior to propellant filling and explosive operations carried out after filling with propellant in the prepared motor case. The document must contain approved process conditions and safety instructions to be followed during the operations. The basic process document is further disseminated, suitable for each process and process stations. These documents include process checklists and corresponding safety checklists. This documentation is very important and follow up of the process and safety checklists during the operation leads to the safe completion of the operation. The safety checklists enlist the conditions of processing, the safety appliances to be used by the operators as well as safety precautions to be taken while performing the process steps with check points to various stage.

In large scale propellant processing and solid propellant grain manufacturing, safety and quality go together. Adhering strictly to the process and safety checklists while carrying out the process steps leads to realization of safe and quality product.

Hazard Analysis

The introduction of any new or modified material, process or procedure in propellant development raises the question, 'does this create a hazard'? The hazard question is complicated since most materials processed can be hazardous in a given situation. Hence, the problem becomes one of deciding: a) the degree of hazard, b) whether the precautions to prevent initiation or explosion are sufficient, and c) whether the protection of facilities and personnel is sufficient. All these considerations must be balanced against the original justification for introducing the new material or process, whether it was for reasons of economy, quality or performance improvement. Catastrophic incidents, unnecessary and cost precautions, or the rejection of a material or process because of the inability to properly assess suspected hazards places any developmental programme in an untenable position.

The most common approach to hazards assessment is sensitivity testing of sub-scale samples to establish the tendency of a material to initiate or explode. The common initiation tests have been thermal, impact, friction, electrostatic and shock. To establish properties after initiation, tests have been made to determine tendencies toward transition to explosion and propagation of an explosive reaction. The data generated by such tests are usually abstract and are expressed in statistical terms such as 50% probability of initiation or explosion occurring. These data are useful for comparing one compound with another. Thus, the probable presence of hazards has traditionally been predicted on the basis of experience which cannot be applied to all new situations. This severely restricts the ability to assess the practicality, from the standpoint of economy and safety, of using new materials, procedures and equipment. Consequently, the need arises for a method of applying sensitivity data directly to process conditions. The System Engineering Approach to Hazard Analysis meets this requirement. This technique involves the following phases of analysis: i) Process survey, ii) Sensitivity testing, iii) support studies, and iv) Hazard evaluation resulting in the degree of hazard, required modifications and subsequent margin. The concept mainly permits a decision concerning margin of safety, feasibility of modification of process or material, the extent or method of modification to be made on a quantitative basis rather than on the basis of opinion, experience, abstract data, or some empirical tests. Hazard analysis techniques such as fault free analysis and logical tree analysis, which were earlier adopted for assessment of process hazards in chemical and photochemical industry, are being applied to propellant processing.

5. AGING STUDIES

Propellant Ageing and Shelf Life

Ageing phenomena could be due to chemical, physical or mechanical ageing. In the case of polybutadiene class of propellants, chemical ageing is mainly due to the oxidative cross-linking of double bonds present in the polymer backbone, the temperature dependence of which is well described by Arrhenius' equation. Physical ageing is mainly due to environment which is discussed separately. Mechanical ageing is the time dependent mechanical degradation which results from long term storage and sustained or cyclic application of loads.

The ageing characteristics and storability of a propellant are evaluated by accelerated ageing tests at elevated temperatures, e.g. 50 or 60°C for different time intervals spreading over 150 to 200 days and measurement of mechanical properties. It is then possible to estimate and predict the useful shelf or storage life of a rocket motor. Mechanical properties such as tensile strength, elongation and modulus are measured periodically at regular intervals of time.

Layton/Christianson's ageing model:

$$P = P_0 + K \log (t/t_0)$$

Where P = property at time t
   P_0 = initial property at time t_0
   K = rate of change of property P
   t_0 = time at the end of cure (zero time)
   t = ageing time.

This model equation could be used to describe the dependence of mechanical properties with time and temperature and has been used to extrapolate long term mechanical properties from thermally accelerated ageing results. Knowing the K values, the propellant properties of any ageing period can be computed using the above equation. Lower the K values, better the ageing characteristics of the propellant. Also, knowing the specification of mechanical properties, shelf life or useful time of the propellant can be computed. If time permits, natural ageing study of propellant samples along with the motor can give a realistic picture.

The life of a particular motor depends on the particular propellant, the frequency and magnitude of the loads or strains imposed, the design and other factors. Typical life values range from 5 to 25 years. Shelf life can be increased by increasing the physical strength of the propellant, selecting chemically compatible, stable ingredients with minimum long term degradation or by minimizing vibration loads, temperature limits and transport environment.

Environmental Effects on Propellants and Storage Conditions

Physical ageing is largely due to environment, i.e., the ingression of moisture into the propellant which is reversible. Propellant properties at different` RH conditions and recovery of properties on re-desiccation can be evaluated to establish the critical humidity for propellant storage. A case study of HTPB propellant shows that propellant samples stored at RH 50% for long duration of 600 days do not show significant change in mechanical properties while propellant samples exposed to RH 90% for even 7 days showed drastic reduction in mechanical properties. This type of changes in propellant properties is due to the migration of moisture under high humid conditions leading to de-wetting of oxidizer-binder bonding. However, the original properties can be re-gained by heating or desiccation at RH 50%.

This necessitates proper storage conditions for keeping propellant or rocket motor. Some options are listed below:

1. Rocket motors/propellant grains or samples are to be stored in humidity controlled building (RH 50–55%).

2. Suitable containers/bags with nitrogen purging provision can be used to avoid moisture ingression.
3. Sealing the motor with suitable plugs will help to some extent to reduce moisture contact.
4. Segments with bottom support and rotation facility, or horizontal mode with rotating facility can prevent slumping characteristics to some extent.

6. DISPOSAL OF WASTE AND AGED PROPELLANTS

Disposal of waste propellants and aged propellants have assumed great importance in modern time due to the large requirements of advanced solid rocket motors of sizes ranging from 75 mm diameter sounding rockets weighing 5 kg propellants to 2.8 meter diameter PSLV solid booster of having 139 tonnes of propellant in five segments and subsequent growth to S-200 boosters of GSLV of 3.2 meter diameter having 207 tonnes of solid propellants in three segments. Waste propellant management is also given adequate attention due to enhanced sensitivity of the propellant because of finely divided nature and also due to contamination with gritty materials. With large size rocket motors, the quantity of waste propellant also increases. The main objective of waste disposal is to avoid accumulation of explosives and chemicals which are hazardous to human life, environment and buildings, thereby minimizing the exposure to toxic and hazardous materials. This also helps in controlling the explosives limit in an explosive building conforming to explosive regulatory rules.

Sometimes sidelined propellant grains can accumulate in the storage magazine, thus blocking the magazine space temporarily, thereby imposing a production constraint. Water jet cutting of big rejected motors for quality or life expired one is to be planned.

The safe operation of propellant disposal calls for a thorough knowledge and understanding of the materials to be disposed off. Some of the studies envisaged are:

1. Physical and chemical characteristics of the waste
2. Explosive characteristics
3. Toxicology of waste materials before and after disposal
4. Explosive limit for each disposal operation
5. Specific area for disposal of explosives
6. Ensure safety in waste disposal and availability of fire fighting force, etc.

Origin of Propellant Waste

Propellant wastes are originated mainly during propellant processing, machining, and batch rejection and aged propellants.

Propellant Casting

Since most of the solid motors are processed by well proven vacuum casting technique, wastage of propellant slurry is minimum. In other casting methods like bayonet casting or pressure casting, propellant wastage is slightly more. Rejection of slurry batches due to non-conformity creates more wastage. The overall slurry wastage using a 300 gallon vertical mixer with a maximum mix capacity of 2.7 tonnes is found to be about 5–10%.

Propellant Machining and Sample Preparation

The ends of the motor/segments are to be trimmed/machined to the required dimensions before inhibition/segment bonding. Sample preparation for mechanical properties, peel strength, tensile bond/shear bond, burn rate evaluation, ageing studies, etc. create cured propellant wastes. In all these cases, propellant waste is in the form of chips, pieces or powder.

Propellant Waste from Test Labs

Propellant wastes from physical, chemical and mechanical testing Labs and also from rheological characterization labs are found to be of meager quantity.

Propellant Wastage due to Defects

Major defects in solid motors are due to blow holes, voids, cracks, de-bonds, soft curing and poor mechanical and interface properties. When these defects are beyond the acceptable quality limits, the motors are to be rejected and kept aside for disposal. The propellant wastes from rejected motors thus become a major concern for disposal.

Propellant Waste from Aged Motors

Propellant wastes arising from aged propellant blocks, segments and motors are of major share to the waste disposal. Solid propellant motors which cross the shelf life and non-conforming to quality acceptance standards are to be disposed off.

Handling of Propellant Wastes

Propellant wastes are to be categorized and packed in suitable containers or bags and stored in identified explosive area. Free standing long grains or bigger blocks can be partitioned for easy disposal. Assembled motors are to be dismantled and separated for further process. Explosive limits are to be strictly followed for waste storage in explosive buildings and subsequent transportation. Proper earthing is to be provided for the building and waste propellants for discharging any static electricity produced. Transportation of propellant for waste disposal requires explosive van accompanied with fire fighting force to face any critical situation.

Methods of Waste Disposal

Many methods are available for disposal of propellant wastes. Some of them are in the R&D level and some are operational in the lab level/plant level only. Some of the methods attempted is listed below.

1. Chemical methods
2. Molten salt oxidation
3. Bio-degradation
4. Reclamation/recycling method
5. Waste burning pit method
6. Un-instrumented motor test
7. As fertilizer
8. As commercial explosives.

Chemical Methods

Chemical methods use hydrolysis, chemical neutralization, oxidation, photolysis, etc. It is a slow process and produces small amounts of gaseous products. RDX can be disposed off chemically using a 5% solution of NaOH which can decompose it. Small quantities can be disposed off by this method and is not practicable for large scale disposal.

Molten Salt Oxidation (MSO)

Molten Salt Oxidation (MSO) method is applicable for disposal of composite propellants containing AP, Al, polymeric binder, etc., double base propellants containing NG, NC and high explosives and pyrotechnic wastes. Here, a molten salt bed is maintained at about 800–1000°C. The propellant wastes are fed with oxygen below the surface of the molten salt. In this process, carbon is converted to CO_2, hydrogen to water and nitrogen is released as gaseous nitrogen. Any halogen present is converted to sodium chloride and potassium chloride and metals to their oxides in the salt bed. No gaseous chemical is found. However, the solid content is very high in this process.

Efficiency in waste destruction is very high (99%) in this method and is evaluated in laboratory level only.

Bio-Degradation

Bio-degradation method is an eco-friendly method using micro organisms to degrade organic materials into less hazardous compounds. Nitrocellulose based propellants can be degraded by anoxic (de-nitrification) and aerobic biological treatment. High explosives like RDX and HMX can be degraded into water, nitrogen and CO_2 under anaerobic conditions. Using micro-aerobic reactor, lower concentrations (less than 1.5%) of perchlorates, nitrates, hydrolysates, etc. can be reduced to non-detectable concentrations. The major disadvantages of bio-degradation process are: a) very slow process, and b) it works at lower concentration of ingredients.

Reclamation/Recycling Method

In this process, the components in a propellant waste are separated by various techniques such as dissolution, distillation, absorption, adsorption., membrane separation, sedimentation, precipitation, floatation, etc. Some cases, combination of techniques are to be employed. Single base propellants containing Nitrocellulose (NC) can be reprocessed to make lacquer NC. Re-cycling is a profitable method for recovery of high cost explosives. High explosive HMX can be separated from HMX-TNT mixture by dissolving in toluene, followed by filtration of HMX. The recovered HMX is washed with water and steam and then re-crystallized from acetone. Waste TNT is disposed off by decomposing it with sodium sulphide.

Composite solid propellants contain 70–75% Ammonium Perchlorate (AP) as oxidizer. Solubility of AP in water is 23.5%. Suitable dissolution technique, followed by concentration, crystallization and re-crystallization can recover a major portion of AP. However, the process may not be economically viable in plant level.

Waste Burning Pit Method

Small blocks, pieces, chips, slurry waste, etc. can be burned off in a waste burning pit protected by thick concrete walls located in a remote place. The burning is carried out using pyrotechnic squibs by remote control. The spreading of fire can be controlled by fire fighting water jets.

Defective propellants from bigger motors or aged motors or defective segments can be removed as small pieces by water jet cutting method. Water jets of pressure ranging from 50–100 MPa can effectively cut and remove solid propellants as pieces. Thus, the motor case can be salvaged and the propellant pieces can be burned off in the burning pit. The pollution created during waste burning are due to release of various toxic gases, oxides of nitrogen, chlorine compounds, etc. Aluminum oxide is found to be non-toxic. This is a conventional method with quantity limitation and of course is not an environment friendly method. However, with quantity limitation and keeping the toxic gases with in safe limits, this method is an easy route for waste disposal.

Glycidyl Azide Polymer (GAP) is an explosive material (hazard division of 1.3) and its shelf life is about 2 years. A method has been devised by HEMRL for disposal of life expired GAP by spreading GAP on kerosene soaked saw dust and burning it with all safety measures. The toxic gases such as CO, HCN, and particulate matter are found to be in acceptable limits.

Un-Instrumented Motor Test

Aged motors and defective motors can be disposed off using un-instrumented motor static test in the test stand. Here, the propellant is burned off and recovered the motor case.

As Fertilizer

NC, NG, RDX, HMX, etc. can be converted to nitrite and nitrate ingredients by alkaline hydrolysis. These nitrite and nitrate materials can be used as fertilizers.

As Commercial Explosives

Another viable method is the incorporation of waste propellants and explosive ingredients in commercial explosives. Commercial explosives are used for rock blasting, demolition of buildings, fireworks, etc. In collaboration with various explosive industries in the country, waste propellants and explosive materials generated by production agencies can be converted to commercial explosives.

Even though many methods are listed, an eco-friendly and economically viable method for disposal of propellant and explosive wastes in plant level is yet to come.

7. QUALITY AND QUALITY CONTROL

Manufacture of solid propellants involves many critical operations. Solid propellants especially based on Hydroxyl Terminated Polybutadiene (HTPB) resins are the current state of the art propellants for launch vehicles. Advanced space missions project large scale requirements of high performance solid propellants. Solid propellant technology in India has grown from 75 mm diameter sounding rockets weighing 5 kg propellants to 2.8 meter diameter PSLV solid booster of having 139 tonnes of propellant in five segments and subsequent growth to S-200 boosters of GSLV of 3.2 meter diameter having 207 tonnes of solid propellants in three segments.

Processing of such booster motors demand many critical operations such as storage, handling and qualification of raw materials in large quantities, multiple number of propellant mixing batches in vertical/horizontal mixers, propellant casting operations under vacuum, curing, de-coring, machining, non-destructive testing, inhibition, assembly and testing. Quality control on each operation plays a major role in realizing a superior quality large solid booster motor.

Quality

Quality of a product is characterized by its compliance with the specifications. The term conformity is used for fulfillment of requirements. Thus, the quality of a product means its conformance to the requirements. Corrective actions are taken then and there to eliminate a detected non-conformity and further recurrence of the non-conformity is eliminated by preventive actions.

Quality is defined in several ways like "fitness for use" or "conformance to requirements." The quality comes from a process. Any process is the transformation of a set of inputs, which can include actions, methods and operations into the desired output in the form of products, information, services or results. To process an output which meets users/customers requirements, it is essential to define, monitor and control the input as well as the process.

In solid propellant production area, desired operational characteristics of a propellant can be ascertained only if quality control is ensured for all ingredients used and each stage of manufacturing. Chemical formulation, purity of every ingredients, moisture content, particle size and its distribution have vital role on performance of these high energy materials. Hence, rigid quality control over each stage is essential to keep the quality of the finished product to the desired standard to achieve desired performance. Quality control of raw materials is generally done by either wet chemical analysis method or by instrumental methods. Quality control during solid propellant processing needs several sophisticated equipments. The different equipments used in quality control of ingredients and in-process finished products are particle size analyzer, moisture content analyzer, volatile matter determinator, viscometer, densitomer, bomb calorimeter, Crawford bomb for burning rate measurement, sensitivity tester includes impact, friction, shock, jolt, vibration and other external stimuli, thermal analyzer, mechanical property testing, radiography or non destructive testing, chromatographic technique, spectrometry and ballistic evaluation.

The product quality relies on the measurements of propellant characteristics such as weight, dimensions, burning rate, mechanical properties of propellant, etc., as a means to ensure that a good quality product reaches the customer. The concept is synonymous with quality control concept wherein the product characteristics are measured at the end of each stage of the process and this is used as a feed back to monitor the performance or correct the process, if necessary.

The process quality, on the other hand, relies on control of process characteristics such as speed of mixer, temperature of slurry, vacuum level in the mould, relative humidity, etc., to ensure product quality. The basic philosophy is "if process is right, the product will be right." This concept is synonymous with the quality assurance concept wherein process control ensures suitability of design, selection and use of right process, raw materials, control of product and delivery process, etc., to ensure the right quality product is obtained. The ISO 9000 series of standards are oriented towards process quality as a focus of ensuring the quality of overall management system to ensure product quality.

Process based Quality Management System (QMS) conforming to the requirements of ISO-9001:2000 defines and establishes an organization's quality policy and objectives. It also allows an organization to document and implement the procedures needed to attain these goals. A properly implemented QMS ensures that the procedures are carried out consistently, that the problem can be identified and resolved and that the organization can continuously review and improve its procedures, products and services. It is a mechanism for maintaining and improving the quality of products and services so that they consistently meet customer's stated or implied needs and fulfill their objectives.

Based on user's qualitative requirements, the propellant grain requirements are determined and reviewed. Design and development plan is prepared. Every manufacturing process requires specific data for reproducing quality characteristics. The experimental data is generated from small scale processing of propellant grain. The target end properties of the product are achieved in these trials and propellant formulation is finalized for large scale processing. Design review team is constituted with members from project group, service group, user's group and quality assurance division. This team reviews the raw materials quality, large scale processing parameters, process facilities, other resource requirements, quality assurance and qualification testing plan and readiness, etc. The team clears large scale processing if the review is satisfactory, otherwise recommends guidelines for project group to generate some more data to establish confidence on the processing parameters to realize the quality product.

Process Control

The manufacture of solid propellants involves number of stages—raw material processing, mixing, casting, curing, machining and inhibition. During the process, process evaluations tell us whether activities and quality control checks are conducted from input to output according to the approved quality plan. Composition control ensures the input materials meet the qualitative and quantitative requirements of the overall quality scheme.

This is achieved by following various analytical checks at every stage of raw material processing. Moisture content of all ingredients should be within tolerable limits as moisture reacts with other ingredients like TDI (in Polyurethane Propellants) and MAPO (in PBAN propellants) and bonding agents (in HTPB propellants) to produce carbon dioxide which can generate voids and cracks or lead to soft curing. The average particle size and its distribution of ingredients play an important role in the rheology of the slurry, mechanical and ballistic properties of cured propellant. The composition control task requires the application of sensitive and accurate weighing balances for various ingredients. Proper and periodic calibration of balances and equipments are necessary for repeat results.

Proper sequential addition of raw material ingredients is a must to achieve uniform wetting of ingredients. For this, coarse ingredients are added first in steps to the binder mix in the mixing bowl, followed by fine ingredients in steps. Temperature control of mixer by circulating hot water through the jackets is necessary after addition of solids to binder and after addition of curative (TDI). The required vacuum level during mixing is necessary to remove the volatiles and unwanted gaseous reaction products formed during the mixing. Clean mould, free from any contaminants is used to receive the propellant slurry. The vacuum level, humidity and temperature, viscosity build-up and casting rate are to be watched during casting. The casting time is to be monitored to take care of pot-life of the propellant slurry. During curing, the temperature of the oven is monitored to avoid thermal stresses in the propellant.

After curing, the propellant is ready for use and is tested for density, calorimetric value, mechanical properties and chemical composition, besides visual inspection. Along with main propellant grains of a lot, carton and BEM grains are always cast. Ballistic evaluation using Ballistic Evaluation Motor (BEM) is carried out to ascertain ballistic competence of the propellant grain produced. X-ray is mandatory for propellant grains before evaluation. For BEM grains, 450 KV X-ray unit is sufficient but for large motors and case bonded grains, minimum of 4.5 MeV X-ray linear accelerator machines are required for radiography for detection of voids, cracks and flaws. Presence of voids and flaws can increase the burn rate and may cause bursting of the rocket motor. Absence of voids in inhibition layer and bonding between propellant and inhibition layer are ensured by radiography. Case bonding introduces more loads on X-ray machines in addition to grain radiography. Exposures of interface become more important and number of shots for clearance of motor has increased tremendously. Once cleared by non-destructive testing, the propellant grains/motors are ready for end trimming and inhibition.

Samples from top and bottom of the main grain are tested for density, calorimetric value and chemical analysis to check the homogeneity of the grain. Dimensions of the grain and composition of the inhibition and the bond strength between propellant and inhibition are checked.

Compressive and tensile properties of the propellant are tested from carton samples. Therefore, strict quality control is essential during the determination of these properties. Specimen shape and dimensions, Instron testing machine parameters are standardized and kept constant for reproducible results. The burn rate, burn rate coefficients and specific impulse are evaluated using BEM firing at ambient, hot and cold temperatures.

Quality Control

Quality control is for improving the quality of the product during the process of its manufacture. Quality control and inspection are not same. Inspection means checking of a product or component or material for the acceptance or rejection of it. However, inspection is part of quality control. Documented procedures are to be adopted for the various operations to control the quality of system. The following are the quality control steps for realizing a quality propellant.

1. *In-process control for propellant processing are:*

(a) propellant raw materials specifications
(b) process operations—mixing, casting, curing, cooling, de-coring, machining, inhibition, assembly and testing as per documented procedures.

2. *Pre-casting acceptance tests are:*

(a) Viscosity measurement of the slurry for the specified duration
(b) Determination of slurry burn rate and density
(c) Homogeneity of critical components.

3. *After casting acceptance tests are:*

(a) Mechanical properties of the cured propellant
(b) Burn rate and burn rate law of the cured propellant
(c) Evaluation of interface properties
(d) Detailed characterization listed under chapter-V as per project requirement.

4. *Inspection and testing of the motor*

Inspection of the motor is for the rejection or acceptance of the product. It is done through dimensional inspection and radiography as per specified procedure. The accepted motors are tested as per well defined test procedures.

Many pre-view and review mechanisms are adopted for multi-level quality control and surveillance of solid propellant motors. Some of them are listed below:

1. *Raw Material Specification Committee (RMSC):* RMSC will look into the critical specifications of propellant raw materials and allied materials.
2. *Local Salvage Committee (LSC):* LSC will look into the minor deviations in the test results which can be salvaged for use.

3. *Calibration and Test Standards Committee (CTSC):* CTSC will look into the calibration methods and test standards to be followed for instruments, equipments, machines and their periodicity in use.

4. *Non-Conformance Review Board (NCRB):* NCRB will look into the various non-conformances of various components which go into the system and for clearance.

5. *Sub-System Development Review Board for Propellants (SDRC-P):* SDRC-P will look into the progress of development of sub-systems consisting of various components and for necessary corrective actions.

6. *Facility Fitness Review Board (FFRB):* FFRB will look into the facility readiness for various operations such as motor lining, mixing operations, casting facility, motor curing, trimming, inhibition, assembly and testing.

7. *Non-Destructive Testing (NDT) Committee:* NDT committee will look into the NDT standards to be adopted for various size motors.

8. *Test Article Review Board (TARB):* TARB will look in totality the readiness of the motor for static testing or flight.

8. RELIABILITY AND MARGIN OF SAFETY

Reliability

Reliability is the probability that a system will perform successfully under given conditions for a specific period of time. Reliability is only one of the tools of the management which is supported by other tools like quality control and design of experiments. Quality control maintains the quality of the product and hence, affects reliability. Quality control is meant for a short period of manufacture of a product where as reliability is entirely a separate function associated with quality over a long period.

Reliability = Quality Now + Quality Later

Some of the factors associated with reliability are specifications, intended function, life and environmental conditions. The five effective areas for achieving reliability of a product are: design, production, measurement and testing, maintenance and field operation. Reliability tests are conducted to verify the working of a product satisfactorily for a time period. Reliability tests consist of functional test, environmental test and life testing.

Product Assurance

Product assurance ensures that failure, hazard and degradation aspects of design and software are properly considered in launch vehicle engineering. To achieve system effectiveness, it is necessary to understand the dependency of design and hardware on product assurance and vice versa. The best design in the world is useless if it cannot be: i) analyzed for its weakness, ii) manufactured and maintained, and iii) tested. The objective of product assurance is to ensure that the consequences of failures and hazards do not affect life, the space vehicle mission, or the space vehicle itself. An essential element in all product assurance disciplines is the feedback cycle, in which the results of design, analysis, manufacturing, development, test and operation are immediately fed back to all involved areas. There is a continuous interchange and updating of all data, and a measure of significance and status. Disciplines which collectively constitute product assurance are reliability, quality, safety, configuration control, parts (electronic, etc.) and materials and process evaluation.

The aims of product assurance are as follows:

1. To protect human life, the investment, the environment, public and private property and the mission.
2. To establish and implement parametric derating criteria to ensure that no electrical, mechanical or chemical overstressing occurs during the mission.
3. To define the probability that the system will perform successfully according to the specification.
4. To identify and control critical elements within the system such that the success of the mission is not jeopardized.
5. To verify that all elements of design, hardware, manufacturing and testing are of an adequate standard, and are consistent and correlatable during all phases of the programme.
6. To ensure that all failures, hazards and non-conformances are identified, their total effects understood and adequate rectification and retest validation carried out.

A general policy of prevention rather than cure is applied.

Since its inception, the rocket propulsion industry has concentrated primarily on designs that maximize payload capability in the context of specified, usually well-defined mission and performance considerations, with costs and reliability of secondary consideration. This is not to assume that reliability was not considered to be important. It is often a critical factor. It is much more difficult to predict the performance and weight, especially at the design stage. Hence, reliability evaluation was based on full system testing and operational outcomes.

The reliability levels that the industry has succeeded in achieving have come about through the use of large margins (safety factors) in part design and of a build-test-fail-fix approach. The success of design as far as reliability was concerned thus depended on factors which were determined by engineering judgment from the lab and the engineer's background, past experience, historical data, statistics and similar information sources.

Reliability analyses that were performed were usually based on simplified models and assumptions. In addition, process and operations induced failures were seldom adequately addressed. As a result of these and many other difficulties, past reliability predictions have almost overestimated the reliability of an operational system.

This situation has been changing considerably since 90s, primarily as a result of the reliability and cost requirements specified for various missions. For space transportation engine, 0.99 reliability be achieved with 90% confidence. This stimulated interest in analytical efforts towards quantification of reliability leading to recognition of the need for careful and thorough identification of failure modes as well as the need for development of more realistic engineering and statistical models of failure mechanisms and their likelihood of occurrence.

A close relationship exits between quality and reliability. A good quality product without reliability is useless. But a reliable product is quite likely to be of good overall quality.

Margin of Safety

A knowledge of the various loads such as thermal strains during the cooling phase, slumping on storage, internal pressurization at the port, vibration/acceleration loads, ageing conditions, humidity effects, etc. of the motor are essential for determining the margin of safety and structural integrity. Quality control tests such as raw material acceptance tests, pre-casting acceptance tests such as propellant mix viscosity, propellant mix homogeneity tests, density to check the composition, slurry burn rate, etc. and cured propellant acceptance tests such as evaluation of mechanical properties, interface properties, density, burn rate, etc. of solid propellants are strictly followed to improve the system reliability.

Margin of Safety (MS) = SF – 1, where SF is the factor of safety which is capability/induced load. As a general rule, a minimum safety factor of 2, which gives a margin of safety of 1 is essential.

Thus, detailed characterization of solid propellants helps in predicting the margin of safety and structural integrity of a motor.

Typical Questions

1. What are the main hazards involved in the processing of solid propellants?
2. Explain the safety tests for solid propellants?
3. Explain a shock sensitivity test for explosives?
4. What is shelf life? How is it evaluated?
5. How are the ageing characteristics of a solid propellant evaluated?
6. How environmental conditions influence the propellant properties?
7. Explain the methods of handling waste propellants and explosives? How are they disposed off?

8. How is a safety culture induced in a work environment?

9. Explain an eco-friendly method for the disposal of propellant waste?

10. What is the difference between quality control and inspection?

11. What are the quality control steps to be looked into during the processing of a solid propellant?

12. What are the analytical tools required for quality control of solid propellant?

13. What are the control review mechanisms for maintaining a quality product?

14. What are the aims of product assurance? What are the disciplines which collectively constitute product assurance?

15. What is reliability? What are the critical areas to be looked into for achieving good reliability of a product?

16. Explain the importance of quality policy in an organization?

17. What is the importance of designing in achieving better reliability?

18. What do you understand by the terms:

    (a) Structural Integrity and (b) Margin of Safety.

REFERENCES

[1] Agrawal, J.P. and Hodgson, R.D., Organic chemistry of explosives, John Willey and Sons Ltd., The Atrium, Chichester, England (2007).

[2] Baker, P. and Mellor, A., "Critical Impact Iitiation Energies for Three HTPB Propellants," AIAA 90–2196, *26th Joint Propulsion Conference*, July 16–18, 1990.

[3] Biagioni, J.K., "Resourcs Recovery System for Solid Rocket Propellants," *23rd ICT, Karlsruhe*, June 30–July 3, 1992, pp. 16/1–11/6.

[4] Christiansen, A.G., Layton, L.H. and Carpenter, R.L., *Journal of Spacecraft*, 18 (3), (1981) 211.

[5] Claus, H., Bausinger, T., Lehmler, I., Perret, N., Fels, G., Dehner, U., Preub, J. and Konig, H., *Biodegradation*, 18 (2007), 27–35.

[6] Dutta, Shantanu K., Upadhyay, V.P. and Sridharan, U., Environmental Management of Industrial Hazardous Wastes in India. *Journal of Environ. Science & Engg.*, Vol. 48, April 2006.

[7] Feigenbaum, Anand V., Total Quality Control, 111 Ed., McGraw Hill Publications, 1986.

[8] Hancox, R.I. and Bentley, J.R., "A New tool for Explosive Ordenance Disposal," *18th International Pyrotechnics Seminar*, Colorudo, 13–17 July 1992, pp. 359–365.

[9] Kanakaraju, P. and Athithan, S.K., Quality Control and Surveillance in the Manufacture of Solid Propellant, *Proc. International Conference on Propellants, Explosives, Rockets and Guns*, IIT, Chennai, 1998, pp. 455–459.

[10] Kihumba, F. and Carroll, William, Open Burning of Wastes, Guidance by Source Category. *International Council of Chemical Associations*, April 2004.

[11] Kulkarni, K.S., Soman, R.R., *et al.*, Environmental Friendly and Safety Aspects during Disposal of Life Expired GAP, *Proc. 9th International High Energy Materials Conference*, Feb. 13–14, 2014, p. 1016, Trivandrum, 2014.

[12] Kumar, Amar Jeet, Gopal, Anish and Kapoor, K.G., Contemporary Safety Issues during Processing and Disposal of Solid Propellants—Application of "Beyond Zero", Safety Management Tool in Military, *Proc. 9th International High Energy Materials Conference*, Feb. 13–14, Trivandrum, 2014.

[13] Machcek, O., *et al.*, Waste Propellants and Smokeless Powders as Ingredients in Commercial Explosives, *23rd International Conf. of ICT, Karlsruhe,* June 30–July 3, 1992, pp. 11/1–11/4.

[14] Mahajan, M., Book on Statistical Quality Control, Published by Dhanpat Rai & CO. (p) Ltd, 1999.

[15] Mary Celin, S., *et al.*, Microbial Remediation: An Eco-friendly Technology for Explosive Waste Management, *Proc. 9th International High Energy Materials Conference,* Feb. 13–14, 2014, p. 1052, Trivandrum, 2014.

[16] Mehilal, Dhabbe, K.I., Kumari, Anjali, Manoj, V., Singh, P.P. and Bhattacharya, B., Development of an Eco-Friendly Method to Convert Life Expired Composite Propellant into Liquid Fertilizer. *Journal of Hazardous Materials.* Elsevier B.V. January 2012.

[17] Mellor, A.M., "Hazard Initiation in Solid Rocket and Gun Propellants and Explosives," *Progress Energy Combustion Science,* Vol. 14, 1988, pp. 213–244.

[18] Meyer, Rudolf, Kohler Josef and Homburg, Axel, "Explosives," Revised Edition, 2007.

[19] Mezhanov, A.G. and Abramov, V.G., "Thermal Explosions of Explosives and Propellants— A Review," *Propellants and Explosives,* Vol. 6, 1981, pp. 130–148.

[20] Nahlovsky, D., Boris and Wong, Micheal K., Recovery of Aluminum and Hydrocarbon Values from Composite Energetic Compositions. Aero Jet-General Corporation, May 1994.

[21] National Ambient Air Quality Standards, Central Pollution Control Board Notification, New Delhi (2009).

[22] Rajagopal, Chitra, "Explosive Safety Management: Challenges and CFEES Initiatives," *Proc. 9th International High Energy Materials Conference,* Feb. 13–14, 2014, pp. 1007–8, Trivandrum, 2014.

[23] Ramu and Krishnamurthy, V.N., "Hazard Characteristics of AP based HTPB Propellants," *Proc. International Conference on Propellants, Explosives, Rockets and Guns, IIT,* Chennai, 1998, pp. 455–459.

[24] Richard, Simpson L., Foltz and Frances, M., "LLNL Small-Scale Drop—Hammer Impact Sensitivity Test," Report Jan 1995, Lawrence Livermore National Laboratory, Livermore, CA.

[25] Sanghavi, R.R., *et al.* (2001). "Studies on Thermoplastic Elastomers based RDX Propellant Compositions," *Journal of Energetic Materials,* 19: 79–91.

[26] Scott, Geller, "The Psychology of Safety Handbook," CRC Press, LLC, Washington DC, 2001.

[27] Shaw, R.W., Research in the Destruction of Energetic Materials, *23rd International Conference of ICT,* Karlsruhe, June 30–July 3, 1992, 5/1–5/6.

[28] Simson, G.M., Waste Disposal, *14th International Pyrotechnics Seminar,* U.K, 18–22, Sept. 1989, pp. 463–466.

[29] Singh, R.B., Viswanathan, K., *et al.*, "Integrated Safety Plan for Static Testing of Large Size Solid Propellant Rocket Motors," *Proc. of the 7th National Seminar on High Energy Matewrials,* Feb. 1994, pp. 105–106.

[30] Singh, Haridwar, "Disposal of Munitions and Explosives: A Eco-friendly Approach," *Proc. International Conf. on Propellants, Explosives, Rockets and Guns,* IIT Chennai, 1998, ISBN 81-7023-885-4, pp. 50–55.

[31] Sundaram, S.T. *et al.*, "Bio-degradation of Nitrocellulose," *J. of Energetic Materials,* No. 3&4, Vol. 13, 1995, pp. 283–298.

[32] Veeramani, M. and Sivapirakasam, S.P., "Safe Storage and Handling of Pyrotechnic Chemicals," *Proceeding of Indian Chemical Engineering Congress,* 2002, pp. 126–128.

[33] Veit, P.W., Landuk, L.G., Simpson, J.W. Jr. and Svob, G.J., Evolution of an Ageing Programme—Minuteman Stage II Solid Rocket Motor, AIAA-88-3328, July 1988.

CHAPTER 9

Homogeneous Propellants[*]

1. INTRODUCTION

Modern solid propellants can be classified into two categories viz. homogeneous and composite. The major homogeneous propellant ingredients contain oxidizer and fuel components in the same molecule, whereas in composite propellants they are in distinct molecules and that makes the fundamental difference between the two categories. In the race for higher energy levels, composite propellants outshine the homogeneous variety. However, homogeneous propellants are found better suited for certain missions on account of their superior mechanical properties, ruggedness, high level of reliability and reproducibility, comparatively low smoke level in the exhaust plume, long storage life, etc.

Smokeless and double-base propellants have been widely used as the primary source of ordinance ammunition. They are more energetic than black powder. For rocket application, however, double-base propellants have two severe drawbacks: i) During storage, it tends to shrink leading to separation of the grain from the case. This leads to exposure of excess surface and production of large amount of gas during burning, exceeding the burst pressure of rocket case. ii) The combustion rate is sensitive to temperature. This leads to variation in thrust with temperature.

NC, Nitrocellulose $[C_6H_7O_2(ONO_2)_3]_n$ invented in 1845, is the basic ingredient in homogeneous propellants. NC, more exactly to be called cellulose hexa nitrate, is a solid amorphous polymeric material. Nitrocellulose is mainly responsible for the structural properties of homogeneous solid propellants. Nitrocellulose based smoke less powder was first introduced as a propellant for use in guns by Vieille in 1886. Nitroglycerine or glycerol trinitrate made in 1846 is the other major ingredient. Homogeneous propellants, in addition to nitrocellulose, contain other ingredients such as gelatinisers, burn rate additives, opacifiers, plateaunising agents, flash suppressors, smoke suppressors, phlegmatizing agents (compounds used for desensitizing explosives), surface moderators, lubricants, etc. required to impart specific properties to the propellant.

---

[*]Co-written with Shri. A.J. Kurien, Former Scientist, Vikram Sarabhai Space Centre, Thiruvananda Puram–695022, India.

Cellulose, $[C_6H_7O_2(OH)_3]_n$ is a polymeric material made up of large number of highly oriented anhydro glucose monomer units. Nitrocellulose, produced by the nitration of cellulose, is a nitric ester. It is called single base propellant. Single base propellants also contain minor amounts of non explosive plasticizers, stabilizers, burning rate modifiers, etc.

When Nitrocellulose (NC) is plasticized with (NG) Nitroglycerine $C_3H_5(ONO_2)_3$, one gets double base propellants. Glycol dinitrate, ethylene glycol dinitrate, metriol trinitrate, etc. are also liquid high energy plasticizers suitable for use as plasticizing agents for nitrocellulose. These high energy plasticizers also carry both oxygen and fuel components in the same molecule. Nitroglycerine is the most popular high energy plasticizer world over. Triple base propellants, in addition to NC and NG, contain another ingredient called nitro guanidine or picrite (an aliphatic nitramine compound $C(NH_2)_2N–NO_2$) which is used for making cooler burning propellants for use in guns to prevent barrel erosion.

Nitrocellulose can be dissolved in many organic solvents, e.g. acetone, acetic esters, alcohol, etc. Solutions of nitrocellulose are colloidal in nature with very high viscosity. Colloidal Nitrocellulose can be made with non volatile solvents called gelatinizers/plasticizers, and such colloidal mixture can function as a binder for all the constituents in a propellant formulation. This technique is used in the preparation of all homogeneous solid propellants. Hence, these propellants are also called colloidal propellants. Molecular weight of NC ranges beyond ten lakhs and it can be suitably gelled with plasticizers in the propellant formulation to produce a plastic like homogeneous structure and hence, the name homogeneous propellant. This tough gelled nature accounts for the superior mechanical properties of NC based propellants.

Homogeneous propellants are thermoplastic in nature i.e. they soften with increase in temperature. Unlike the case with composite propellants, curing of these propellants is a physical phenomenon. They are treated as smokeless propellants as the combustion products contain only negligible amounts of solids, soot, colored gases, liquids or condensable gases. The combustion products are mostly water vapor, nitrogen and oxides of carbon.

Unlike nitroglycerine, NC is not a simple compound, but a high polymer. If we assume that all the three hydroxyl groups in the monomer units in cellulose macro molecules have undergone nitration, such high nitrogen NC can have a maximum nitrogen content of 14.15% by mass. Fully nitrated cellulose with such high a nitrogen content is unstable due to steric effects and undergoes slow decomposition with nitrogen content falling below 13.9%. Generally, propellant grade nitrocellulose has nitrogen content varying from 12.2 to 13.4%. Low nitrogen NC with nitrogen content less than 12.2% is used in the plastic industry. When we think of oxygen balanced combustion, we find that propellant grade NC is oxygen deficient approximately by 30% and that accounts for the low energy level of single base propellants.

Nitroglycerine is oxygen surplus by 3.5%. If we aim at an oxygen balanced double base propellant formulation for the exploitation of maximum combustion energy, such a propellant must contain NG 8.6 times that of NC. But such a mixture will remain only as jelly and cannot have the desired physical properties. In optimizing a solid propellant formulation one cannot aim at a perfect oxygen balance due to other considerations such as the need for good mechanical properties of the propellant. More over in such situations, one end up in combustion products like CO_2 in place of CO and H_2O in place of H_2, which directly offsets the advantage of low molecular weight combustion products, high flame temperature and high heat of combustion. As specific impulse is directly proportional to flame temperature and inversely proportional to the mean molecular weight of combustion products, the specific impulse of single base propellant is low, may be of the order of 180 seconds. Double base propellants offer I_{sp} in the range of 215–220 seconds, whereas composite propellants offer practical I_{sp} level of 245 seconds.

The energy level of homogeneous can be increased by supplementing oxygen i.e., by incorporating a crystalline inorganic oxidizer like ammonium perchlorate in the double base formulation. Significant amounts of a metallic fuel viz. aluminum powder also can be added to exploit optimum energy from the filled double base variety and such high energy version is called Composite Modified Double Base (CMDB) propellant. This filled system does not fit well with the definition for homogeneous propellants. However, they also offer very good mechanical properties suitable for cartridge loading. While composite modified double base propellants can offer energy level equal to that of high energy composite propellants, we also find certain short comings in this class of propellants compared to straight double base variety. Ammonium perchlorate by its decomposition produces large amounts of HCl vapor and aluminum powder pollutes the atmosphere with ultrafine particles of aluminum oxide. The CMDB variety does not offer a smoke free exhaust but is less polluting as compared to the composites.

Energy enhancement in homogeneous double base propellants is also possible by adding high energy—high explosive ingredients in the propellant formulation. Two such high energy ingredients are RDX and HMX. Both are cyclic nitramine compounds and are comparatively new comers in the explosive industry. RDX or Royal Detonating Explosive is also known as cyclonite or cyclo trimethylene trinitramine. It has the chemical nomenclature 1,3,5 trinitro-1,3,5 triaza cyclo-hexane. HMX or Her Majesty's Explosive, also known as octogen has the chemical nomenclature 1,3,5,7 tetra nitro-tetra aza cyclo octane or cyclo tetramethylene tetranitramine. HMX or RDX incorporated propellants are generally called nitramine propellants. RDX and HMX have positive heats of formation 16.7 k cal/mole and 17.2 k cal per mole respectively and hence, can impart marginal energy increase to the propellant on this account. Both HMX and RDX are crystalline compounds;

they cannot be plasticized but can function as filler in the double base matrix. Both are oxygen deficient by 21.6%. They can impart higher energy to the propellant formulations by release of large volume of low molecular weight combustion products and high heat of combustion. When incorporated in small quantities in the double base matrix, the density of packing of the crystals gets reduced and the high detonation velocity of these compounds also gets tuned in tandem with the required deflagration ranges in a homogeneous propellant formulation.

World over, large quantities of homogeneous propellants are used in defense applications. The first sounding rocket launched from Thumba, Equatorial Rocket Launching Station near Thiruvananthapuram by name Nicke Apache was of American origin and the two stage rocket employed homogeneous double base propellant in both stages. The first indigenous sounding rocket launched from Thumba also employed double base propellant processed by solventless processing. More than one thousand numbers of two stage Russian M 100 rockets launched from Thumba during 1970 to 1993 period were of extruded double base category.

Homogenous propellants find application in sounding rockets, missiles, jet assisted air craft take off systems, ship engine starters, aircraft seat ejection cartridges, blasting compositions (generally known as gelatin sticks), gas generators as well as for various industrial applications.

2. HOMOGENEOUS PROPELLANTS—INGREDIENTS AND THEIR FUNCTIONS

Homogeneous propellant formulations contain in addition to nitrocellulose, explosive plasticizers, inert plasticizers, stabilizers, ballistic modifiers, opacifiers, high energy ingredients, combustion instability suppressors, plateaunising agents, flash and smoke suppressors, etc.

Nitrocellulose

Cellulose is the most abundant natural fiber and is the condensation product of anhydro glucose units represented by a general chemical equation given below:

$$nC_6H_{12}O_6 \rightarrow (C_6H_{10}O_5)_n + nH_2O$$
$$\text{Glucose} \qquad\qquad \text{Cellulose}$$

Cellulose has D glucose units in its molecule which has 3 hydroxyl groups free for esterification. The raw material for nitrocellulose is purified cotton linters, though wood pulp after special treatment for removal of lignin has also been used in UK. In unbleached cotton, the degree of polymerization is to the tune of 10,000 and above and the molecular weight is of the order 15,00,000 and beyond. This figure varies depending on the source of cellulose.

NC is prepared by the nitration of cellulose with a nitrating mixture consisting of HNO_3 and H_2SO_4. It is the nitric ester of cellulose with its characteristic $-O-NO_2$ groups replacing the $-OH$ groups. The nitrated cellulose is pulped to reduce the molecular weight to the desired level and washed thoroughly to remove the residual acids and stabilized with the addition of a suitable stabilizer. The final product is a white amorphous, fibrous/powdery material with a density 1.6 g/cc. The density also shows very minor variation depending on the nitrogen content of cellulose. The molecular structure of nitrocellulose, trinitrate/hexa nitrate is depicted below:

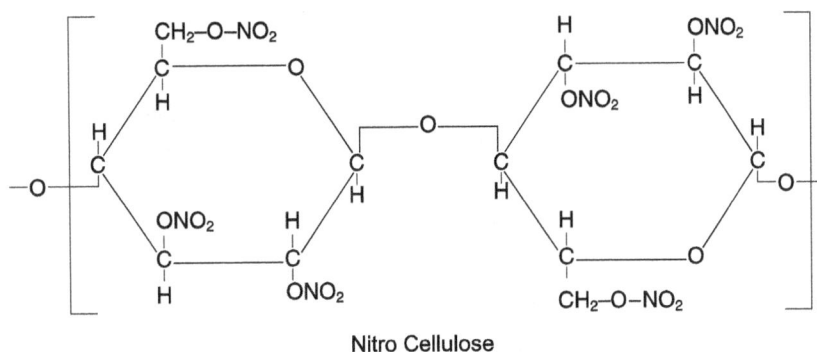

Nitro Cellulose

The solubility of nitrocellulose in a typical process solvent mixture of ether and alcohol in 1:1 ratio depends on the nitrogen content of the nitrocellulose. NC with nitrogen content of 10.5 to 12.2% is highly soluble. The solubility of NC decreases drastically to less than 10% level as the nitrogen content varies (increase or decrease) from this range. The soluble NC is used in making celluloid films, paints, plastics, varnishes, etc. The nitrogen content lies between 12.2 and 13.4 for the propellant grade NC. The mechanical and ballistic properties of the propellant are very much affected by the nitrogen content. Hence, suitable blends of different grades of nitrocellulose within the propellant grade itself are made for getting the desired properties. Properties of two different grades of nitrocellulose Type A and Type B, and their blend in 20:80 ratio determined experimentally are given in Table 1.

Table 1: Properties of Two Different Grades of Nitro
Cellulose Varying in Nitrogen Content

Nitrogen Content of NC Used (%)	13.17 (Type B)	12.4 (Type A)	A Blend of Type A 20 and Type B 80
Ether-alcohol solubility (%)	11.6	99	23.65
B & J test at 70°C (N_2/g)	0.95 mg	0.05 mg	0.84 mg
Viscosity C/s	6.3	2.9	5.4

Nitrocellulose, like nitroglycerine, is a powerful explosive with very high velocity of detonation and very sensitive to heat, friction and impact. The explosive properties of nitrocellulose viz., heat of explosion, volume of gases on explosion, temperature of explosion, velocity of detonation, etc. mainly depend on the nitrogen content of NC. Explosive properties of two different grades of nitrocellulose varying in nitrogen content are given in Table 2.

Table 2: Explosive Properties of Two Different Grades of Nitrocellulose

Nitrogen content of NC (%)	12.6	13.1
Volume of gases cc/g	900	874
Heat of explosion cal/g	973	1046
Temperature of explosion °C	2840	3100
Heat of combustion cal/g	2400	2200

The heat of explosion and explosion temperature increases as the nitrogen content is increased. Low nitrogen NC has a higher heat of combustion compared to high nitrogen NC, since the former is comparatively more under oxidized.

Nitroglycerine

Nitroglycerine (NG) is the nitric ester of glycerol, also called glycerol trinitrate. NG is highly sensitive to shock and friction and is a powerful explosive. Hence, it is transported in the solid form by absorbing in kiesulghur and is known as dynamite. It is a colorless and transparent oily liquid with a specific gravity of 1.6 at 15°C. The commercial product is straw colored or yellowish. On cooling, nitroglycerine crystallizes to form stable bi pyramidal crystals at around 13°C and to labile triclinic crystals at 2°C. NG is readily miscible with most organic solvents and itself is a good solvent. NG is able to dissolve NC only partially since the latter is a high polymer. The solubility again depends on the nitrogen content of Nitrocellulose. NG plasticizes the high molecular NC and the plasticized NC is less hygroscopic.

NG starts boiling at 180°C with exothermic decomposition and explodes at 215–218°C. The explosive decomposition of nitroglycerine is generally expressed by the equation:

$$4C_3H_5(ONO_2)_3 \rightarrow 12CO_2 + 10H_2O + 6N_2 + O_2$$

Typical explosive properties of Nitroglycerine are presented in Table 3.

Table 3: Explosive Properties Nitroglycerine

Temperature of explosion	3185°C
Heat of explosion	1455 K cal/kg
Heat of combustion	1623 K cal/kg
Heat of formation	85 K cal/mole
Volume of gases	715 l/kg
Volume of gases after condensation of water	469 l/kg
Specific energy	1139 kJ/kg

Nitroguanidine

Nitroguanidine is an aliphatic nitramine, existing in two tautomeric forms:

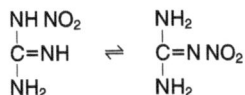

$$
\begin{array}{ccc}
NH\ NO_2 & & NH_2 \\
| & & | \\
C{=}NH & \rightleftharpoons & C{=}N\ NO_2 \\
| & & | \\
NH_2 & & NH_2
\end{array}
$$

Molecular structure of nitroguanidine—tautomeric forms

Though it does not detonate easily, nitroguanidine is a powerful explosive. It is also known as picrite and exists in two crystalline forms ά and ß. The ß form is produced either alone or together with some ά compound by the nitration of a mixture of guanidine sulphate and ammonium sulphate. ά form is produced by precipitation of the product with water. The solubility of nitro guanidine in water as well as in organic solvents is limited. Only finely powdered form of nitroguanidine is suitable for incorporation into colloidal propellants. Nitroguanidine has an apparent density of 0.96 g/cc. Nitro guanidine is used as a cooler burning, high energy ingredient in homogeneous propellants, since the gases from the decomposition of nitro guanidine are less erosive on account of its comparatively low temperature of explosion. The reported explosive properties of nitro guanidine are given in Table 4.

Table 4: Explosive Properties of Nitro Guanidine

Temperature of explosion	2098°C
Heat of explosion	717 k cal/kg
Heat of combustion	1995 k cal/kg
Gas volume	1077 l/kg

Certain guanidine derivatives such as tri amino guanidine, nitro amino guanidine and trinitroethyl-guanidine also find use as suitable cooler burning high energy ingredient in triple base propellant.

RDX (Cyclotrimethylene Trinitramine)

RDX, first synthesized in 1899 is a very powerful explosive.

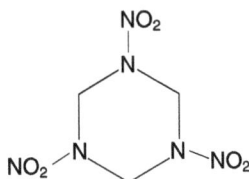

RDX

Typical reported explosive properties of RDX are given in Table 5.

Table 5: Explosive Properties of RDX

Heat of explosion	1390 k cal/kg
Heat of combustion	2261 k cal/kg
Gas volume	780 l/kg
Detonation velocity	8380 m/sec at density 1.7 g/cc
Specific energy	1394 kJ/kg

HMX (Cyclotetra Methylene Tetranitramine)

The molecular structure of HMX is depicted as:

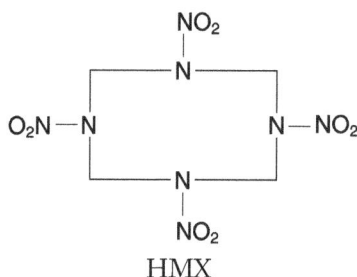

$$O_2N - N \quad \begin{matrix} NO_2 \\ | \\ N \\ \\ \\ N \\ | \\ NO_2 \end{matrix} \quad N - NO_2$$

HMX

HMX is a white crystalline powder, having four different polymorphic forms distinguished as ɑ, ß, γ and δ, differing from one another in specific gravity and sensitiveness to impact. The most stable ß form is used in propellant processing. Thermal ignition of RDX occurs at 260°C, whereas HMX gets ignited at 335°C. It is on account of its high melting point (276°C), HMX is also called high melting explosive. The explosive properties of HMX are given in Table 6.

Table 6: Explosive Properties of HMX

Heat of explosion	1355 k cal/kg
Heat of combustion	2253 k cal/kg
Gas volume	755 l/kg
Detonation velocity	9100 m/s at a density of 1.9 g/cc
Specific energy	1387 kJ/kg

RDX has an edge over HMX in terms of specific energy. HMX has higher density of 1.96 g/cc as compared to the density value 1.82 g/cc for RDX. Hence, the energy per unit volume is marginally higher for HMX.

Non-Explosive Plasticizers

Non-explosive plasticizers are generally high boiling and high molecular weight esters, compatible with nitrocellulose and nitroglycerine. Sometimes, camphor, centralite, dinitro toluene, etc. are also used as solid inert plasticizers. The inert plasticizers are used to improve the mechanical properties of the propellant and to reduce the sensitivity of NG and to adjust the energy level and burn rate or to aid the processing/extrusion characteristics of the propellant. High melting candellila wax in small quantities is used to impart flexibility/plasticity to solventless propellant formulations at the elevated extrusion temperature. Mineral jelly or Vaseline is used as a plasticizer as well as a phelgmatizing agent. A commonly used inert plasticizer is triacetin or glycerol triacetate which is an acetic ester of glycerol and has structural compatibility with nitroglycerine. Other compatible esters include diethyl phthalate, dibutyl phthalate, tricresyl phosphate, etc. Plasticizer molecules are able to permeate into the twisted, bundled fibrous nitrocellulose polymer chains and reduce the cohesive forces between them and impart segmental mobility to the macromolecules.

Methyl methacrylate is at times used in cast propellant formulations to improve the mechanical properties of the propellant as it polymerizes on curing at elevated temperature and increases the modulus of elasticity. GAP or glycidyl azide polymer is a new energetic binder that finds application as a plasticizer in double base propellants and also as an energetic ingredient which can enhance the burn rate.

Stabilizers

Nitric ester (NC/NG) based propellants on storage undergo slow decomposition releasing oxides of nitrogen which accelerate further decomposition of nitrocellulose. These decomposition products exert a detrimental effect on the stability of the propellant. The presence of moisture also accelerates the ageing process. Stabilizers retard the rate of auto catalytic decomposition of nitrocellulose and such nitric esters by chemically combining with the oxides of nitrogen formed. The commonly used stabilizers are diphenyl amine, ortho nitro diphenyl amine, symmetric dimethyl diphenyl urea, also known as centralite-I, symmetric diethyl biphenyl urea known as centralite-II and carbazole (it resembles biphenyl amine). Di phenyl benzamide, nitronapthaline, aniline, etc. are also capable of retarding the decomposition of nitric esters.

Nitrocellulose is stabilized during the manufacturing stage by adding a minor amount of calcium oxide/calcium carbonate or magnesium carbonate. Non volatile solvents such as dibutyl phthalate and camphor also have stabilizing effect. Due to their basic properties, they tend to abstract the oxides of nitrogen. Diphenylamine gets slowly converted into N-nitroso diphenylamine, dinitro diphenylamine and trinitro diphenylamine by absorption of oxides of nitrogen.

The typical stabilization reactions undergone by diphenyl amine while absorbing the oxides of nitrogen are shown in Table 7.

Table 7: Typical Stabilization Reactions of Diphenyl Amine

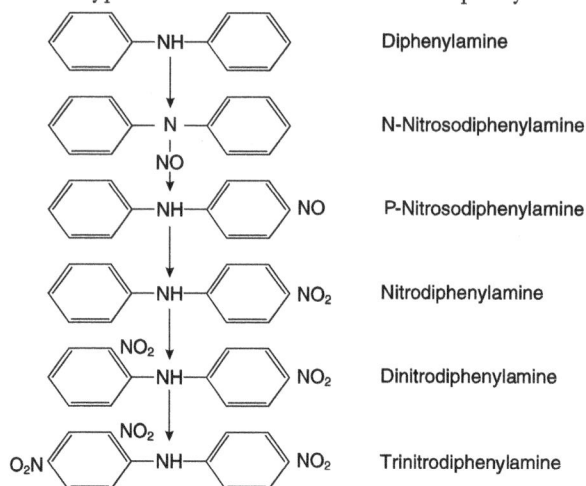

	Diphenylamine
	N-Nitrosodiphenylamine
NO	P-Nitrosodiphenylamine
NO_2	Nitrodiphenylamine
NO_2	Dinitrodiphenylamine
NO_2	Trinitrodiphenylamine

Darkening Agents

Darkening agents such as carbon black in minor percentages [0.1 to 0.5%] are added to make the propellant opaque. This helps in preventing the uncontrolled burning of the propellant below the desired burning surface by absorbing radiant heat from the hot combustion gases.

Burn Rate Modifiers

The burn rate of double base propellants generally vary from 5 to 20 mm/s under standard conditions at 70 ksc pressure and is also a function of its energy level/calorific value and flame temperature. Hence, double base propellants give a higher burn rate compared to single base and triple base. In CMDB formulations, the surplus oxygen from the decomposition of ammonium perchlorate interacts with the under oxidized reactive species of the double base matrix resulting in higher flame temperature and higher burn rate in the range of 10 to 20 mm/s. Ultra fine AP of 5 microns size is used to enhance the propellant burn rate. Carbon black in finely divided form in concentrations up to 1% is also used as a burn rate additive. Addition of potassium perchlorate can enhance burn rates. However, the energy level will be lower since high molecular weight reaction products and solids are formed by the decomposition of potassium perchlorate.

Flash Suppressers

In tactical warfare, either in guns or in rockets, propellant flash is undesirable. During night hours, glaring flash from guns easily discloses their location. A secondary flash a few meters away from the muzzle end can occur if the out coming gases from the gun barrel form an inflammable mix with air. Such secondary flames from high caliber guns may be visible even at 50 km distance. Hence, flash suppressors are added in the propellant formulations. Dinitro toluene liberates considerably less heat compared to nitrocellulose. It reduces the flame temperature and flash and also acts as a non explosive plasticizer for nitrocellulose. Dinitro toluene also reduces the hygroscopicity of nitrocellulose. Potassium salts such as potassium nitrate, potassium cryolite (potassium aluminum fluoride), potassium bi tartrate, etc. are used as flash suppressers. Potassium sulphate is used in propellant formulations to suppress muzzle flash by preventing the ignition of fuel rich combustion gases that are expelled out of the gun barrels. Potassium cryolite is helpful in suppressing the pressure index of burn rate to a small extent and also helps in avoiding cessation of propellant burning. However, potassium compounds significantly increase the smoke level in the exhaust plume.

Plateaunising Agents

Solid propellant burn rate is dependent on chamber pressure. Burn rate-pressure relationship is shown by St. Robert's law $r = a\,P_c^{\,n}$, where r is the burn rate in cm/s; a is a constant for a given propellant and is characteristic of propellant composition, P_c is the chamber pressure in ksc and n is the pressure index. In general, for homogeneous propellants, n lies between 0.5 and 0.8, whereas in composite propellants, the value lies between 0.2 and 0.5. As n approaches 1, the burn rate becomes linearly dependent on the pressure. A high value of pressure index can cause rapid increase in chamber pressure and failure of the rocket motor. At lower values of n, the burn rate is less dependent on pressure and when $n = 0$, the burn rate becomes independent of chamber pressure. This burning pattern is called plateau burning. Ballistic agents used to reduce the pressure index are known as plateaunising agents. Certain lead compounds, typically lead stearate, lead salicylate, lead ethyl hexoate, etc. are used as plateaunising agents in nitric ester based propellants. They tend to enhance the burn rate at low pressure regions and suppress the burn rate at high pressure.

3. METHODS OF HANDLING NITROCELLULOSE

Nitrocellulose for propellant processing is handled by various techniques such as dry gun cotton (a common term for high nitrogen NC), alcohol dehydrated NC, and water wet NC, nitrocellulose-water slurry and dense NC or spheroidal/globular nitrocellulose. A convenient form is chosen based on the propellant processing method selected.

Dry NC and NC-NG Paste

The earlier practice was mainly to use dry nitrocellulose for propellant mixing. Here, water wet NC with water content around 30%, received from the nitrocellulose processing unit is dried in a current of hot air in separate compartments with lead flooring (Lead flooring is provided to prevent static electricity and reduce friction). The drying is continued at around 45°C for a few days so as to get rid of the absorbed moisture. After bringing the moisture content of NC to the specified minimum value (usually below 0.5%), measured quantity of NG of known weight is poured on a pre-weighed quantity of nitrocellulose and mixed at random in a lead trough provided with a chute. This NC-NG mix called paste is transferred into rubberized bags and stored for a few days to allow free absorption of NG by NC. The NC-NG paste already quantified is taken for dough mixing. The fully dried NC and the paste is highly susceptible to static charge accumulation and the attendant fire/explosion risk and requires meticulous care in handling and transportation. This practice of using dry NC-NG paste in large quantities is now almost discontinued.

Water Wet Nitrocellulose

Nitro cellulose in a uniformly water wet condition with 30% water content is safe for handling, storage and transportation. Water wet NC is packed in polythene bags and stored in 200 liter capacity steel drums and transported over long distances in specially designed explosive vans with proper danger signals displayed.

Alcohol Dehydrated Nitrocellulose

Nitrocellulose in the alcohol dehydrated form is a convenient and safe method to use. Water wet nitrocellulose received from the plant contains 25 to 30% water. Special types of hydraulic presses with four vertical cylinders rotating around a shaft are used for alcohol dehydration of nitrocellulose. Each cylinder consists of a thick walled basket made of stainless steel. A brass mesh and a linen cloth are placed at the bottom of the cylinder. About 20 kg of water wet nitrocellulose is charged into the cylinder. While the charged mass is supported by another bottom piston, the NC is compacted by ramming by I piston. Around 25 to 50 ksc pressure is applied for compacting.

At the end, the brass mesh and linen cloth adhere to the bottom of the compacted mass. Now about 20 liters of 95% alcohol is poured on the compacted NC bed and the alcohol is forced through the NC layers at 50–100 ksc pressure using the II piston. Water, dilute alcohol and finally less dilute alcohol flows out through conduits provided in the bottom support for the cylinder. In the third position, the upper piston applies 200–300 ksc pressure and squeezes out the alcohol reducing the alcohol content to 30–35%. Here also, the cylinder is properly supported and equipped with conduits to drain the squeezed alcohol. Now the cylinder moves to its fourth position, when the upper piston pushes the dehydrated NC cake down. The brass mesh and linen are removed and the cake is broken into small lumps and the alcohol content of NC is determined by a simple gravimetric analysis.

This process of washing nitrocellulose with alcohol and alcohol dehydration also serves another purpose. It successfully washes out the degradation products of nitrocellulose which otherwise on storage tend to destabilize nitrocellulose. Further processing (dough mixing) is completed after adding required quantity of ether or acetone which is good solvents for nitrocellulose.

Nitrocellulose-Water Slurry

Nitrocellulose-water slurry, keeping the NC content at around 10% by mass is prepared in a slurrying vat/large container fitted with a stirrer. The concentration of NC in the slurry is determined accurately after continuously and uniformly agitating the slurry for not less than an hour and drawing samples from the slurry which is still under uniform agitation. Measured quantity of NC-water slurry is drawn and used in wet mixing process for preparation of NC-NG wet paste. The mass of NC used for paste mixing is calculated based on the concentration and volume of the slurry.

4. TESTING OF NITROCELLULOSE

Nitrogen content of nitrocellulose is determined using Lunge's nitro meter by Lunge Sederholm's method. Here, nitrocellulose is treated with an excess quantity of sulphuric acid in presence of mercury when nitric oxide is quantitatively evolved. Weight percent of nitrogen is calculated from the quantity of nitric oxide evolved from a known weight of the sample.

Other important parameters are: i) Percentage solubility of NC in ether-alcohol solvent, ii) Percentage of acetone insoluble matter, iii) Percentage of ash content after igniting the NC sample in a platinum crucible, and iv) alkalinity/acidity of distilled water extract of NC samples.

Stability of nitrocellulose and NC powders is another critical parameter from safety and storage point of view. A number of devastating accidents had occurred in the early days of NC manufacture on account of poor stability of nitrocellulose. Stability test at 134.5°C with standard methyl violet test paper and Abel heat test are the standard methods for testing the stability of nitrocellulose as well as nitric ester based propellant samples.

In modified Abel heat test, samples are heated at 80°C in a test tube immersed to a depth of approximately 7.5 cm in a water bath. The test tube is closed with a cork carrying a specially prepared starch potassium iodide indicator paper suspended at a standard height above the sample with the help of a platinum wire. The platinum wire is conveniently hung from the bottom of the cork. A tiny drop of glycerine-water mixture is placed at the centre of the paper by a pointed glass tube. Oxides of nitrogen evolved are detected by the indicator paper by the appearance of dark line between the moistened and un-moistened parts of the paper. The time elapsed in minutes from the moment heating is started till the detection of oxides of nitrogen is noted as test result. The Abel heat test values vary depending on the nitrogen content of nitrocellulose used. Typical values are 18 minutes for high nitrogen NC, 20 minutes for double base propellants with high nitrogen NC, 25 minutes for double base propellants with low nitrogen NC and 15 minutes for dynamite.

Determination of residual stabilizer is also used as a convenient method for assessing the stability and storage life of NC and NC powders.

5. GLOBULAR/SPHEROIDAL NITROCELLULOSE

Fibrous nitrocellulose can be converted to globular/spheroidal nitrocellulose which is also known as dense NC. Globular NC can be conveniently used as a single base casting powder for making cast charges through the slurry cast process. Figure 1 gives a process flow chart for processing globular nitrocellulose.

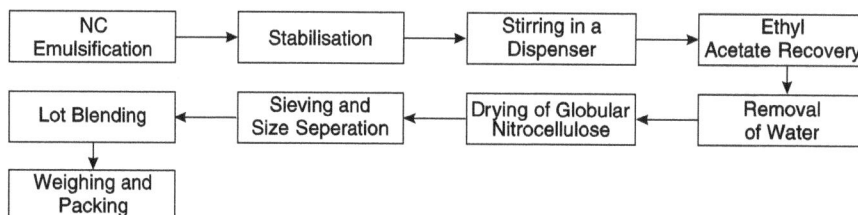

Fig. 1: Process Flow Diagram for Globular Nitrocellulose

Soluble nitrocellulose of 12.2 to 12.4% N_2 content is suitable for making globular nitrocellulose. The fibrous nitrocellulose is dissolved in ethyl acetate to get an emulsion. Diphenylamine is added to this emulsion as stabilizer and mixed uniformly (A portion of the required quantity of nitroglycerine can be added here to make double base ball powder by this process). A protective colloid, typically a gel made by boiling corn starch or gum Arabic, in a small amount of water is added into the lacquer and stirred vigorously in a disperser with an additional amount of water. The lacquer quickly breaks into small globules during stirring. After formation of the globules has been completed, the solvent is evaporated at a controlled rate. Finally, water is removed and the product is dried to moisture level below 0.5%. The globule size is strictly controlled to 0.2–1 mm by adopting process control methods like design and speed (RPM of rotation) of the disperser, concentration of the emulsion/lacquer, rate of evaporation of the solvent, etc. The dense NC is sieved through proper sieves to remove over size and under size granules and lot blended.

6. PROCESSING OF HOMOGENEOUS PROPELLANTS

In general, two different process techniques, viz. extrusion and casting are adopted for processing homogeneous propellant charges. Extrusion is done by solvent process or by solventless process, depending on the use of a volatile solvent for achieving the extrudable (plastic like) consistency for the propellant mix. Two different routes are followed for casting viz. slurry casting process and interstitial casting process. The slurry casting process (also called ball powder process) makes use of globular nitro cellulose. Interstitial casting process makes use of casting powders specially prepared by solvent extrusion process. The important considerations in the selection of a process are the charge size and grain geometry as well as the ingredients composition.

The extrusion process gives grains exhibiting better homogeneity and reproduceble performance characteristics. The process is highly suitable for mass production. The disadvantages are restrictions in propellant mass, size and configuration due to limited capacity of the extrusion press. The casting technique permits the processing of propellant grains virtually having no limitation as far as shape, size and configuretion is concerned. It allows fabrication of bi-composition as well as free standing or case bonded grains.

Solvent Extrusion Process for Propellant Manufacture

Propellant grains of thin webs are generally called propellant powders. Single base powders, double base powders and triple base propellant powders (in general they are called smokeless powders) as well as high energy casting powders which contain inorganic oxidizers like ammonium perchlorate, metallic fuels like aluminium powder, nitramine compounds like RDX/HMX, etc. are processed by solvent extrusion process with the required process modifications dictated by the powder composition. Since removal of the residual solvent becomes difficult with the increase in web thickness, propellant charges with less than 30 mm web thickness only are processed by solvent extrusion process. Moreover, grain shrinkage due to removal of residual solvent also becomes problematic as the web thickness increases. The early solvent extruded double base propellant was in the form of chords and hence, they were called cordite. The first cordite propellant called cordite MK1 contained 58% nitroglycerine, 37% nitrocellulose (12.9 to 13.1% N_2) and 5% Vaseline.

Today solvent extruded propellants are made in different forms-chord like, tubular, multi-tubular, slotted tubular, etc., to conform to different grain designs, to suit the ballistic requirements of burn time, surface area of burning, pressure in the gun barrel, projectile velocity, etc. and weapon system in which the propellant system is used. The charge length may vary from less than 0.5 mm to a few centimeters. While smokeless powders are used in small arms and high caliber guns, casting powders are used for making massive cast propellant charges weighing even hundreds of kilograms per single grain.

Nitrocellulose of different grades is selectively used or sometimes they are blended in suitable proportions to achieve the desired burn rates, processability, mechanical properties, etc. The solvents used are acetone, acetone-water mixture in nearly 92:7 ratio or ether-alcohol mixture in which the ether content may vary from 50–65%. Acetone-water mixer in 92:7 ratios has higher solubility for nitrocellulose compared to anhydrous acetone as solvent. Hence, such mixtures are preferred in the processing. As the water content increases beyond this optimum level, the solubility drastically decreases and at 12% concentration of water, the nitrocellulose gel starts precipitating. A mixture of 55% by volume of ether with 45% by volume of alcohol has much higher solubility for nitrocellulose than ethyl alcohol. In an alcohol-ether mixture as the alcohol content is increased beyond the optimum level, the solubility decreases and at 75% alcohol level the nitrocellulose gel starts precipitating. Because of its poor solubility for nitrocellulose, ether alone is used to extract nitroglycerine from double base samples.

Nitrocellulose can be extracted with ethanol partially, depending on the nitrogen content. The residue from the first extract can be further extracted with ether-alcohol mixture which has a higher solubility for nitrocellulose. Still there may be an insoluble residue of high nitrogen NC. Acetone-alcohol mixtures in suitable ratio are used in the processing of homogeneous propellant powders. The molecular weight and degree of polymerization of nitrocellulose has a significant effect on solubility of nitrocellulose. In general, higher the molecular weight, lower the solubility. Consequently, nitrated cotton shows higher viscosity in process solvents compared to nitrated wood pulp. Higher the molecular weight of cellulose used for nitration, higher will be the viscosity of NC solutions. These general factors are to be borne in mind in formulating the propellant compositions and process conditions for any end use.

Processing of Single Base Propellant

Alcohol dehydrated nitrocellulose with known alcohol content is used in the processing of single base propellant. A process flow diagram for single base propellant is given in Figure 2.

Weighed quantity of alcohol dehydrated nitrocellulose of suitable nitrogen content is gelatinized in a kneading machine where it is mixed with the additional solvent ether, stabilizers and other additives to achieve an adequate consistency for extrusion. Here, ether-alcohol mixture forms a suitable binary solvent for nitro-cellulose. A small percent of potassium nitrate as leaching agent is also added in the mix which later gets leached out in the water treatment process. The dough well formed by kneading is unloaded and kept warm for maturing. The matured dough is used for extrusion. The dough is filled in the cylinder of a vertical extrusion press provided with a multiple die assembly at the bottom to shape the emerging propellant strands in the desired form. The plunger applies controlled pressure on the dough and the speed of the ram is monitored to match the consistency of the mass under extrusion. The propellant strands extruded through the dies are initially cut into long lengths of approximately 2 meters and tied as bundles. These bundles are partially dried in drying cabinets by exposing to a current of injected warm air. The solvents evaporate out leaving a small amount of alcohol still remaining in the propellant. The solvent vapors from the mixing room and extrusion and pre-drying rooms are sucked out into the absorbers where the solvents are absorbed in activated charcoal and subsequently recovered for reuse.

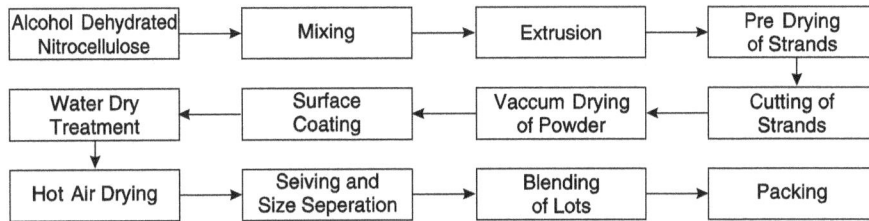

```
┌──────────────────┐    ┌──────────┐    ┌──────────┐    ┌──────────────┐
│Alcohol Dehydrated│ →  │  Mixing  │ →  │Extrusion │ →  │  Pre Drying  │
│  Nitrocellulose  │    │          │    │          │    │  of Strands  │
└──────────────────┘    └──────────┘    └──────────┘    └──────────────┘
                                                                │
┌──────────────┐    ┌──────────┐    ┌──────────────┐    ┌──────────────┐
│  Water Dry   │ ←  │ Surface  │ ←  │Vaccum Drying │ ←  │ Cutting of   │
│  Treatment   │    │ Coating  │    │  of Powder   │    │   Strands    │
└──────────────┘    └──────────┘    └──────────────┘    └──────────────┘
       │
┌──────────────┐    ┌──────────────┐    ┌──────────┐    ┌──────────┐
│ Hot Air Drying│ → │  Seiving and │ →  │ Blending │ →  │ Packing  │
│              │    │Size Seperation│    │ of Lots  │    │          │
└──────────────┘    └──────────────┘    └──────────┘    └──────────┘
```

Fig. 2: Process Flow Diagram of Single Base Propellants

The pre-dried strands are cut into granules in a cutting machine and the cut powder is sieved through sieving plates. The sieved material is spread in aluminium trays and loaded in shelves heated by hot water in vacuum drying ovens. In the surface coating process, the propellant powder is coated with fine graphite powder and the coating is done in rotating drums similar to sweetie pans used for sugar coating tablets in pharmaceutical industry.

The surface coated propellant is then filled in cotton bags and immersed in hot water for two or more days in order to remove the final portion of the solvents. The leaching agent dissolves in the water rendering the propellant surface porous. This technique is made use of to achieve the desired burn rate for single base powders. The duration of this water treatment is varied depending on the web thickness of the propellant charge. The bags containing water treated propellant powder are placed on a screen in the drying house and dried in a blast of hot air to remove the moisture. The finally dried powder is passed through a continuously operating screening drum to remove the dust. It is then passed through a pan sifting machine to remove larger/out of size grains. Batches of sifted propellant are blended in a blending apparatus to ensure homogeneity and then packed in suitable containers.

Single base powders for pistols, 7.62 mm rifles, 30 mm anti aircraft guns, 40 mm high caliber guns, etc. are processed by this method.

Typical composition of single base propellant grains by solvent extrusion process is shown in Table 8.

Table 8: Typical Composition of Single Base Propellant

Ingredient	Mix Composition (%)	Final Composition (%)
Nitrocellulose	50–55	90–95
Stabilizer diphenyl amine	1–2	2–4
Solvent ether	30–35	Nil
Solvent alcohol	15–25	1–2
Leaching agent (KNO₃)	1–2	0–0.5

Various quality control measures are adopted to maintain the quality of the propellant powder such as determination of density, bulk density, granule size, calorific value, etc. Ballistic evaluation of the gun propellant is done in proof ranges. Important ballistic parameters such as cartridge base pressure on firing each round is used as a quality control measure. Projectile velocity at muzzle end and between the muzzle end and projectile target are also measured with the help of counter chronometers for proving casting powder lots

Processing of Double Base Propellant Powders by Solvent Extrusion Technique

Double base propellant processing by solvent extrusion process can make use of either dry NC-NG paste or NC-NG paste made by wet mixing process, after bringing the moisture level below 1%. Alcohol wet nitrocellulose with necessary process steps for addition of nitroglycerine also can be made use of.

Figure 3 shows a process flow chart for double base propellant processing using alcohol wet nitrocellulose by solvent extrusion process.

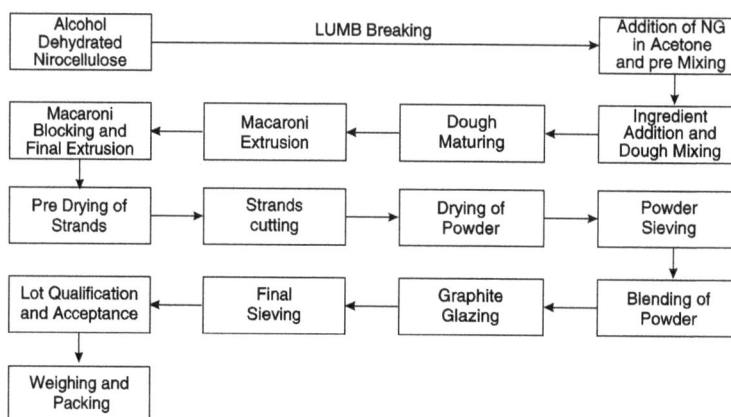

Fig. 3: Process Flow Diagram for DB Propellant by Solvent Extrusion Process

Nitroglycerine in acetone is mixed with alcohol dehydrated cellulose in a Schrader or low shear mixing step called pre-mixing. The premix is then loaded in the final mixer where all the additives are also incorporated. Alcohol-acetone mixture forms a suitable solvent when high nitrated cellulose is used. The solvent quantity is adjusted for getting a suitable consistency of the mix for extrusion and to adjust the charge weight to any specified grain length. The dough mixing is done in standard kneading machines with low clearance between the impeller edge and the kneader bowl. The two impellers in the dough mixer rotate at controlled low rpm in opposite directions, one feeding the other. Certain additives are incorporated at the dough mixing stage after initial gelling of the paste. Initial circulation of warm water in the gland at the bottom of the kneading machine helps the solvation of NC with the solvent as well as plasticizers. Once the solvation is started, heat of dissolution may tend to raise the dough temperature.

During mixing care is taken to see that the dough temperature is maintained as desired between 35 to 40°C by circulating hot/cold water. In CMDB formulations, the high energy ingredients HMX, AP or aluminum are added with special precautions in small installments, after a gelatinized matrix has been formed. The usual mixing time for preparing homogeneous dough with suitable consistency for extrusion varies between 4 to 5 hours. At the end of the mixing, the dough is broken into smaller lumps by reversing the direction of rotation of the impellers and unloaded into rubberized bags to prevent loss of solvent. The dough is aged in warm compartments for a day or two prior to extrusion.

The extrusion is carried out as a two step operation. The first step is called macaroni extrusion. In the macaroni extrusion stage, the dough passes through suitable strainer or filter mesh placed at the bottom of the extrusion basket and then gets extruded through a set of large dies. This step helps to further improve gelation of NC by the plasticizer on account of the mechanical work done by subjecting the dough to 50–100 ksc pressure and also serves as an initial trial extrusion. By this process, large nitrocellulose fibers which resist gelation are filtered off with the help of the strainers. The macaroni strands are collected and compacted in the final extrusion basket and extruded through properly contoured dies with accurate die dimensions. The final extrusion applies higher specific pressure on the dough for better compaction of the mass under extrusion. Typical composition of double base propellant is given in Table 9.

The extruded strands are collected in a suitable container and these strands after pre-drying attains a suitable consistency for cutting into smaller grains. In the next step, these strands are horizontally fed to cutting machine equipped with a circular cutting head, rotating in a vertical plane. The rotating disc cuts the strands into grains of uniform length. The cut length can be varied depending upon the specific end use. This process is used for making double base casting powders for casting large size propellant charges also. In the case of casting powders, usually the diameter to length ratio of the granule is maintained close to unity. The cut propellant granules still retains a certain percent of solvent. The solvent is removed by subjecting the granules to a draught of warm air and the solvent is recovered. Graphite glazing is done depending upon the requirements to facilitate easy flow of the grains and also to discharge static electricity. Lot blending is done to even out batch to batch variation in chemical, physical and ballistic properties. Depending upon the charge weight requirements, bigger size grains are processed and made into bundles for use in high caliber guns.

Table 9: Typical Composition of Double Base (Solvent Extrusion) Propellant

Ingredient	Approximate Mix Composition (%)	Final Composition (%)
Nitrocellulose (13.1 N_2)	40–45	60–75
Nitroglycerine	15–20	20–30
Stabilizer ethyl centralite/orthonitro diphenyl amine	1–3	2–5
Plasticizer–dinitro toluene or mineral jelly	05–07	07–1.0
Plateaunising agent (Lead stearate)	0–2	0–3
Flash suppressor (Potassium sulphate)	0.5–1.0	1–.1.5
Solvent (Acetone/acetone water mixture)	35–40	0.5–2.0

The process steps are varied to suit the requirements. In NG containing propellant grains, the residual solvent after drying is comparatively less. Moreover, the propellant surface is not moisture permeable. Hence, water dry treatment is dispensed with. When water soluble additives like AP are present in the formulation, water dry treatment is not advisable.

The propellant dough for extrusion can be prepared using NC-NG paste, prepared by wet mixing process also. Here, the final moisture content in the paste is brought down to less than 1% level by prolonging the drying time. (The earlier practice was to make use of dry NC-NG Paste. Here, NC-NG dry paste is taken directly for dough mixing using acetone-water mixture or acetone-alcohol mixture as the process solvent. In this case, the initial Schrader mixing is avoided since the NC-NG paste is already mixed at random and matured). The paste is taken for dough mixing using acetone-water or acetone-alcohol mixture as process solvent. The necessary ingredients like plasticizers and stabilizers are incorporated at the dough mixing stage. Using the dough, solvent extrusion process is continued as in the case with alcohol dehydrated NC as the starting material.

Any modification in process operations must take into account the compositional details, type and nitrogen content of nitrocellulose as well as the safety aspects. In triple base propellants, the nitro guanidine is incorporated at the premixing stage and the process steps are continued.

Solventless Process for Homogeneous Propellant Manufacture

Solventless process is preferred when the propellant grains with webs greater than 30 mm is required. Figure 4 gives a process flow diagram for solventless extrusion.

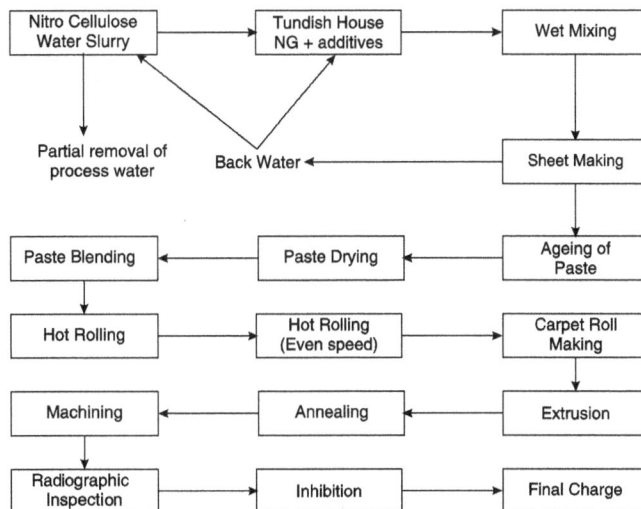

Fig. 4: Process Flow Diagram for Solventless Extrusion Process

Nitrocellulose with 12.2–12.6% nitrogen content is suitable for solventless extrusion. Nitrocellulose-water slurry is prepared in a large vessel, called NC vat. The concentration of the slurry under agitation is properly determined and a measured quantity of the slurry is pumped with the help of a diaphragm pump and received in a special container. Here, the process water is removed and the slurry is mixed with back water which contains traces of dissolved nitroglycerine from the previous mixings. The slurry is stirred for approximately 10 minutes with back water and pumped to the tundish. The tundish is an open coiled gutter with a tangential outlet built on a cone shaped base. As the slurry now containing back water passes through the tundish, a measured quantity of NG from a lead burette is sprayed on the slurry. The entire quantity of NC slurry with NG sprayed in it flows by gravity through stainless steel pipe to the mixing tank. After completing the spraying of NG, weighed quantity of additives (carbamite slurry in water, precipitated chalk, candelilla wax, etc.) are added through the tundish which is again washed down with back water. Thus, the entire mix is received in the mixing tank where it is stirred for 30 minutes. Absorption of NG in NC takes place in the mixing tank. The slurry is now pumped with the help of special diaphragm pump to a sheeting table.

The sheeting table is essentially a continuous reel cloth band moving over a series of rollers of 1 to 2 inch in diameter and vacuum tubes of similar size and finally passing between two heavy rollers at one end. On the sheeting table, water from the mix drains out and also gets sucked through the reel (filter) cloth with the help of vacuum tubes. As the water is removed, the mix appears like a cake spread on the reel cloth and moves to the edge of the table where it passes between the two rollers. The squeezed paste emerges through the gap in between the heavy rollers and gets pressed into a compact sheet of approximately 1–2 centimeter thickness. Following the rollers, there is guillotine which cuts the sheet into suitable length so that rectangular sheets of paste can be collected one by one. Depending on the width of the reel clothe, the compacted cut sheets may have 50 to 75 cm length and 30 to 40 cm width and approximately 1 to 2 cm thickness. These paste sheets with approximately 30 to 40% moisture content are collected in rectangular aluminum trays of matching dimensions so as to hold the paste without spillage. (An alternate route for reducing the water content in the paste is centrifuging). These trays are kept in drying trucks which can hold up to 24 trays in two racks with air gap in between. The paste is matured for 24 hours. Afterwards, the truck is taken to the drying house where air heated to 45–50°C is blown through the trays for 24 hours so as to bring down the moisture content to 10% level. Water soluble ingredients are added in a separate Schrader mixer.

The paste which is now in powdery form is converted into a tough plastic sheet by working on differential rolling mill. The rolling mill has two parallel horizontal cylindrical rollers made of stainless steel of approximately 30 cm diameter and is heated by hot water line passing through the central bores. These cylindrical rollers rotate at different circumferential speeds. The paste gets gelatinized by hot rolling, i.e. by passing between the rollers maintained at 45–50°C for a specified number of times. The gap between the rollers can be adjusted for 2 to 5 mm thickness as desired to suit the thickness of the gelatinized sheet. The elevated temperature of the rollers causes the moisture content to be reduced to 0.5%. After completion of the differential speed rolling, the sheet is given a number of passes through the even speed rollers. A typical sheet is 500 cm long 75 cm wide and 3 cm thick. The sheet is now cut into10 to 12 cm wide strips. These strips are inspected for foreign material inclusions, ungelatinized NC particles, etc. and defective portions are removed. These strips are then wound by hand or on a carpet rolling machine to form a roll of cordite strips. These rolls, called cheese, are kept in steam jacketed ovens maintained at 50 ± 5°C so as to maintain the required flexibility for extrusion.

The rolled strips are finally converted into propellant grains by batch extrusion process in a horizontal hydraulic extrusion press. Here, the heated carpet rolls are loaded into the preheated press container fitted with a properly designed die. The diameter of the wound carpet rolls are adjusted to match with the inner diameter of the extrusion press container. The diameter of the press basket may be around 30 to 40 cm which can accommodate 3 or 4 carpet rolls at a time. The overall processing capacity of the press can be still higher. Having loaded the required number of carpet rolls, the press is set in motion. As the plunger of the press just enters the container, the press is stopped by a pre-set/automatic arrangement. The annular gap between the plunger and the container is now closed with a flanged flexible sealing ring fitted on the plunger head. Now the container is fully evacuated through a vacuum line opening to it. This is an important safety precaution to avoid adiabatic compression of the entrapped air which can cause accidental ignition of the propellant during extrusion. When the evacuation is completed, the press is again set in motion and the plunger head engages the heated cheese at a pre-set speed. The propellant cheese gets well consolidated on account of the contoured entrance region of the die and finally emerges with a smooth finish. The extrusion pressure is of the order of 200 ksc.

The extruded cordite moves on roller conveyor kept in front of the die. After completion of the extrusion, the propellant is cut with a guillotine cutting machine and removed. The plunger is reversed after disengaging the sealing ring. In a standard extrusion press, up to 50 kg propellant (Cheese) can be loaded in the container which is the size limiting factor for an extruded charge. The diameter of the extruded grain can vary up to 200 mm, beyond which it is difficult to extrude, due to the requirement of massive extrusion presses and safety installations. Typical composition of double base propellant by solventless extrusion process is given in Table 10.

Table 10: Typical Composition of Double Base Propellant by Solventless Extrusion Process

Ingredients Used	Composition (%)
Nitro cellulose-12.2–12.6% N_2	45–55
Nitroglycerine	40–45
Stabilizer-ethyl centralite II	3–5
Precipitated chalk	0–0.5
Flash suppresser-potassium cryolite	1–2
Die lubricant candelilla wax	0–1
Darkening agent carbon black	0–0.5

Die temperature, charge temperature, extrusion pressure and propellant composition and consistency are the variables which affect the quality of extrusion. The extruded propellant is annealed to relieve stress concentration and brought to room temperature. The semi transparent solventless extruded propellant has smooth finish and very high mechanical properties not less than 30 ksc tensile strength and 500 ksc tensile modulus. Depending on the die design, the charges can have desired geometrical configurations—tubular, cylindrical, cruciform, etc. The propellant charges are trimmed to the required length with a hand operated pneumatic cutting machine and are examined for visual flaws and charge weight as well as for dimensional accuracy—charge length and diameter using "go" and "no go" gauges with specified tolerance limits and finally radiographed for internal flaws. On account of the thermoplastic nature of the propellant, the rejected charges can be heated and sliced into thin slices and used for reprocessing, starting with rolling and cheese making process

The propellant is characterized by various methods. The most commonly used one is the determination of tensile properties at constant strain rate at different temperatures. Typical mechanical properties of double base propellant prepared by solventless extrusion process and its dependence on temperature are given in Table 11.

Table 11: Mechanical Properties of Double Base Propellant Prepared by Solventless Extrusion Process at Various Temperatures

Test temperature (°C)	27	18	0	−10
Tensile strength (ksc)	31	39.3	105	181
Elongation (%)	11	10	8.2	8.0
Tensile Modulus (ksc)	975	1230	3335	6160

The typical burn rates of a double base propellant processed by solventless process range between 6 and 10 mm/second under standard conditions of operating pressure (70 ksc). The burn rate pressure index falls in the region 0.6 to 0.65. Various catalysts have been developed to enhance the burn rate as well as to reduce the pressure index to a still safer region.

The burning surface of the charge is restricted by applying a suitable inhibitor to conform to the thrust requirements of the rocket. The conventional inhibitor was made with ethyl cellulose. Migration of nitroglycerine as well as other liquid plasticizers to the inhibitor layer during storage of inhibited double based propellant charges leads to problems of de-bonding as well as the tendency of inhibitor itself to become inflammable and burn with the propellant. In the case with composite propellants, the migration is less to inhibitor since the liquid component is less in cured propellant. Moreover, the functional groups of the polymeric binder in the propellant can be made to crosslink with inhibitor ensuring high tensile bond strength and shear bond strength. In a dedicated effort to develop a suitable inhibitor, a highly cross-linked polyurethane type inhibitor for double base propellant which could withstand stringent requirements, climatic variations on storage and high temperature and low temperature test conditions have been developed.

Solventless extruded propellants are used in short range missiles, sounding rockets, in pressure actuated devices like air craft seat ejection cartridges, ship engine starters and also in gas generators.

Processing of Cast Homogeneous Propellants

Two types of casting process, i.e. interstitial casting process and slurry casting process have been developed as versatile techniques for complicated grain geometries and for incorporating high energy ingredients viz. AP and aluminium powder as well as the nitramines (RDX and HMX). These compositions can be suitably processed for non-case bonded and case bonded applications, the latter demanding low modulus of elasticity and high strain capability in preference to high strength. The non-case bonded charges call for very high modulus of elasticity to the tune of 500 ksc. Improvements in tensile properties are achieved by chemical cross linked formulations in the slurry casting process and special grafting techniques in interstitial casting process.

The cast double base process was developed during the second world war period to meet the requirements of rocket propellant charges significantly larger than those made by extrusion process.

Interstitial Casting Process

In interstitial casting process, casting powders and casting solvents are prepared separately. The casting powder (granule) of about 1 mm length × 1 mm diameter contains nitrocellulose, all the required solid ingredients and a fraction of the plasticizers and stabilizers intended for final formulation. The casting powders are made by solvent extrusion process. Figure 5 shows the process flow diagram for casting powders.

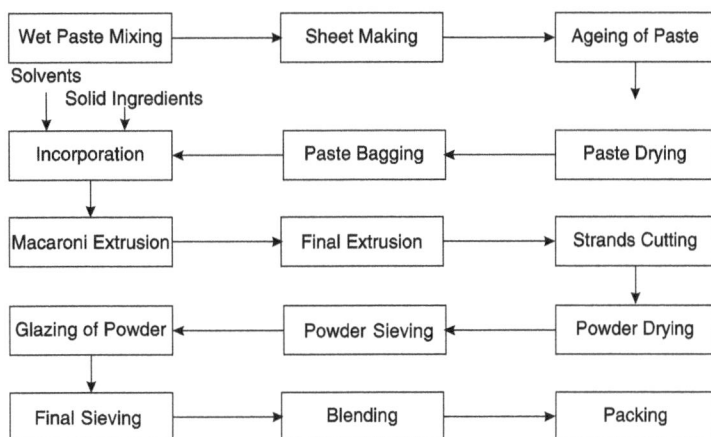

Fig. 5: Process Flow Diagram for Processing Casting Powders

The convenient method is wet mixing process using NC-NG paste containing stabilizers. Here, wet paste is made in the form of felted sheet as described earlier under solventless extrusion process. The sheet/paste are dried by hot air to maintain moisture content less than 1%. The paste with additives is taken for mixing in a kneading machine. The kneader is initially wetted with a small quantity of the volatile process solvent and the paste is mixed with the solvent to form gelatinized dough. The kneader is provided with a water jacket to maintain a moderate temperature preferably around 35 to 40°C. After dough is formed depending upon the type of casting powder formulation base/nitramine, solid/crystalline additives such as HMX/RDX, ammonium perchlorate and aluminum powder are added sequentially in instalments. This step is done with due safety precautions taking into account the sensitiveness of the mix. During gelation, temperature of the mix goes up automatically due to the heat of dissolution. This is helpful in proper dough formation, but precautions must be taken to control the dough temperature not to exceed 40–42°C. Additional quantities of the solvent are also added in one or two instalments and the mixing is continued for the optimum mixing time. After a day's dough maturing, the dough is taken for extrusion and the powder processing is completed on similar lines as in the case of double base propellant powders.

Characterization of the Casting Powder

The casting powder is to be characterized for maintaining quality of the final propellant. An excellent gauge of quality of casting powder is to achieve 97 to 98% of theoretical density. A high screen loading density or bulk density is another important requirement. This is achieved by maintaining the L/D ratio of the granule close to one and by avoiding the irregular or slanted cut surfaces.

The safety characteristics of the powder in terms of friction and impact sensitivity and auto ignition temperature are determined by standard test procedures. In general, the casting powder ignites at 160 to 170°C, when heated at a rate of 5 to 7 degrees centigrade per minute. The friction and impact values depend on the explosive plasticizer level as well as the high energy fillers. The crystalline additives render the powders more sensitive and are to be handled with due precautions. The stability of the casting powder is determined by estimating the stabilizer content and by heat test determinations. Risky operations with dry powder such as glazing, sieving, sifting, pouring are to be done with special precautions to discharge static charge. The heat of combustion (calorific value) of the casting powder determined in a bomb calorimeter gives an indication of the energy level and its value depends on the powder composition. In general, calorific value of double base formulations vary from 800 to 1200 calories per gram, whereas, with high energy casting powders containing AP and Al powder, the calorific values beyond 2000 calories per gram have been achieved. However, high calorific value alone cannot be considered as a proper measure of the energy level of a propellant.

A mixture of explosive plasticizers like nitroglycerine or DEGDN and inert plasticizers like glycerol triacetate, dibutyl phthalate, etc. in suitable proportions is used as the casting solvent.

The non-explosive plasticizers are essentially added to desensitise the explosive plasticizer. Certain inert plasticizers like glycidyl methacrylate or methyl methacrylate are used to improve the mechanical properties by grafting on to the cellulose macro molecules. Figure 6 shows the process flow chart for the interstitial casting process.

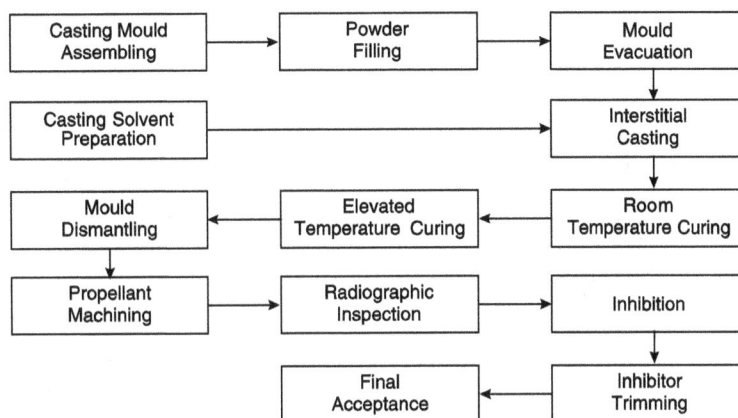

```
 ┌──────────────┐     ┌──────────────┐     ┌──────────────┐
 │ Casting Mould│────▶│   Powder     │────▶│    Mould     │
 │  Assembling  │     │   Filling    │     │  Evacuation  │
 └──────────────┘     └──────────────┘     └──────────────┘
                                                   │
 ┌──────────────┐                          ┌──────────────┐
 │Casting Solvent│─────────────────────────▶│  Interstitial│
 │  Preparation │                          │   Casting    │
 └──────────────┘                          └──────────────┘
                                                   │
 ┌──────────────┐    ┌──────────────┐      ┌──────────────┐
 │    Mould     │◀───│   Elevated   │◀─────│     Room     │
 │  Dismantling │    │Temperature Curing│  │Temperature Curing│
 └──────────────┘    └──────────────┘      └──────────────┘
        │
 ┌──────────────┐    ┌──────────────┐      ┌──────────────┐
 │  Propellant  │───▶│ Radiographic │─────▶│  Inhibition  │
 │  Machining   │    │  Inspection  │      │              │
 └──────────────┘    └──────────────┘      └──────────────┘
                                                   │
                     ┌──────────────┐      ┌──────────────┐
                     │    Final     │◀─────│  Inhibitor   │
                     │  Acceptance  │      │   Trimming   │
                     └──────────────┘      └──────────────┘
```

Fig. 6: Process Flow Chart for Interstitial Casting

Casting Process

In the interstitial casting process, the casting powder is filled approximately 70% volume in the mould with a suitable core which decides the shape and dimensions of the port of the cast propellant charge. A perforated disc is provided at the bottom of the mould which supports the powder bed and also permits the casting solvent during the casting process. The casting solvent is fed through the bottom of the casting mould. Teflon coated mould is used for easy extraction of the free standing cast grain after completion of curing. Alternately, a suitable inhibitor lined casting mould which acts as the motor case can also be used. Developing a suitable inhibitor for double base propellant is difficult because of nitroglycerine migration into the inhibitor on storage and hence, double base propellants are mostly used as cartridge loaded.

Before casting, the powder filled mould is continuously evacuated for a few hours to remove the air entrapped in the interstitial space in the powder bed and then the casting solvent is admitted to the powder bed. The powder bed gets flooded with the casting solvent and the excess solvent overflows to a collector vessel while the powder is retained by a fine strainer placed at the top of the powder bed. The casting powder is allowed to freely absorb the casting solvent for some time, preferably 24 hours. Afterwards, the mould is detached from the joints and sealed. Figure 7 shows schematic representation of interstitial casting process. Figure 8 shows photograph of laboratory set up for interstitial casting process.

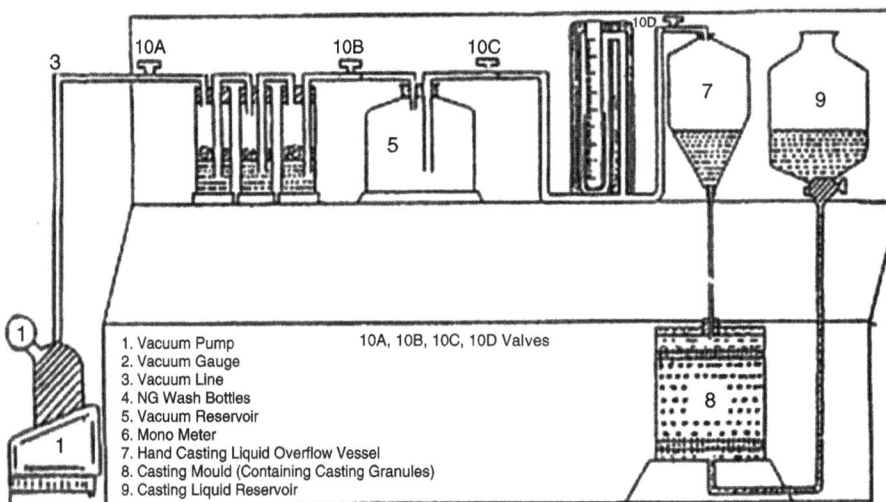

1. Vacuum Pump
2. Vacuum Gauge
3. Vacuum Line
4. NG Wash Bottles
5. Vacuum Reservoir
6. Mono Meter
7. Hand Casting Liquid Overflow Vessel
8. Casting Mould (Containing Casting Granules)
9. Casting Liquid Reservoir

10A, 10B, 10C, 10D Valves

Fig. 7: A Schematic Process Diagram for Interstitial Casting Process

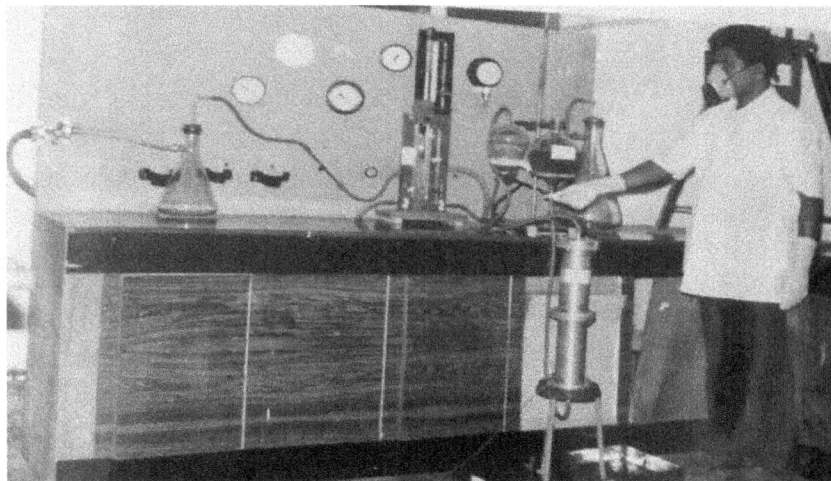

Fig. 8: Laboratory Set-up for Interstitial Casting Process

The curing is carried out in two distinct phases—an ambient rest period during which the cast mould is stored undisturbed for a day or two which is followed by curing at elevated temperature of approximately 60°C for 5 to 10 days. During the curing process, mutual diffusion of the nitrocellulose macromolecules and the plasticizer takes place with the result the granules coalesce and their boundaries disintegrate and a macroscopically homogeneous cast propellant is formed.

The mechanical properties at the end of optimum cure depends on the propellant formulation, the nitrocellulose content, the type and grade of nitrocellulose as well as the plasticizer and the grafting/cross linking techniques employed by different producers. Granule size and screen loading density of the casting powder are very important process parameters which decide the casting powder to casting liquid ratio. These parameters have to be fine tuned for optimizing the ballistic and physical properties of the propellant.

Slurry Casting Process

The very high rate of gelation of nitrocellulose with nitroglycerine is a major disadvantage for casting large propellant charges by slurry cast process. This is overcome by using granular nitrocellulose, also known as densified nitrocellulose in the slurry casting process. The slurry casting process can also avoid the lengthy process of making casting powder through solvent extrusion process. The process of making granular nitrocellulose has already been described. The propellant ingredients can be homogenized in one simple mixing operation. Here, the plasticizer mix is poured into a mixing pot and the solid ingredients including granular nitrocellulose are added. The mix is stirred for a short duration to achieve proper homogeneity of the ingredients and poured into the mould or rocket case and cured. This method is adopted to make large propellant charges and for processing the CMDB variety by incorporating high energy ingredients—AP, Al powder, RDX/HMX and different explosive and inert plasticizers.

The mechanical properties of slurry cast propellants depend on various factors such as nitrocellulose content, solid loading, chemical cross-linking effected, etc. Chemical cross-linking technique utilizes the residual hydroxyl groups in the polymeric binder which can react with a cross-linker, typically toluene diisocyanate.

7. NITRAMINE PROPELLANTS

Nitramine propellants offer energy levels 3 to 5% more than high energy double base propellants. Table 11 shows theoretical specific impulse and chamber temperature of HMX/RDX containing propellant formulations with a systematic variation in NC content versus NG in the formulation. The theoretical specific impulse was calculated using NASA software program employing extended Fortran IV language based on NASA/SP/273 report. Performance characteristics like specific impulse, flame temperature, gamma, mean molecular weight, etc. were evaluated using this program. Theoretical specific impulse and flame temperature of 15% HMX/RDX containing propellant formulations by varying NG vs. NC content is given in Table 12.

As the NG content is increased at the expense of nitrocellulose, theoretical specific impulse as well as chamber temperature shows a corresponding increase. RDX on account of the higher specific energy has an edge over HMX in this respect.

Table 12: Theoretical Specific Impulse of 15% RDX/HMX Containing Propellant Formulations

Propellant Ingredient	Percentage Composition of HMX Containing Formulation				Percentage Composition of RDX Containing Formulation			
Nitrocellulose	50	45	40	35	50	45	40	35
Nitroglycerine	25	30	35	40	25	30	35	40
Plasticizer (DBP)	10	10	10	10	10	10	10	10
Nitramine	15	15	15	15	15	15	15	15
Theoretical Performance Evaluation Data								
Theoretical specific Impulse, sec.	225.9	229.9	232.5	235.6	228.1	231.4	234.6	237.7
Chamber temperature, K	2396	2481	2565	2647	2442	2526	2609	2691

Typical Combustion Products of Nitramine Propellant

The combustion products of a typical nitramine propellant as theoretically computed consist mainly of carbon monoxide, carbon dioxide, nitrogen, hydrogen, water vapor and a small percentage of elemental carbon. Table 13 shows the mole fractions of different combustion products at the nozzle exhaust of nitramine propellant formulation containing 10% HMX and the change in gas composition effected by systematic introduction of an inorganic oxidizer ammonium perchlorate in small percentages in place of HMX. Ammonium perchlorate brings in hydrogen chloride, which condenses in humid atmosphere and leaves a detectable trail. Yet, it is helpful in eliminating elemental carbon in the exhaust which also leaves a signature. On introduction of an additional oxidizer, as expected, under oxidized combustion products decreases and fully combusted products like CO_2 and H_2O shows an increasing trend.

RDX/HMX incorporated nitramine propellants have been processed by solventless extrusion technique as well as by slurry cast and by interstitial casting process. The process sequence for preparing casting powder shown in Figure 6 can be followed for preparing HMX/RDX incorporated advanced casting powder. The required quantity of HMX/RDX can be incorporated in 30% alcohol wet form. Safety considerations recommend the use of ß HMX in preference to the other three forms. Cast nitramine propellant can be processed by interstitial casting process shown in Figure 7. Typical ballistic data from ballistic evaluation of interstitially cast nitramine propellant with approximately 9% HMX in the formulation in circumferentially inhibited 3.4 kg size ballistic evaluation motor is presented in Table 14.

Table 13: Typical Combustion Products of Nitramine Propellant at the Nozzle Exhaust

Propellant Ingredients	Percentage Composition					
Nitrocellulose	45	45	45	45	45	45
Nitroglycerine	35	35	35	35	35	35
Triacetine	10	10	10	10	10	10
HMX	10	9	8	7	6	5
Ammonium perchlorate	0.00	1	2	3	4	5
Combustion products	Mole fractions					
Carbon (s)	.0279	.0253	.0226	.0199	.0171	.0145
CH_4	.0137	.0138	.0139	.0140	.0141	.0142
CO	.2906	.2898	.2891	.2883	.2875	.2867
CO_2	.1849	.1869	.1890	.1911	.1938	.1953
H_2	.2632	.2633	.2633	.2634	.2634	.2635
H_2O	.0885	.0894	.0903	.0912	.0923	.0930
NH_3	.00004	.00004	.00004	.00004	.00004	.00004
N_2	.1313	.1297	.1280	.1263	.1247	.1230
HCl	–	.0019	.0039	.0058	.0078	.00980

Table 14: Ballistic Data from Ballistic Evaluation of 3.4 kg Size Nitramine Propellant Charges

Ballistic Parameter	BE Motor No PED/1934	BE Motor No PED/1935
Bare grain weight (kg)	3.392	3.388
P_{max} on ignition (ksc)	64.9	65.01
Ignition delay (seconds)	0.0085	0.0087
Burning time (seconds)	17.18	16.7
Average pressure (ksc)	43.1	44.1
Pressure integral (ksc × seconds)	740.5	740.8
Discharge coefficient (per second)	6.683×10^{-3}	6.674×10^{-3}
Characteristic exhaust velocity (m/second)	1468	1470
Specific impulse (seconds)	231.1	231.3
Burn rate at web average pressure (mm/second)	11.0	11.1
Burn rate at 50 ksc pressure (mm/second)	11.2	11.4

The pressure time graph of the above tests is presented in Figure 9.

The ballistic data indicates higher level of energy as compared to conventional double base propellant with a high level of reproducibility.

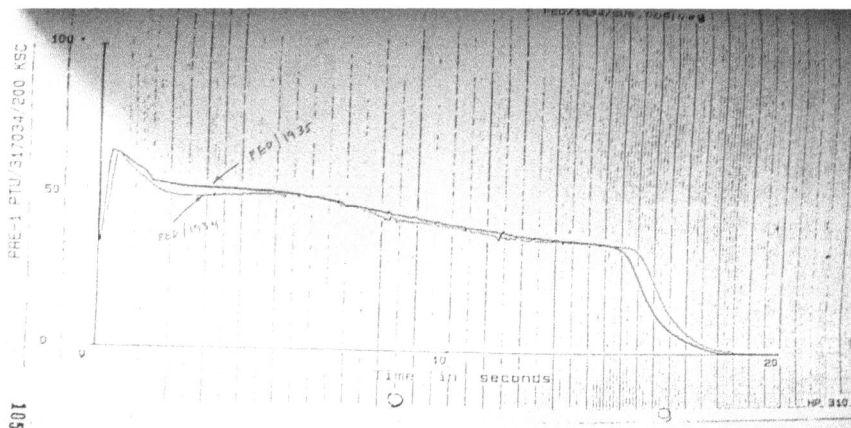

Fig. 9: Pressure-Time Graph of Ballistic Evaluation Motor PED 1934 and PED 1935

8. MECHANISM OF HOMOGENEOUS PROPELLANT BURNING

Research workers Crawford, Huggett and Mc Brady developed some simplified theory of burning. They consider that the burning of homogeneous propellant occurs in three physically distinct zones identified as foam zone, fizz zone, and flame zone sequentially and the burning progresses in a direction perpendicular to the burning surface. The regressing condensed phase is called foam zone. In the foam zone, some liquefaction of the propellant ingredients takes place by absorption of radiant heat from the flame and fizz zones. Though self sustained exothermic reactions are initiated in the foam zone, (the propellant has very low thermal conductivity) the layer below the foam zone does not directly partake in the burning process. Liquefaction and decomposition of the propellant ingredients occurs at the foam zone with evolution gaseous products. The diffusion of NG in the temperature range of 90 to 180°C involving cleavage of $RO–NO_2$ bonds is the rate determining step in the decomposition of double base propellants. The surface temperature of the foam zone is about 300°C.

The fizz zone reactions above the foam zone involve continuous decomposition of larger molecules releasing energy. Maximum energy is liberated in the flame zone reactions involving various molecular rearrangements. The temperature of luminous flame zone is in the range 2800 to 3000°C. At high pressures above 1000 psi, the fizz zone disappears and the high temperature flame zone nearly grazes the burning surface resulting in an increase of burning rate. This explains the high pressure dependence of burn rate in homogeneous propellants.

Compounds like lead stearate reduces the burn rate pressure index of nitric ester as well as nitramine propellants. When a lead compound is added to a double base propellant, the burn rate is increased in the lower pressure regions. As the pressure increases, the rate falls to that of a comparable unleaded composition, or even below it leading to plateau region and a rapid fall of burn rate gives a mesa curve. The mechanism of combustion in double base propellants is not changed fundamentally when lead compounds are added to propellants. The basic burning process includes the formation of a limited amount of carbon from condensed phase and this carbon, projecting into the gas phase is capable of catalyzing certain exothermic reactions involving the reduction of NO which otherwise occur only comparatively slowly. Thus, the rate of burning is higher in presence of carbon. Reduction of nitric oxide is highly exothermic. In presence of lead compounds, the carbon is far more active and is more readily oxidized by NO. In the low pressure region, under the influence of lead, the carbon gets oxidized by NO imparting higher burning rate. As the pressure increases, the surface carbon is consumed as fast as it is produced and the burning rate falls since there is little carbon on the surface to catalyze. Today, compilations are available on the type of lead compounds suitable for a given level of burning rate, together with the range over which the plateau will occur.

The overall initial decomposition process of nitramines is depicted as below:

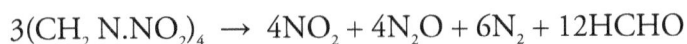

$$3(CH_2 N.NO_2)_4 \rightarrow 4NO_2 + 4N_2O + 6N_2 + 12HCHO$$

Since both RDX and HMX are oxygen deficient similar to the double base matrix, the flame structure of nitramine propellant is similar to that of double base propellants. It is reported that the burn rate of nitramine propellant is independent of the particle size of the nitramines but is dependent on its concentration. Increase in niramine concentration results in enhancement of burn rate. Possibly, on account of its higher heat of explosion, the burn rate of RDX containing nitramine propellant is marginally higher than of that of HMX containing nitramine propellant under identical concentration. Nitramine concentration also marginally enhances the pressure index of burn rate.

9. COMPOSITE MODIFIED DOUBLE BASE PROPELLANT

Composite modified propellants can be considered as a cross breed of double base and composite propellants. The double base matrix functions as binder for the energetic ingredients like ammonium perchlorate and aluminum powder. Various combinations of the major ingredients viz. nitrocellulose, nitroglycerine, ammonium perchlorate and aluminum powder can offer similar energy levels. In general, the matrix together with the inert components has to be of the order of 50 to 60% by mass so as to maintain the required level of mechanical properties. The burning rate of these class of propellants range between 12 to 20 mm/sec at standard conditions and the pressure index falls in the range 0.5 to 0.7. Decomposition of nitric ester is initiated in the temperature region of 160 to 170°C where as the decomposition of ammonium perchlorate takes place only at an elevated temperature of above 240°C.

The enhanced burning rate is considered to be due to the secondary combustion of partially oxidized deflagration products of the double base matrix imparting higher level of heat transfer to the burning surface. A maximum theoretical specific impulse of 267 seconds at 70 ksc has been computed for this class which is at par with the high energy composites.

Processing of CMDB propellant can be done either by interstitial casting process or by slurry casting process. In the interstitial casting process, all the solid ingredients including ammonium perchlorate and aluminum powder and a portion of the nitroglycerine required are incorporated in the formulation at the dough mixing stage. The casting solvent consists of nitroglycerine and an inert plasticizer like dibutyl phthalate or triacetin, and stabilizers for nitroglycerine. A polymerizing monomer like methyl methacrylate is at times used to reduce the sensitivity of the casting solvent and also to improve the mechanical properties of the cured propellant. The casting powder to casting solvent ratio, which again depends on the mutual diffusion of the casting solvent and casting powder during the casting and curing stages, determines the final composition of the formulation; and hence, trial and error experiments are essential in evolving the final formulation.

In the slurry casting process, granular nitrocellulose is made use of in place of fibrous NC. The slurry consists of all the ingredients in the required ratio. The slurry is cast in the moulds under vacuum. After completion of the curing cycle, the mould is dismantled and the propellant is machined to the specified dimensions. The particle size and chemical specifications of the propellant ingredients have a stringent role to play in determining the burn rate and mechanical properties of the propellant. In general, ammonium perchlorate of 60 microns average particle size and aluminum powder of 20 microns average particle size is recommended for both processes.

The specific advantage of CMDB is a combination high energy level par with high energy composites, better mechanical properties—hardly achieved by the composites and higher density level not achievable by the straight double base species. The higher energy level is achieved on account of 34% surplus oxygen available from the decomposition of ammonium perchlorate according to the equation:

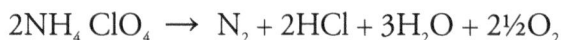

$$2NH_4 ClO_4 \rightarrow N_2 + 2HCl + 3H_2O + 2\tfrac{1}{2}O_2$$

This surplus oxygen is utilized in the combustion of the high energy metallic fuel and the secondary combustion of the partially under oxidized deflagration products of the double base matrix.

The CMDB variety was very specifically chosen in American space vehicles like Scout and Vanguard, ICBM Minuteman under water/sea launched missile Polaris and interceptor missile Sprint in some of their stage motors. Indian Akash missile made use of CMDB variety processed by slurry cast process in its booster stage.

10. CONCLUSION

Homogeneous and as well as composite modified double base propellants still continue to render the modern solid propellant profile versatile and attractive, offering a wide range of ballistic and mechanical properties to suit different mission requirements.

Typical Questions

1. Explain the following:
   i) Single base propellants
   ii) Double base propellants
   iii) Triple base propellants
2. What are the features of Composite modified double base propellants?
3. What are Nitramine compounds? Explain the explosive characteristics of 2 nitramine Compounds?
4. Explain the methods of handing Nitrocellulose?
5. What is dynamite? How is Nitroglycerine extracted? What are the safety precautions to be adopted?
6. Explain the stabilizing reactions of diphenyl amine on double base propellants?
7. What are: i) plateaunising agents, and ii) Flash suppressors in double base propellants?
8. Explain the extrusion process for single base propellants?
9. Explain the solvent less process for extrusion of double base propellants?
10. Explain the typical compositions of single base and double base propellants?
11. What is interstitial casting? How is it used for casting of CMDB and Nitramine propellants?
12. How do HMX and RDX influence the performance of double base propellants?
13. What is globular nitrocellulose? How is it prepared?
14. Explain the P-T curve of a nitramine propellant motor?
15. Explain the combustion mechanism of homogeneous propellants? How does it differ from the combustion mechanism of composite propellants?
16. What are the safety precautions to be adopted during the processing of double base and nitramine propellants?

REFERENCES

[1] Allen, Davenas, Book on Solid rocket propulsion technology edition 1993, pp. 1–10, 35–61, 215–225.
[2] Badgujar, D.M. and Mahulikar, P.P., Advances in science and technology of modern energetic materials: An overview, *J. Haz. Mat.*, 151, 2008, 289.

[3] Beakstead, M.W., "Solid propellant combustion mechanism and flame structure," *Pure & Applied Chemistry*, 65 (1993), 297–300.

[4] Benreuven, M., Caveny, L.H., Vichnevetsky, R.J. and Summerfield, M., "Flame zone and sub-surface reaction model for deflagrating RDX," Symposium (International) on Combustion, [Proceedings], 16 (1977),1223–33.

[5] Cohen-Nir, E., "Combustion characteristucs of advanced nitramine based propellants," Symposium (International) on Combustion [Proceedings], 18 (1981), 195–206.

[6] Diwaar, R.P., Pandey, R.K. *et al.*, Studies on Preparation of Ultrafine RDX by Economic and Effective Method, *Proc. 9ᵗʰ International High Energy Materials Conference*, Feb. 13–14, Trivandrum, 2014.

[7] Fischer, N., Klapötke, T.M., Matecic Musanic, S., Stierstorfer, J. and Suceska, M. TKX-50, New Trends in Research of Energetic Materials, Part II, Czech Republic, 2013, 574–585.

[8] Gallagher, P.M., Kmkonis, V.J. and Coffey, M.P., Gas anti-solvent recrystallization and application for the separation and subsequent processing of RDX and HMX, US Patent 5389263, Phasex Corporation, Massachusetts, USA (1995).

[9] George, P. Sutton, "Book on Rocket Propulsion Element," Vol. 7, edition 2001, pp. 27–36, 46–84, 417–453, 474–511.

[10] Gould, R.F., Propellants Manufacture, Hazards, and Testing, Advances in Chemistry, No. 88, American Chemical Society, Washington, DC, 1969.

[11] Kurian, A.J., "Homogeneous Solid Propellants" in Propellants and explosives technology, edited by Krishnan, S., Chakravarthy, S.R. and Athithan, S.K., ISBN 81-7023-884-6, Allied Publishers Limited, India, 1998.

[12] Kuwahara, Takua, "Combustion mechanism of nitramine/AP composite propellants," Kogyo Kayaku, 47 (1986), 137–43.

[13] Lengelle, G., Duterque, J. and Trubert, J.F., Combustion of solid propellants, ONERA, May 2002.

[14] Manfred, A. Bohn and Mueller, Dietmar, "Insensitivity aspects of NC bonded and DNDA plasticizer containing gun propellants." Fraunhofer Institute for Chemical Technology (ICT), Pfinztal, Germany 37ᵗʰ ICT on Energetic Materials, 27–30ᵗʰ June 2006.

[15] Muthiah, R.M., Varghese, T.L., Rao, S.S., Ninan, K.N. and Krishnamurthy, V.N., Realization of an eco-friendly solid propellant based on HTPB-HMX-AP system for launch vehicle application, *Propellants, Explosives, Pyrotechnics*, Vol. 23, 1998, pp. 90–93.

[16] Miya, Hiroshi and Tanaka, Shinichiro, "Nitramine propellant," *Jpn Kokai Tokkyo Koho* (2006), JP2006151791 A 20060615.

[17] Muellar, Dietmar, "Low Temperature Coefficient (LTC) Gun Propellants," Fraunhofer Institute of Chemical Technology (ICT), D-76327, Pfinztal, Germany, 29ᵗʰ ICT, Westminster, Colorodo, July 2002, Hosted by International Pyrotechnics Seminar.

[18] Nakastsuka, Kanji; Sumikawa, Kenji; Suzuki, Naohisa and Fukuda, Takaaki., "Nitramine system composite propellants with low pressure index," Jpn. Kokai Tokkyo Koho (1985), JP 60235787 A 19861122.

[19] Pillai, A.G.S., Sanghavi, R.R., Dayanandan, C.R., Joshi, M.M., Velapure, S.P. and Singh, Amarjit, "Studies on Effect of RDX Particle Size on LOVA Gun Propellant Formulations," *Propellants, Explosives, Pyrotechnics*, Vol. 26, Dec. 2001, 226–228.

[20] Pillai, A.G.S., Sanghavi, R.R., Khire, V.H., Bombe, P.D. and Karir, J.S., "Processm Technology development for LOVA gun propellant," *Indian Journal of Chemical Technology*, Vol. 7, May 2000, 100–104.

[21] Rekha, Sangtyania; Manoj, Gupta, *et al.*, Effect of Particle Size of Nitramines on Physical, Mechanical and Ballistics Properties of Composite Propellant, *Proc. 9th International High Energy Materials Conference,* Feb. 13–14, Trivandrum, 2014.

[22] Simpson, L. Richard; Foltz, M. Frances, "LLNL Small-Scale Drop-Hammer Impact Sensitivity Test," Report Jan 1995, Lawrence Livermore National Laboratory, Livermore, CA.

[23] Stojanovic, Radislav and Filipovic, Milos., "Combustion of non-catalyzed cyclotrimethy-lenetrinitramine-coposite modified double-base propellants (Part II)," *Scientific Technical Review,* 53 (2003), 3–8.

[24] Urbanski, T., Chemistry and Technology of Explosives, Vols. 2 and 3, Pergamon Press, New York, 1985.

[25] Walter, Langlotz and Dietmar, Mueller, US Patent No. US 2001/000 3295 A1 dt. 14th June 2001.

[26] Wang, Belhai, "Effect of additive HMX upon burning rate behaviour of AP/HTPB composite propellants," *International Annual conference of ICT 19th (Combust, Detonation Phenom),* (1988) 57/1–57/10.

[27] Washburn, E.B., Beckstead, M.W., Hencker, W.C., Howe, J. and Waroquet, C., "Modeling condensed phase kinetics and physical properties in nitramines: Effect on burning rate temperature sensitivity," *JANNAF 37th Combustion Subcommiittee Meetings,* 1 (2000), 309–318.

[28] Yano, Y. and Kubota, N., "Combustion of HMX-CMDB propellants (II)," *Ropellants, Explosives and Pyrotechnics,* 11 (1986), 1–5.

[29] Zhao, Baochang and Zhao, Zhijian, "High pressure combustion characterisation of RDX based propellants," Symposium (International) on combustion [Proceedings], 22 (1989), 1835–42.

[30] Zimmer-Galler, Roswitha, "Correlations between deflagration characteristics and surface properties of nitramine-based propellants," *AIAA Journal,* 6 (1968), 2107–10.

The Evolution of Solid Propellants in ISRO

1. INTRODUCTION

The successful launching of the two stage *Nike-Apache* rocket on 21st November 1963 around 6.30 pm from the shores of Thumba, a sleepy and fishing village near Trivandrum, excited the scientists in India to develop rockets for space exploration. The *Nike-Apache* rocket used solid propellants in its propulsive units—booster and sustainer. These solid propellants belong to the class of double base propellants and are generally used in sounding rockets and missiles. Composite propellants and composite modified double base propellants are the other classes of solid propellants.

Solid propellants are energetic materials and constitute the major and integral part of solid rocket motors. A solid propellant is a complex and stable mixture of an oxidizer and a fuel, which react chemically on ignition to liberate large amounts of energy in the form of heat. If the oxidizer and the fuel are in the same molecule, as they are in double base propellants, used in *Nike-Apache* rocket, it is classified as homogeneous propellants. If tiny, discrete particles of oxidizer is uniformly dispersed throughout the fuel matrix, then classified as composite propellants. Rockets based on solid propellants are by far simplest to operate which burns with a predetermined mass burning rate till all the propellant is consumed. Solid propellant rockets, therefore, have no on-off capability and the control of trajectory during the flight has to be achieved by means of auxiliary rockets or flexible nozzles.

A solid rocket motor assembly basically consists of a motor case, which contains the propellant grain and also acts as a high pressure combustion chamber, a convergent-divergent nozzle for expanding the hot gasses generated inside the motor, insulation for protecting the case and nozzle from high operating temperature of the gases in the chamber and an igniter for starting the burning of the solid propellant grain. The criterion for the choice of a propellant for a particular mission may widely vary depending on the energy content and the physicochemical and operating factors. The energy producing capacity of a particular propellant system is commonly referred to as specific impulse (I_{sp}), which is defined as impulse (thrust × time) per unit weight of the propellant. In all advanced countries, space application was considered a spin-off of their defense efforts. Being strategic in nature, propellant technology has to be developed in-house. Accordingly, chemically cross-linked composite propellants, for use in sounding rockets and launch vehicles programme, had its origin in Propellant Engineering Division (PED) in the end of 1967.

The dedication of Thumba Equatorial Rocket Launching Station (TERLS) to the United Nations on 2nd February 1968 by the then Indian Prime Minister Mrs. Indira Gandhi further catalyzed the enthusiasm of Indian scientists. Indian sounding rocket programme, thus, called for development of solid propellants at Thumba, to have Indian rockets launched from this site for exploring space. The Propellant and Igniter Engineering Division (PIED) and Rocket Propellant Plant (RPP) started the development of solid propellants for sounding rocket programmes.

2. PROPELLANTS FOR SOUNDING ROCKETS

The RPP was set up by the end of 1968 under license from Sud Aviation, France to manufacture the *Centaur*—a two stage rocket in India. Both stages of *Centaur rocket* used solid propellants based on Poly Vinyl Chloride (PVC). This propellant, called *plastolite propellant*, used mainly Ammonium Perchlorate (AP), PVC, Dioctyl Phthalate (DOP) and Lithium Fluoride (LiF). Initially, imported raw materials were used to make non-aluminized propellant grains, which had a solid loading of 74%. The hardware was fabricated from imported 15CDV6 sheets at the central workshop of Department of Atomic Energy. The first launch of the *Centaur rocket* made at RPP was successfully flown on 26th February 1969 with indigenous scientific payload.

Concurrently, RPP carried out the indigenization of raw materials for *Centaur* propellant. For AP, WIMCO Mumbai was persuaded, as they were manufacturing regularly chlorates for safety matches. The AP production unit was set up at Ambernath, near Mumbai, by WIMCO, based on Swiss know-how, using platinum anodes and steel cathodes. The plant had a capacity of 50 tonnes per annum. The PVC made by Chemplast, Chennai was selected after extensive trials. The DOP plasticizer used in the formulation was made by Indo Nippon, Mumbai which met all the French specifications. The burn rate modifier LiF was required in small quantities for the propellant manufacture. Mafatlal Chemicals came forward to manufacture this chemical for RPP. All the chemicals from indigenous sources were qualified by making propellants and static firing the grains of 200 mm diameter and 1000 mm long, known as *mimosa grains* (currently known as *Agni grains*) in proof hardware. The first static test of propellant grain, from indigenous raw materials, was carried out on 2nd March 1969.

The restrictor tube used for centaur propellants is also based on PVC containing Titanium dioxide filler and lead salts as stabilizers to withstand the high temperature involved in the curing/gelling process of the propellant. This was indigenized by RPP using the facility available at M/S Bingam Plastics, Mumbai and with the in-house raw materials. These sheets were then welded into tubes of required diameter at RPP for propellant casting. The propellant grains for Centaur first and second stages, known as Venus and Vega respectively, were then made with indigenous raw materials and static tested to ensure that they meet the specifications (total impulse, density, burning rate, mechanical properties, etc.) of imported grains, laid down by the French. The first flight using indigenous raw materials based propellant was successfully flown on 7th December 1969 from Thumba. Though, the flight was successful, it had higher range due to higher burning time of the sustainer Vega grain, probably due to the indigenous LiF.

The propellant grains were charged into steel chamber by chilling the grains below 2°C for 24 hours and pushing them in. During the cooling time, the grain contracts and goes into the chamber. In case of some difficulty, even after cooling, the restrictor tube was shaved partially to push fit the grain. The annular space between steel chamber and PVC restrictor tube containing propellant was filled with unsaturated polyester resin cured with peroxide curative and cobalt naphthenate. The resin on gelling holds the grain in position inside the chamber. This type of casting propellant grain separately and then inserting the grain into the metallic chamber is known as *free standing grains* and are suitable only for small diameter sounding rockets and missiles. The other advantage of this technique is that it is independent of the availability of metallic chambers for casting. Also, if the propellant grain is found to be defective, it can be rejected and thereby saving the hardware. In missiles, the aged grains can be removed easily and save the hardware. The other charging chemicals and polymers used in the Centaur rocket were also indigenized by RPP and PED. In 1970, the PVC technology was also used to develop other sounding rockets, like *RH-100*, *RH-125*, *MONEX* and *RH-560*.

The PVC propellant made for Centaur rocket was upgraded for energetics by incorporation of aluminum powder (20 µ) from Metal Powder Company, Madurai. The aluminized PVC propellant formulation containing 12 to 15% aluminum, called as *RPP-II*, was used in a two stage *RH-560 rocket* (RH stands for Rohini and 560 the propellant diameter in millimeters). The rocket is very similar to the French Dragon sounding rockets. *RPP-II* propellant had a solid loading of 78.8%. This rocket weighing more than 1000 kg could reach 350 km altitude with 150 kg of payload. The first successful flight of *RH-560* (with a 560 mm diameter booster and 300 mm diameter sustainer) was carried out from Sriharikota/SHAR range on 27th January 1973.

The imported centaur technology and its indigenization was a major milestone in the development of sounding rockets. This process helped to understand the centaur propellant system thereby establishing the future growth of rocketry in India. The indigenisation programme has helped ISRO scientists to understand better composite propellant technology though the PVC technology is not suitable for building launch vehicles since this propellant is not amenable for *case bonding technology*. Other short comings of this technology include the following: (i) high temperature (175°C) required for curing this propellant, which may cause safety hazard with large case bonded propellants, and (ii) The energy available from this propellant, even after aluminum powder addition is low. Hence, the PVC propellants if used in booster motors may not lift the launch vehicle or may call for huge sizes that are difficult to handle.

The time between *Nike Apache* launch in 1963 and the RPP commissioning in the beginning of 1969 saw a number of sounding rockets development under the banner of Rohini series, using double base propellants available from ordinance factories. The first project was *RH-75* rocket. The main objective of the project was to design a 75 mm diameter solid propellant rocket so as to enable engineers and technicians to get experience in the design of rockets. The experience demanded includes development of various methods to predict the performance characteristics—internal ballistics of the motor, trajectory, stability, drag, etc. The rocket is also used for flight testing of various electronic instruments such as transmitter, accelerometer, etc. In the first phase of the project, *SUK cordite propellant* blocks from Aravankadu Cordite factory was used after conducting many static firings, both in proof and flight hardware, at Thumba for their suitability and acceptability. These propellant grains were made by extrusion technique at Aravankadu Cordite factory. The rocket was flight tested on 20th November 1967, four years after the first flight (*Nike Apache* rocket) from Thumba. This is the first indigenous rocket in true sense. However, the propellant was not made in Trivandrum. The cordite propellant used had a length of 1100 mm and a diameter of 75 mm and weighed 5.02 kg. The total rocket weighed 10.8 kg including nosecone and fins and reached an altitude of 8 to 9 km.

A shorter version of this *RH-75*, known as test rocket, was developed as an operational rocket for training personnel in open flight tracking and also to prove the indigenous propellant in flight conditions. This rocket had a propellant length of 600 mm with a propellant weight of 2.4 kg and overall weight of 6.9 kg. This test rocket also used the SUK cordite propellant from Cordite factory. After successful static firings, the rocket was flight tested on 21st January 1970. Four rockets were flight tested on this day followed by another six rockets on 29th January 1970. In one of the flights, the igniter fell off to the ground and the flight was aborted. The motor was afterwards static tested successfully. In another flight, the rocket chuffed on the launcher and after an ignition delay of 15 seconds, the rocket took off normally and the performance was normal. All flights had 1.1 kg dummy payload and reached an altitude of around 6 km when launched at an angle of 65°.

The development phase of *RH-75* TR project was completed in 1969–70, using cordite propellant. Simultaneously, *RH-100* and *RH-125* projects were approved and rockets were flown with the same SUK cordite propellant blocks. The use of Cordite or double base propellant was only the first step in these programmes. Cordite blocks from Aravankadu factory had only circular central port while the projects in Trivandrum wanted star shaped central port. The Aravankadu factory was persuaded to make star shaped grains and the Thumba projects supplied the die for their extrusion. The other draw backs of extruded double base propellants are their limitation in larger diameters and length. For required length, these blocks had to be bonded even for 100 and 125 mm diameter blocks. These double base propellants could not be easily case bonded, though it was one of the prime requirements for launch vehicles. This class of propellants have low elongation and hence, not amenable for case bonding technique. Above all, these propellants have low density and energy i.e., their specific impulse is of low order of 195 to 200 seconds. If these propellants are used in launch vehicles like SLV-3, the size and weight will be so enormous that the satellite launch vehicle may not lift off. Further the cartridge loading technique adopted for charging reduces the loading density and subsequently its performance.

Solid propellants are broadly categorized into three types—double base, composite and composite modified double base. The double base propellant is a homogenous propellant with nitrocellulose and nitroglycerine as major constituents and is used in sounding rockets and missiles. On the other hand, a composite propellant contains an oxidizer, a metallic fuel, a polymeric binder and some additives like plasticizer, burn rate modifier, bonding agents and process aids. These propellants, known by the polymeric binder used in their formulations, are used in launch vehicles, sounding rockets and missiles. The Propellant Engineering Division (PED) concentrated on the development of composite propellants. The composite modified double base propellant is a blend of composite and double base propellants. It is used in missiles and launch vehicles. The energy of a propellant available from the combustion of the propellants is measured by its specific impulse. Double base propellants have a specific impulse of around 200 to 205 seconds while composites have a specific impulse of 245 seconds under normal conditions. The specific impulse depends on the solid loading (the amount of aluminum and AP in the propellant formulation), carbon-hydrogen ratio in the polymeric binder, the molecular weight of the exhaust combustion gases and combustion temperature.

Unlike RPP which got the French knowhow for making PVC propellants, the PED started from scratch, with commercially available unsaturated polyester resin (Hylam, Hyderabad)—HSR 8521 and HSR 8411 as the polymeric binder. The other ingredients used are AP from Central Electrochemical Research Institute

(CECRI), Karaikudi and Aluminum powder from Metal Powder Company, Madurai. The work started right earnest in August 1967 in an asbestos shed with plywood partitions for various operations of propellant making. Initially, when there was no mixer, the propellant formulation was made in a beaker with a PVC rod as a stirrer. The perchlorate was ground using a porcelain mortar and pestle and sieved using a kitchen sieve. The sieved perchlorate and aluminum powder were added in small quantities to the resin taken in a beaker and mixed. About 70% of AP and aluminum was mixed till dough like consistency was achieved. The curative Methyl Ethyl Ketone Peroxide (MEKP) and cobalt naphthenate catalyst supplied by the firm was added to the dough and mixed for ten minutes to ensure uniform mixing. The dough was transferred to a steel mould and cured in a hot air oven. This study was carried out to fix the curative amount and the catalyst concentration, since "use as per" the firm's recommendation gave an immediate gel on mixing the ingredients and the catalyst. After finalizing the quantities, the mixing was carried out in a two kilogram stainless steel indigenous mixer. The polyester propellant from the mixer was fed into steel chambers of 50 mm diameter and 175 mm long tubes. PVC rods or Teflon coated steel rods were used as mandrel to get a port of 20 mm diameter and was removed in open after curing. The ends of the propellant were covered with a commercial polyester (HSR 8111) mixed with titanium dioxide and curatives. The mix was poured on to the top and bottom ends, one side at a time. These steel chambers were then fitted with head and nozzle flanges with bolts. Graphite throat inserts were used and ignited with a squib and black powder. These motors, called *zero-zero motors*, were static tested in the Veli static test facility operated by propulsion group of Space Science and Technology Centre (SSTC). These propellants were called ABC (a) propellant or Veli-1(a) propellant.

PED's efforts continued to improve the processibility or flow characteristics of the propellant by adding different amounts of plasticizers. One such formulation contained 6% nitroglycerine as plasticizer and 67% solid loading. This formulation was mixed and the propellant was cast in *zero-zero motors* and static tested after curing and end inhibition. After five successful static firings, the propellant was cast in mild steel chambers of *RH-75* test rocket and static tested. After two consecutive successful firing, the propellant was cast in aluminum chambers of the test rocket and static tested, initially uninstrumented. Following this, the propellant in aluminum chambers was static tested with instrumentation for pressure and thrust. Two such successful static tests are the minimum requirement for flight testing the propellant. The propellant, called *'Mrinal'* after Sarabhai's wife Mrinalini, was flight tested on 21st February 1969 around 8.05 am in the morning.

The flight was successful and reached an altitude of 4.5 km. The flight preparations were made by RED group of TERLS under Shri. Abdul Kalam. The test rocket with Mrinal propellant was named as Dynamic Test Vehicle (DTV). On 21st February 1970, the first anniversary of the first composite propellant launching in *RH-75* test rocket was celebrated. All members of PED and RED assembled at Kovalam palace lawns. The members recalled the efforts put by the divisions and decided to celebrate the day as PED day every year and arrange a family get together on that day. On that occasion, a poem was read and an honorary degree "Master Booster" was conferred on Gowariker. The poem is given under annexure. The teams rededicated themselves to work together for future goals of space programme. It was also explored to celebrate the day as National Polymers Day, similar to National Science Day.

The first flight of composite propellant in India was thus carried out in the midst of fierce competition by PED. Muthunayagam's group at Propulsion Division (PSN) was also working on the development of composite propellants based on liquid rubber, a commercial product. A liquid rubber propellant was successfully flight tested the next day, 22nd February 1969. Sarabhai's encouragement of competition paid rich dividend in India by having two composite propellants for use in a short time. Sarabhai did not believe in territories or areas and encouraged competitions initially, as competition brings out creativity and hard work in achieving the goal in a short time. In December 1968, however, the chairman of ISRO assigned the task of propellant development exclusively to PED. It was further clarified by him in August 1970 that the development of technology of casting large size propellant blocks (up to 1000 mm) is part of the recognized activity of PED and that the development of propellant blocks for SLV-3 would be dovetailed with this R&D activity in due course.

As the commercial resins were not consistent in quality (wide tolerances), PED started making the unsaturated polyester resin in the laboratory. The polyester resins are condensation products of dicarboxylic acids with dihydroxy alcohols and are considered suitable as propellant binders. One of the advantages of polyesters is that they can be cured without application of pressure and are known for their strength. PED started with polyester resins based on phthalic anhydride, maleic anhydride and propylene glycol. This unsaturated polyester resin was cross-linked with styrene or diallyl phthalate using peroxide curative and a cobalt complex catalyst. The resin was named as VI-1 resin. This resin system was used to make propellants for the projects like *RH-100*, *RH-125* and *Menaka*. The author recollects one memorable static test of one of the VI-1 polyester resin based propellants in *RH-100* motor at the test stand near the present polymer complex. The motor after ignition got detached from the test stand. The head end also got detached from the chamber and the gases started coming out profusely from the head end. Soon the rocket went up in the air after hitting the roof and flew zigzag through the air across the Veli Lake. The motor fell on the banks of the lake beyond the boat club where a herd of cattle was grazing and a heap of dry grasses was stored. Luckily everything ended up without any other disastrous incident.

Two polyester resins based 125 mm rockets in fiber glass casing were flown for the Menaka project on 6th and 25th June 1969. High cure exotherms and lower mechanical properties were the draw backs of Mrinal type propellants. During the developmental phase, scientists and engineers used to work on long hours to meet the deadline, as most of them were bachelors at that time. Gowariker along with his wife will visit the division at nights with lot of eatables from railway canteen to ensure the scientists and engineers have the food and continue the work. This act of his was morale booster as he will discuss during such visits further plans and request for views. Sometimes, static tests were arranged in the midnight to meet the deadline. In absence of vehicles in the night, the rockets were carried on shoulders to static test stand.

To meet the requirements of the projects, in collaboration with binder development laboratory of PED, the polyester resin was scaled up in the pilot plant section. The binder development group also made other kinds of polyesters by using itaconic anhydride in place of maleic anhydride and called the resin VI-1(M) and a boron containing polyester by using boric acid instead of maleic anhydride in the polyester production. Propellants were also made with these polyester resins and offered to the projects. The resin production was shifted to a building in Karamana area to meet the demands. PED was housed in an asbestos shed at Veli where all the activities of the division from propellant processing, inhibition of propellant grains, motor assembly for static testing and propellant curing were going on apart from engineers' sitting place. As the operations are hazardous, PED got 5 buildings built in the first phase and another 7 buildings in the second phase. In addition, a magazine for storing of the explosive materials was also got ready. These buildings were built on the slopes of Veli hills as per STEC regulations. The city cell, as it is called at that time, met the requirements of resins and chemicals needed for propellant making in view of space shortage at PED complex. The cell had a pilot plant facility with multipurpose reactor of 50 kg capacity and a planetary mixer for blending the resin batches. The new buildings were ready in early 1970.

The first building was occupied by Chemical Analysis Laboratory which was inaugurated by Sarabhai in one of his visits by performing a titration. The laboratory qualitatively and quantitatively analyses raw materials and finished products used by PED for propellant making. The laboratory also played a vital role in the import substitution of chemicals and paints used in Centaur charging. The second building catered to the needs of propellant processing including scaling up of R&D groups' developments and meeting of the project requirements. The third building met the needs of the inhibition requirements of all developmental and project motors in addition to refurbishing the static tested motors. The fourth building housed the propellant physical and mechanical testing laboratory. This laboratory was also inaugurated by Sarabhai by pressing a switch to ignite a propellant strand inside a Crawford bomb, used in the determination of burning rate of propellants as part of quality control test. The fifth building was meant to carry out machining operations on propellants meant for development and projects.

The second phase buildings that came up by middle of 1971 eased the operations of PED and accelerated the work on propellant development and facility buildup. The binder development laboratory continued to be housed in the original shed on the Veli hills. The new buildings mainly catered to propellant processing and curing bays where the ovens heated with either hot air or water meant for the curing operations of propellants were housed. The ovens have to run continuously for days and also at different temperatures for development work. In one building, remotely operated mixers were erected. To overcome the extrusion process in double base propellants and to cast bigger diameter double base propellants, nitroglycerine extraction from dynamite and interstitial casting technique for making double base propellants were established separately in the next two process buildings. The development of double base propellants is essential in view of large requirements of double base propellants by defense sector and to complete the propellant activities under one umbrella.

PED thus, established the minimum facilities required for carrying out the major task of formulating new and energetic propellants. The propellant development pattern followed was the same as United States. Encouraged by the success of the first polyester resin based propellant in flight, a two pronged attack in the development was pursued. One group worked on polysulphide resins based on imported Thiokol's LP 33 polysulphide resin. Polysulphide was the first polymeric binder used by the Jet Propulsion laboratory, Pasadena in 1946. Two formulations were made and small motors were static tested. These propellants, designated as Veli-3(a) and Veli-23, had no metallic powder in its formulation in view of water elimination during the curing of the resin with either lead dioxide or para-quinone dioxime. Though the propellant had excellent mechanical properties and case bonding capability, it had a low specific impulse. Polysulphides are also used as curing agents for end inhibition of motors with epoxy resins. It is reported that the polysulphide based propellants were used in key operations like moon missions, to ensure the firing of escape motors from the moon atmosphere. During one *zero-zero* motor test firing, the mild steel chamber was converted into a sheet of steel at the stroke of zero with a big bang. So powerful was the propellant based on polysulphide! Gowariker used to ask after every static firing, whether the propellant was powerful meaning the noise level was high since he felt sound level was an indication of propellant energy.

The other group focused on the development of polyurethane propellants. Polyurethane can be made either with hydroxyl terminated polyester or polyether. On reacting this diol with isocyanate gives an isocyanate terminated prepolymer which can be further cross-linked with a triol or a mixture of diol and a triol. Initially, PED chose the easily and indigenously available hydroxyl terminated polyether, polypropylene glycol 2000 (PPG 2000). Since large quantities of this polymer are used in foam industries, this chemical was easily available. Also, the hydroxyl groups are less reactive as these are secondary in nature. The polymer also has low viscosity and hence, can take higher solid loading than polyester or PVC and therefore higher specific impulse. The curing takes place around 60 to 75°C. These prepolymers are named as pedathane resins by PED. Pedathane-22 was used for making propellants with 83% solid loading. These propellants, christened as Veli-22, were characterized by static testing 2 kg motors. In addition, another formulation, called pedathane *P-10* was also developed with good processibility. This propellant was offered for *RH-300* project that was created in early 70s to replace the centaur rocket. The single stage rocket with high energy propellants replaced the two stage PVC based centaur rocket, under the project direction of Mathur. The first static testing of PP-10 propellant in *RH-300* motor was fired in presence of Sarabhai in June 1970. After the test, Sarabhai wanted the propellant to be flown in the *RH-300* rocket by the year end.

The isocyanate terminated polyurethane prepolymer is sensitive to moisture and hence, had to be made and used within two days and also had to be stored under dry nitrogen blanket. The resin used to gel on the surface by reaction with moisture in the nitrogen or ingression of moisture during storage. Removal of scum from the surface was thought as a solution but the resin properties were different and propellant properties achieved were not acceptable. To avoid this, one has to find out the availability of *bone dry nitrogen* from Indian Oxygen ltd. Chennai. After repeated enquiries, we could get *bone dry nitrogen* in cylinders. By the use of this nitrogen, the resin can be stored for 3 to 4 days. This made us work round the clock in pilot plant cell in the city to make the required amount of resins for the propellant casting of the project motors.

The resin, though gave good mechanical properties and specific impulse, the propellant had to be made under dry conditions or Relative Humidity (RH) less than 50%. This was really a problem when the mixing was done either in Veli or anywhere in the coastal areas. To overcome this problem, we had to resort to air-conditioning of the propellant mixing and casting areas in PED. It is known that propellants based on AP are susceptible to moisture embrittlement, but the effect is more pronounced in case of the pedathane resins based propellants. This is because of the prepolymers containing propylene oxide moiety are capable of dissolving large quantities of AP in presence of water. This leads to lowering of propellant mechanical properties, especially the percentage elongation or strain capability. In other words, the propellant becomes stiff and loses its elongation. This made us store the propellant in air-conditioned magazines or keep the propellant port closed with silica bags to prevent exposure to moisture and to avoid further deterioration in propellant properties. Drying the propellant restored the mechanical properties to some extent in small propellant samples but, not possible with propellant grains or case bonded motors. This moisture embrittlement is not of concern in other polyurethane propellants based on ISRO Polyol or Hydroxyl Terminated Polybutadienes (HTPB) as they are less polar and hence, absorb less moisture.

Bonding agents in small quantities were tried to improve the mechanical properties of Pedathane propellants by improving the adhesion of oxidizer particles and the binder matrix in the propellant. These bonding agents, as reported in literature, are polar molecules that bridge the oxidizer and the binder matrix either by chemical links or adsorbed on the oxidizer particles. PED-O-Bond, a reaction product of 12-hydroxy stearic acid and di-ethanolamine, was used in the formulations. We also sometimes found soft-center cure reaction in the *PP-10* propellants. That is, the propellant is hard and well cured on surface but inside was soft. This could be either due to side reactions consuming the curative or insufficient amount of curative or cure time. To improve the processibility, plasticizers were used. Since the prepolymer has a linear aliphatic chain backbone, a linear chain plasticizer was selected. Dioctyl adipate was chosen based on the availability in the market though dioctyl sebacate is preferred.

For cross-linking the pre-polymers, different types of tri-functional triols were available in the market—trimethylol propane, glycerol, triethanolamine and castor oil. Initially, castor oil was used after drying to ensure low moisture content. Castor oil is compatible with Pedathane resins as both contain linear chains. However, being a vegetable product, castor oil had some variations from batch to batch and this was overcome by procuring the material in bulk.

During 1970s, when PED was toying with polyurethane propellants based on PPG 2000, a book entitled "Propellant Manufacture, Hazards, and Testing" was published based on a symposium on propellants for the 153rd American Chemical Society meeting in April 1967. This book was read, reread and discussed thoroughly by the scientists and engineers of the division. This helped the scientists to chalk out further course of action. The team decided to start work on polybutadienes especially with Carboxyl Terminated Polybutadiene (CTPB). These almost bifunctional prepolymers give substantially improved mechanical properties at high solid loadings, especially at low temperatures.

As per the available literature, CTPB prepolymer is synthesized by two methods, either by a free radical or lithium initiated anionic technique. It is available in US from four companies. The free radical based prepolymer is available from B.F. Goodrich Company and Thiokol Chemical Company. As a broad molecular weight distribution is needed in the polymer, free radical based prepolymer technique was chosen for the synthesis of CTPB. Also, the fact that, free radical method is simpler compared to lithium based anionic technique confirmed the chosen method for the synthesis. The requirements of the chemicals for this prepolymer are butadiene gas, free radical initiator and a solvent. The butadiene gas is available either from Synthetic and Chemicals, Bareilly, Uttar Pradesh or from NOCIL, Mumbai. While the source of the former is ethyl alcohol, the source of the latter is petroleum refinery gases. Butadiene gas is transported either in gas cylinders or in tankers as is being done in the case of cooking gas. The initiator used is either glutaric acid peroxide or 4,4'-azo-bis-4-cyanopentanoic acid. As these are not available in India, the more easily available 4,4'-azo-bis-4-cyanopentanoic acid was imported and used in the polymerization reactions. The solvent used is toluene which was distilled and dried over sodium.

To get the gas, the first requirement is fabrication of gas cylinders and the valves. The fabricated gas cylinders had to be certified by explosive directorate at Nagpur after proof pressure testing. Both fabrication and valves were made after great exploration and follow up. Getting certification for gas filling from Nagpur directorate involved considerable correspondence and a couple of visits as this was the first time that such a requirement has been projected. After long correspondence and justification for the objections, 20 cylinders were cleared along with cylinder valves. These gas cylinders, fabricated at Mumbai, were sent to Bareilly to fill the butadiene gas and this was brought to Trivandrum in the southern corner of India. All these hurdles took two more years. Meanwhile, the group procured 200 kg of HC-434 resin from Thiokol, USA and initiated the propellant studies. The CTPB resin is cured with epoxies or aziridines. The epoxide (GY 252) was procured from CIBA Geigy, Mumbai and the tris (2-Methyl Aziridinyl 1-Phosphine Oxide) (MAPO) from Arsynco Chemicals, USA. 92% pure MAPO was available and is required to give cross-linking to the binder matrix because of its trifunctionality. MAPO was preferred over trifunctional epoxies because of its storage problem and reactivity.

The group started working on propellant formulations based on imported CTPB. As per literature, CTPB polymer can take solid loading of about 90%. Hence, to take advantage of the prepolymer, a solid loading of 88% was tried. To achieve such high solid loading, a bimodal distribution of AP was tried. The bimodal distribution contained coarse AP of average particle size 300 microns and fine AP, made by grinding the coarse AP, of average particle size 45 microns. The total AP content in the formulation was 68%. In addition, the formulation also contained 20% aluminum powder. The propellant was mixed in the 2 kg mixer and the curative MAPO was added as suggested in the literature based on the carboxyl value. The propellant slurry was poured in Teflon coated steel moulds and cured at 60°C, as recommended in the Thiokol's CTPB product bulletin. These experiments were conducted to find the optimum curative level and the processibility of the formulation. The basic formulation was further improved by adding plasticizer like Dioctyl Adipate (DOA), upto 3%, in the formulation.

Dr. Sarabhai was a dreamer and recognized the great importance of advanced technologies in space and electronics. The most important practical applications of space research are those related to meteorology, communications and geodesy. This advanced technology is essential for a country like India which is vast and depends on rain for agriculture. This requires weather prediction including cyclone and storm warnings. Sarabhai was dreaming of satellites going round the earth giving this information, since the sounding rocket programme started in India with the successful launching of *RH-75* rocket from Thumba in November 1967. He was also confident that, if India had to play a meaningful role nationally and among the group of countries in the world, we must be second to none in the use of advanced technologies to solve the real problems of man and society, which we find plenty in India. He emphasized the need of our own launch vehicles and satellites from the time we succeeded in developing sounding rockets. This dream to come true, one has to translate it to reality. Gowariker helped in realizing the dream by planning the setting up of space booster plant in early seventies, when the propellant group was exploding *zero-zero* motors on the test stand during the development phase. While R&D chemists and scientists were engaged in formulating high energy propellants, a small group of engineers and scientists prepared a project report for the setting up of a solid propellant manufacturing unit at Sriharikota, about 90 km North of Chennai, at the end of 1969 and submitted to Sarabhai for approval. This SHAR Island, in the east coast, extending up to 25000 acres was selected by Sarabhai for satellite launching station to take advantage of earth's rotation. About 1000 acres of land was identified for the space booster plant considering the safety distances, quantities handled and the future expansion that may need. The plant was envisaged to develop solid booster grains, including allied systems, weighing up to 10 tonnes in a single operation and assembly of segments weighing up to 40 tonnes of propellants. The processing capacity of the plant was around 500 tonnes annually. The capital cost of the plant was projected around ₹ 8 crores with a foreign exchange component of ₹ 80 lakhs. The major reason for concentrating on a solid propellant plant was the simplicity in design and operation of solid motors. The Solid Propellant Space Booster Plant (SPROB) was approved in late 1971 and was expected to go into production by the end of 1975 or early 1976. The entire plant was designed and built with completely indigenous know-how, skills developed at Vikram Sarabhai

Space Centre (VSSC), when India's first launch vehicle was not even on the drawing board when SPROB was planned.

Following the approval of SPROB project report, a small team was formed with Mujumdhar as the leader to plan the details and execute the project so that the plant will be ready for the launch vehicles when needed. The project team studied the literature of plants, where big solid motors such as 260" motor by Douglas that contained 2.4 million pounds of propellant, 156" motor and the 120" motor of Aerojet general were made and their processing facilities available for casting, curing and transporting them. Based on the literature, the plant facilities and processing techniques were planned. The most sophisticated technique adopted is that of segmentation in view of the advantages like ease of handling, transportation, and assembly and the use of relatively smaller equipments for bigger motors. Segmentation provides further flexibility in the grain design and in the choice of internal configuration and perforation. The insulation of joints and seals is, however, a critical factor in segmented motors and unless the insulation is optimized, the motor weight can increase undesirably. Another factor is that a large number of propellants involve stress relieving flaps and more inhibiting. The other advantages of segmented motors are: i) ease of storage, ii) working with limited amount of propellant, iii) casting segments with 10 feet or more in diameter and weighing upto 60 tonnes each, and iv) ability to reject just one part of the block in case of fault rather than rejecting the entire motor as is being done with monolithic motor. The first stage of USAF Titan 3C, which is one of the most standardized motor for space launch vehicles, uses two segmented solid motors of 120" diameter.

The plant layout was based on a three –zone concept, each zone having its own level of hazard and spread-out risk. The proposal also envisaged the deputation of some of the engineers to foreign propellant establishments for study and selection of process equipments required in the proposed plant. A team of engineers would visit some foreign suppliers for initial survey and selection of these equipments. They would also explore the possibilities for training our engineers to work on such machines. Based on the outcome of this mission, a small number of R&D staff would be sent for training in different operations involved in the various stages of large propellant grain development. These trained people will aid in the execution of the programme. This training would also provide essential basis for imbibing safety consciousness commensurate in the staff, who are employed with large scale hazardous operations. As per the project proposal, teams visited countries like Germany, France, UK and USA in the second half of 1971 and the inputs and discussions with various production facilities were incorporated. The civil works and the plant and machinery finalization started by middle of 1972. The project core team which worked from a rented building in city moved to Sriharikota in 1973 to follow up closely the procurement action and to liaise with the construction works.

The plant when ready for trials in second half of 1976, the capacity was reduced to 250 tonnes. Much of the equipments were made within the country, although some critical equipments, such as large mixers of one ton capacity (Guitard mixers, France) and 8 MeV linear accelerator needed for the Non-Destructive Testing (NDT) of finished propellant grains, had to be imported. A remote controlled trimming machine, for example, was designed and fabricated with the help of the Central Machine Tools, Bangalore. A number of trial mixings were done to ensure the smooth functioning of the equipments. ISRO polyol based propellant was used in casting trials in a number of intermediate stages and the ballistic and mechanical properties evaluated before taking up the monolithic motor casting. SPROB procured a vertical mixer and replaced the horizontal mixers because of the advantages like, better safety, remote addition of ingredients, mixer packing glands not immersed in propellant slurry during mixing and the use of mixer bowl itself as casting hopper (which avoids propellant unloading from mixer and unwanted exposure of operators).

A monolithic SLV-3 first stage (one meter diameter and ten meters long) was cast with IPP 10 propellant, based on ISRO polyol. The casting used about fourteen mixings of one ton capacity and the operations took almost 2 days continuously in December 1976. In addition to monolithic motor, two control blocks and cartons were cast from each mix of propellant. The propellant was then cured at 50°C for 20 days in a planned way. The monolithic motor was test fired on 27th March, 1977 in presence of number of VIPs including Chairman ISRO. The motor exploded after one second and the inauguration function of SPROB planned on the beach in the evening on that day was abandoned. A high level committee headed by Muthunayagam was constituted to look into the failure and suggest the reasons for the failure. The committee had many deliberations and submitted its report to Chairman, ISRO. The monolithic motor was not cast afterwards except for the one tested with HTPB propellant in November 1994. Though PSLV uses six strap-on motors of first stage of SLV-3, segmented motors are preferred, as was done in SLV-3 period, in view of better utilization of RPP facilities.

Initially, SPROB was mixing and casting for SLV-3 the third and fourth stage motors though the plant has the capability for processing bigger grains like first and second stage motors. In view of delay in commissioning of the SPROB, RPP was expanded to meet the requirements of SLV-3. Hence, first stage segments and second stage motors were made at RPP and transported to SHAR after end inhibition and NDT, for segment assembly and static testing. After sometime, the first stage motor was cast in SPROB and supplied to SLV and Augmented Satellite Launch Vehicle (ASLV) programme when there was a need for pair casting. The SPROB plant was expanded further, to meet the demands of PSLV. But for SPROB, the flights of PSLV and GSLV would have taken place at much later time.

While PED scientists were busy with developing advanced high energy propellants using imported polymers, the binder development laboratory was busy in synthesizing the resins in the laboratory. The planned availability of butadiene gas and pressure reactors to carry out the polymerization and the constant temperature bath with rocking arrangement with the pressure reaction bottles were available to start the polymerization reaction to get the CTPB resin. The reactions initially gave 50 g of resin with carboxyl value. The resin synthesis was optimized and the process was scaled up first in binder development laboratory. The synthesized resin was cured with CIBA's GY 252 and MAPO. The curing studies were optimized first in gum stock studies and the beaker level propellant mixing.

The scientists were struggling with the synthesis of CTPB resin and city cell was producing the unsaturated polyester resin and pedathane polyurethane resins for the various Rohini sounding rocket programme. During this time, a small group of R&D scientists were preparing a project proposal to manufacture these resins for the sounding rocket programme and to meet the future demands of the propellant production plants and the requirements of proposed Launch vehicle programme (SLV-3). Two proposals, one for Propellant Fuel Complex for manufacturing the resins and another for Binder Development Laboratory or Polymer Complex to carry out research on polymeric binders required for future advanced high energy propellants, were made.

When the potential of a binder is established for possible use in large scale propellants, there is a need for production of these binders in the required quantity. The need of such a plant will be obvious in view of following considerations: a) Commercial binders are meant for different applications like lamination, foams, etc., b) Propellant application calls for good quality control and close tolerances to ensure reproducible properties, c) The commercially available resins are based on polyesters and epoxy types and are exorbitantly costly, d) The imported resins are not only costly but had to be purchased in large quantities and cannot be consumed within the span of its shelf life. In addition, these binders will be required in tons for the satellite launch vehicle along with a number of plasticizers, catalysts and other additives. These have been made in house and are not available in the country. Since various binders have to be tried before a final combination is chosen, the process mechanics have to be changed for scaling up the process to meet the requirements. Based on Chairman's advice, the propellant requirements for different projects were obtained from respective project leaders before the plant size was finalized. The project proposal was made and submitted to the chairman ISRO in January 1970. The proposal contained the production of polyesters, polysuphides, plasticizers, catalysts, phenol formaldehyde, polyurethanes and polybutadienes at a cost of around ₹ 62 lakhs with a production capacity of 200 tonnes per annum, including the allied chemicals in an area of 7 acres of land in Thumba. Sitarama Sastri was the project leader to get the plant in to operation. The Propellant Fuel Complex's (PFC) role is twofold. Any chemical process that works in laboratory scale had to be demonstrated that it works in larger scale before it is taken up for industrial production. PFC will do the needful as there is no other pilot plant available in the country for this purpose. PFC could also meet the requirements of the projects for the resins. Sarabhai appreciated the reasoning and gave approval for the setting up of the PFC. Sarabhai was no more to lay the foundation stone of PFC or see its functioning. The foundation stone was laid in February 1972 by Professor MGK Menon, who held the reins of ISRO after the death of Sarabhai. PFC came into operation in 1974. During erection and commissioning, a reactor fell into the river near Attingal, Trivandrum while transporting it by road from Mumbai, after hitting the height limiting structure. The reactor was removed after some days with the help of local people and boatmen. However, this accident did not delay the commissioning of PFC except the reactor had to be sent back for repairs to the fabricator.

In the end of 1971, side by side with the project proposal of PFC, Chairman ISRO sanctioned a three storied building in Veli hills with self contained facilities for the smooth functioning of the polymer and other allied chemicals developmental activities of PED. The basic resin development group and fine chemicals group were housed along with necessary control laboratories. The polymer complex was furnished as was done in the modular laboratory of Bhabha Atomic Energy Center, Mumbai. The furnishing contract was awarded in a month's time based on the BARC's purchase order. The total area of the complex was over 26,000 square feet and is based on modular lab type construction. Depending on the requirement, there is single-, bi- and tri-modular labs. The scale up facility was housed in the ground floor. The polymer complex started functioning from mid 1974. Viswanathan and Krishnamurthy planned the polymer complex and the facilities in it. The scientists working in polymer complex developed many polymers and chemicals including CTPB, HEF 20, PBAN, HTPB, phenol formaldehyde resins, silicone resins, Poly imide resins, a host of adhesives and sealants, thermal control paints, etc. These resins have been scaled up in PFC. The static test stand meant for small motor static testing, which had seen many successful tests and explosions, was finally dismantled and moved to TERLS area and co-located with the centaur static test facility.

In the end of 1970, the president of French Space Agency, *Centre National d'Etudes Spatiales* (CNES), visited Trivandrum along with Sarabhai and was impressed by the progress made in sounding rockets. He also enquired whether India could develop the upper stage solid motor for their Diamont BC launch vehicle. He also said a French technical team would be visiting the center in January 1971 for further discussions. The team did visit Trivandrum as planned to assess the level of Indian technology. During the meeting, the French team said that it would take three years for India to develop a propellant formulation. To hasten the process,

French team offered technology transfer at a cost of 12 million francs for each formulation. The raw materials from France would have to be imported for three years or till India made alternate arrangements. Sarabhai informed the French delegation that Indian team would develop a propellant formulation suitable for Diamont BC upper stage in the next six months on its own and could discuss further after that. The propellant group developed in the next six months two formulations named Veli 21 and Veli 22 based on imported CTPB resin and pedathane 22 resin respectively. Some twenty four 2 kg motors were made with each of the formulation and static tested on 29th December 1971 and Sarabhai was reviewing the progress in the engineers building conference hall. After the review, Sarabhai was happy and satisfied with the progress, had meetings with others and returned to Kovalam where he had some more visitors. He retired finally around midnight. In the morning of 30th December 1971, when his secretary went to wake him up, there was no response. He was no more. Probably the over joy of the PED team developing two propellant formulations, as he wanted in front of the French team six months ago and fulfilling his dream had done the damage. The rest was history. After two years, the French dropped their Diamont programme and there was no need of any propellant for their programme from India.

3. PROPELLANTS FOR LAUNCH VEHICLES

Prof. Dhawan became chairman of ISRO and the SLV-3 management programme was finalized. The Satellite Launch Vehicle, SLV-3, is a four stage rocket designed to place 40 kg satellite into a near earth orbit at an altitude of 400 km. The major sub-systems constituting SLV-3 are four propulsive stages with inter-stages, heat-shield, control and guidance systems, telemetry, telecommand and tracking systems, ignition system, separation system, destruct system and instrumentation packages. Since SPROB will not be ready to meet the SLV-3 first flight in 1977, RPP was expanded to take care of first and second stage motors. The biggest motor RPP had made at that time was RH-560, weighing around 700 kg. The propellant was aluminized PVC, called RPP-II. RPP started the developmental work for SLV-3 programme with imported PBAN, and PED was entrusted with Stage 3 and 4 propellants. The PBAN resin was cross-linked with epoxide and MAPO, the same system that was used with CTPB. Of the butadiene prepolymers, PBAN is well-known and comparatively cheap. Thus, RPP started the work on case bonded propellants. The developmental work culminated in a propellant with 84% solid loading containing 12% aluminum powder. The propellant, called RCN-01, was scaled up first in S2 size motor before taking up the S1 segments. The scale up first used three fourth (3/4) size S2 before casting full size S2 motor. The S2 motor has 3.15 tonnes of propellant. The S1 motor was a segmented one and had 3 segments each carrying around 3 tonnes of propellant. Though segmented motor had many advantages, the space shuttle Challenger disaster demonstrated the importance of perfect joints between segments. The segmentation technology was first demonstrated in RH-300 size, before going for the S1 motor. The first segmented motor was static tested during 1975–76.

Simultaneously RPP was developing compatible insulation for the propellant. Silica filled nitrile rubber sheets of 1.5 to 2 mm thickness were calendared and hand laid inside the rocket casing, which was sand blasted and solvent cleaned. The unvulcanised rubber sheet was bonded to the cleaned surface with rubber solution and vulcanized in an autoclave. The insulator was called ROCASIN in view of its role as rocket casing protector. The propellant had to be bonded to the ROCASIN sheets laid inside the motor casing. To bond the propellant, polyurethane resin, called liner, was applied as a thin coat either by brushing or spraying before casting the propellant. The big motors are susceptible for cracking or debonding from the case due to cure shrinkage. In big motors to overcome these defects, the loose flap or stress relieving boot was incorporated in addition to adopting stepped cooling. The loose flaps were then filled when the final inhibition of the segment was done. While RPP was working with PBAN for the lower stages, one S2 motor had low mechanical properties, almost half the specified value. A committee went into the structural integrity of the motor and finally cleared the motor for static test and the motor worked satisfactorily.

PED continued the development of CTPB based propellant for the upper stages and these propellants required higher elongation and higher tensile strength in view of harsher environments to which the propellant would be subjected during flight. The propellant was tailored with 86% solid loading with 20% aluminum powder and tested in 40 kg control block levels. The programme had to be abandoned midway due to an embargo in the supply of the prepolymer. Hence, the effort on indigenous prepolymer synthesis was accelerated. The synthetic process optimized in R&D level was passed on to PFC for scale up and productionisation. The problem of finding large suppliers of butadiene gas and transportation in tankers had to be solved before the production at PFC. In view of the delay anticipated for the procurement, PED started with a novel idea of using polybutadiene rubber as the starting material which is easily available in India and that could be transported easily. Simultaneously, PED developed a low density insulator based on Ethylene Propylene Diene Monomer (EPDM) with silica micro-balloons. The EPDM rubber sheets were formulated using Government Rubber works at Chakkai, Trivandrum after ensuring the calendaring mill is free of carbon. The low density insulator initially posed bonding problems to the metallic case. During PSLV time, RPP developed an EPDM based insulator for PS3 motor with cork particles as filler to get the required low density. Both developments did not find use in solid motors in view of inconsistent density of the insulator because of the crushing of the filler in the final sheet.

It was second half of 1973 that R&D engineers in PED started with polybutadiene rubber to get CTPB. This was achieved by oxidative degradation of the rubber. The rubber cut into pieces and dissolved in a solvent like benzene in a glass flask and mixed with calculated quantity of per-benzoic acid under controlled conditions and refluxed. Further, the solution was mixed with calculated quantity of periodic acid in glacial acetic acid and refluxed. At the end of the reaction, the prepolymer in benzene solution was washed with water to remove the acetic acid. The solvent was removed by distillation. The pure polymer is left in the flask. The final product, when analyzed for carboxyl content, showed a low value compared to imported CTPB. The spectral studies confirmed that the material is a lactone, in addition to its having some carboxylic groups. The resin cured like CTPB with MAPO and epoxide. Propellants were made at 85% solid loading and compared with that of CTPB at the same loading. The mechanical properties were much better than CTPB though the unloading viscosity was marginally higher for the new resin. The new resin, lactone terminated polybutadiene, was called High Energy Fuel (HEF-20), the first of its kind in the world. PED has the unique distinction of giving the world a new binder the lactone terminated polybutdiene to the existing list of carboxyl- and hydroxyl terminated polybutadienes. This was further scaled up at PFC in batches of 35 kg of resin and 3 or 4 batches were mixed together and sent as a composite batch to production agencies for propellant processing.

The upper stages of SLV-3 require better mechanical properties (tensile strength more than 10 kg/cm^2 and elongation more than 35%) compared to lower stages (tensile strength not less than 5 kg/cm^2 and elongation ~25%) and PED used HEF-20 for this propellant in view of the non-availability of CTPB and non-suitability of PBAN. The propellant was tailored to ensure that the unloading viscosity was within the castable limits by choosing an 84% solid loading and adjusting the AP coarse to fine particles ratio to 4:1. The propellant, called HEF-20(1) propellant, after standardization at PED, was scaled up at RPP to evaluate the ballistic and mechanical properties. After ensuring that the propellant meets the project requirements, the full size S4 motor was cast. The collapsible Teflon coated mandrel was used for the S4 motor to give the internal port configuration while the internal configuration of S3 was cylindrical which was further machined to the required configuration. After successful proof motor testing, the propellant was cast in glass/Kevlar fiber motors. The S4 motor takes around 280 to 290 kg propellant while the S3 motor takes 1100 kg of propellant. The S4 motor was put through environmental tests like thermal cycling, hot and cold temperature tests, vacuum test, and vibration and spin tests. These tests were required as S4 motor was chosen as the apogee motor to put India's first communication satellite using Ariane launcher under development at European Space Agency. This satellite, known as Ariane Passenger Payload Experiment (APPLE), was successfully launched on 21st June, 1981. The S4 motor of SLV-3 with HEF-20(1) propellant also successfully put India's first satellite, ROHINI on 18th July 1980. The propellant also was successfully used in the 3rd and 4th stages of Augmented Satellite Launch Vehicle (ASLV) to put the SCROSS satellite in a circular orbit. The details of the work on HEF-20 pre-polymer and HEF-20(1) propellant were presented at the AIAA at Propulsion conference at California in September 1975.

The HEF--20 prepolymer was costly and consumed more energy compared to CTPB because butadiene gas has to be polymerized to get a rubber of around two lakhs molecular weight and then oxidatively degraded to get the HEF-20 resin of molecular weight of around 5000. Hence, the energy spent was almost double that needed to make CTPB. The method was adopted in view of non-availability of CTPB by imports and transportation difficulties associated with the butadiene gas from Bareilly in the North to Trivandrum in the Southern tip of India. However, the prepolymer gave propellants with excellent mechanical and ballistic properties to put India in the space club. The other difficulty with the HEF-20(1) propellant is its aging characteristics on storage, which was found to be not more than one year. Unlike these propellants, CTPB based propellants have been reported to last for more than 5 years. This made us to look for alternate binders.

Two propellants, one based on ISRO polyol and another based on indigenous CTPB were developed to substitute a cartridge loaded Russian sustainer grain weighing around 150 kg. The propellant had a tensile strength of 16 to 18 kg/cm², an elongation of around 10% and very high modulus. Indigenous CTPB was chosen for the replacement of the sustainer propellant. The propellant developed in PED was scaled up in RPP and four grains were made from each mix of 850 kg. Initially the grain was cast as case bonded grain with ROCASIN insulation instead of the original tape wound one. The grain after decoring was found cracked and was attributed to the low elongation and high modulus. This made PED to switch over to free standing mode casting and then doing the tape insulation. The grains were cast as free standing grains with circular port. After decoring, the grain was machined to outer dimensions. A tape soaked with epoxy resin was wound over the machined grain and cured. After the curing is complete, it was machined to meet the outer dimensions and the slots required as per the requirement. To ensure propellant performance, the grain was tested both at elevated and cold temperatures before conducting flight trials. The flight was conducted successfully during September 1982 and was appreciated by one and all including chairman ISRO. The flight motor was kept outside in open for 5 years and flight tested successfully, proving the ageing characteristics of CTPB propellants.

Till now the propellant binders required for solid propellants are based on petro-chemicals. For example, poly propylene glycol, CTPB, PBAN or HEF-20 are all based on petro-chemicals. The polyurethane propellant used in RH-300 is based on PPG which is again petro-chemical based. The R&D scientists of PED, mindful of the oil and energy crisis and related problems, have been concerned for some time about substituting PPG from non-petroleum source within the country. They studied the chemical structures of organic chemicals that were chemically akin to PPG and available in the country. During the search, they stumbled on a common non edible oil, Castor oil. This castor oil was converted into a polyol after hydrogenation and modification of the resulting product. This could also be prepared from twelve hydroxyl stearic acid. The polyol could be used in polyurethane propellants in place of PPG.

At the end of 1973, PED developed a new propellant using this polyol, called ISRO-polyol, with 83% solid loading. The propellant, called IPP-10 was cured with toluene di-isocyanate. The propellant had no plasticizer and had better mechanical properties compared to imported PBAN and the PPG based propellants. This propellant was used in the single stage rockets RH-300 and RH-300 MkII developed for replacing the Centaur rocket. The RH-300 with IPP-10 propellant was flown successfully in January 1983. The RH-300 MkII had a dual configuration. The grain had a tubular configuration at the head end followed by a six star configuration for the rest of the length. The first successful flight of this rocket took place in 8th June 1987. The IPP-10 propellant was also used in the first monolithic version SLV-3 first stage motor by SPROB in 1976 and static tested on March 27, 1977, the day on which SPROB was commissioned and the test was a disaster.

Another formulation IIP-20 was developed as a candidate propellant for replacement of Pechora missile sustainer propellant. The propellant had 84% solid loading with 2% plasticizer and gave high tensile strength and high elongation (>25 kg/cm^2 and 25% elongation). Yet another propellant formulation IPP-40, containing 86% solid loading with 1.5% plasticizer, was developed as a candidate propellant for PSLV booster along with HTPB and indigenous PBAN resin based propellants. The propellant was cast in one meter diameter motor and static tested successfully. Due to the saturated backbone, ISRO polyol based propellants have better aging characteristics, compared to polybutadiene based propellants like PBAN or HTPB. Also, the ISRO polyol has a C7 pendant chain which acts as an internal plasticizer. Hence, no additional plasticizer is required for processing. Absence of external plasticizer means no migration of plasticizer from grain to insulator and better aging characteristics, since the migration reduces the interfacial bond properties. The hydroxyl groups present in ISRO polyol are secondary in nature and hence, are less reactive compared to the primary nature of hydroxyls in HTPB. Last but not the least, the polyol is not based on petro-chemicals and hence, cheaper than other prepolymers. The only drawback of this polyol is the non reproducibility of mechanical properties due to the variations in the composition of naturally occurring castor oil.

During 1992–94, PED came up with a new idea of making big grains similar to building a mansion using bricks. The building block technique can be considered as an extension of segmentation technology. The technology was demonstrated in Agni motor, first using 6 segments each of which 2 star blocks. A number of Agni motors were static tested and 2 motors were flight tested. The technique can also be used to repair big grains for defects by removing the portion having the defect and replacing a block in its place. To prove the point, 10 years old PBAN based AS2 motor was selected. Propellant from defective part of the motor was removed by hand trimming. The propellant removed was replaced by 42 numbers of pre-molded ISRO polyol propellant blocks and bonded with propellant slurry. This motor was successfully static tested at STEX. ISRO owns a patent for this technology and Mrs. Lalitha and her team won the NRDC award for the technology development in 1994. Another new technology developed by the PED is the extrusion of composite propellants using the indigenous press made for double base casting powder. These propellants find use in spin rockets and power cartridges.

Composite propellant consists of finely ground oxidizer in a matrix of combustible polymeric fuel binder. The oxidizer is often inorganic and the fuel binder is organic, along with ground metallic powder (like aluminum). The JATO (Jet Assisted Take Off) device developed in 1942 to enable an air plane take-off from much shorter air strip employed a composite propellant, consisting of potassium perchlorate, asphalt and a plasticizer. This development was a pace setter for future developments in the field of composite solid propellants. In ISRO too, PVC propellants were developed for Rocket Aided Take Off (RATO). More than 100 firings were conducted including actual trials with aircrafts at the Bareilly Air Force station in Uttar Pradesh, at the end of 1972. The indigenization of Russian RATO motors which used Double Base propellant was finally shelved by the Air Force, because of the smoke coming from composite propellants.

Oxidizer accounts for 70% by weight of a modern solid propellant and hence, has a major bearing on the propellant. The most commonly used oxidizer is ammonium perchlorate. Till 1978, this was supplied for propellant development and production by WIMCO from their Ambernath Factory, near Mumbai or by CECRI, Karaikudi. The WIMCO plant has an annual production capacity of 50 tonnes and CECRI had only a research cell. The space profile of India in the 1980–90 period envisaged 2500 tonnes of solid propellants to sustain India's rocket and space vehicle programme. This means there is shortage of ammonium perchlorate in the country.

Also, perchlorates are materials of direct military application, their production, distribution and export come under some kind of state control even in advanced countries. Also, this class of materials is too critical an element in our rocketry, to be left to the dictates of unplanned growth of profit motivation. Hence, ISRO proposed a 150 tonnes plant near Aluva with CECRI know-how in 1972 to the Chairman, Space commission. This would make India self reliant and self sufficient in solid propellants along with RPP, PFC and SPROB. The CECRI technology, is an indigenous technology developed by them, which employed graphite substrate lead dioxide anodes and stainless steel cathodes. The WIMCO technology, a Swedish technology, is based on platinum anodes and steel cathodes. Both used, however, sodium chloride as the starting material and both are batch processes. Electricity is the other major input, as it involves the electrochemical-oxidation of sodium chloride to sodium perchlorate, followed by double decomposition with ammonium chloride to get the ammonium perchlorate.

Dr. Brahma Prakash who became the first director of Vikram Sarabhai Space Centre in May 1972, after discussion, informed that a 6000 Ampere industrial modular cell as proposed in the project proposal be established in TERLS area and run for a year to prove the design. An amount of ₹ 10 lakhs was sanctioned for this purpose. The modular cell was established at project complex, TERLS in February 1974 and the first batch of AP was realized in April 1974. The cell was run for two years (1974–76) to generate the design parameters for the 150 tonnes plant. The Quality Assurance (QA) of VSSC certified the product prepared using this cell in the end 1974. The cell had a capacity of 1000 liters and used 36 anodes. The exercise was used to firm up the details for the 150 tonnes plant. A team consisting of Srinivasa shetty, Viswanathan and Krishnamurthy coordinated the activity. While the modular cell was running, about 58 acres of land for the bigger plant was procured through the collector of Ernakulum. The land was chosen based on the criteria of availability of electricity (the major input), water and raw materials and nearness to industrial belt. FACT and chemical industries are located in Aluva on the banks of Periyar river. The buildings for the plant were ready at the end of 1978, and the anode preparation started. There were 16 modular cells made of concrete tanks lined inside with epoxy polymer and each had 32 anodes of 4" diameter graphite rods coated with lead dioxide and stainless steel cathodes. Cooling coils were provided to keep the cell temperature below 60°C. The plant was inaugurated on 5th February 1979, by the then Chief Minister of Kerala. During the first three years, the plant could produce only around 30 tonnes per year because of large number of losses of anodes in every run. The problems were overcome by modifications in the cell and electrode arrangements and cell cooling arrangements. Later the plant capacity was increased by switching over to triple oxide coated Titanium Substrate Lead Dioxide

Anodes (TSLA) in place of graphite substrate anodes. The ammonium perchlorate was also coated with FERT FLOW (know-how from Sindri Fertilizer factory) to improve moisture sensitivity and flow characteristics. The AP is sieved and blended in 2 tonnes level to the propellant requirement and packed in 50 kg polythene containers (containing silica bags) after drying. Currently, the plant has been expanded to produce 1000 tonnes per annum to meet ISRO's requirements for launch vehicles. Today, AP made at APEP is the work horse oxidizer in all solid propellants made by ISRO.

APEP, in addition to making AP, also meets the other requirements of perchlorates. For example, it developed and supplied strontium perchlorate to Secondary Injection Thrust Vector Control (SITVC) systems of SLV-3 and ASLV projects. Small quantities of perchloric acid and magnesium perchlorate were made and supplied for battery development. The plant holds a patent for development of TSLA and given the know-how of AP manufacture to Defense sector through private parties. Even though world over research is going on in developing high energy oxidizers like Ammonium Dinitramide (ADN) and Hydrazinium Nitroformate (HNF), it will be difficult to replace AP from solid propellants for another 50 years. Ammonium nitrate, though cheap and available in plenty cannot replace AP in view of its hygroscopicity, crystal phase transition near room temperature and lower energy available from it.

In 1968, when PED started with composite propellants, it also initiated work on Double Base (DB) propellants to have all types of solid propellants in its R&D purview. Double base propellants mainly consist of Nitrocellulose (NC) and Nitroglycerine (NG) and other additives in small quantities required for processing. While the Ordnance Factories (OFs) concentrated on extruded double base propellants, which have size limitations, PED decided to go for Cast Double Base (CDB) propellants. This method of casting DB propellants is simple and any size can be cast unlike extrusion process. NC and NG are available only from OFs. NC can be transported easily from OFs as it is a low explosive. On the other hand, being a high explosive and sensitive to shock, NG cannot be transported. Hence, NG is absorbed in kieselguhr and transported. In this form it is less sensitive to friction. This mixture is called dynamite, the process of which was developed by Alfred Nobel.

The cast double base propellant manufacture consists of two steps: a) Manufacture of casting powder, which contains all the NC and solid ingredients with a portion of plasticizer, is made in the form of solid cylinder of 1 mm diameter and 1 mm long, b) Casting and curing, where the casting powder is loaded into a mould and interstices between the granules are filled with a casting liquid consisting of the plasticizer and stabilizer. On heating, interdiffusion of polymer (NC) and plasticizer

(NG) occurs which knits the two components into a single system. Three types of casting powders namely: i) single base casting powder, ii) double base casting powder, and iii) advanced casting powder, are generally made for making propellants. The cast double base propellants are used in making large and homogenized propellants for Scout type space vehicles or intermediate range ballistic missiles. The DB propellants are preferred, especially for missiles, because of favorable exhaust. The radar attenuation by the exhaust is also minimum, in addition to excellent physical properties and favorable shelf life. Initially PED got the double base casting powder from OFs at Aravankadu and Bhandara and the dynamite from Aravankadu factory. PED developed a new process to extract the NG from dynamite using water as the extractor unlike the one used elsewhere which used acetone. The PED method of NG extraction consists of floatation of kieselguhr particles using surface active agents. The kieselguhr particles are carried away with the current of air stream leaving behind water and NG. A series of washings are given in the next step to purify the extracted NG and subsequently dried. This method is advantageous over the conventional acetone extraction method with regard to time and ease of operation. As this requires little handling of NG, it is safer. The extraction efficiency is around 95% and the final product contains 0.2% moisture.

The activity of this DB group was extended to Air Force requirements, substitution of imported DB grains at the instance of Sarabhai. The project Kala was the first project and the indigenous single grain made by casting method was static tested to prove the feasibility. Static testing of a cluster of 14 grains, as required by the project, was done successfully in Thumba in 1969. After the successful static testing of the clustered grains, Gowariker put a notice of appreciation on the PED notice board which included among other things........... "*The PED lion has started roaring.*" There were other double base indigenous projects like Caprod. Vaz, etc. The work could not be continued for want of an extrusion press for making the required casting powders and a mixer. To expand further its work in the DB propellant area, PED designed a 50 ton hydraulic extrusion press with other accessories and got it fabricated by Bemco Hydraulics, Belgaum in seventies. This allowed making Composite Modified Double Base (CMDB) propellant by incorporating ammonium perchlorate and aluminum metal powder in the DB matrix. In fact in 1970, a small quantity of advanced casting powder was imported and processed to CMDB grains. This gave experience in handling advanced casting powder and making CMDB grains by casting technique. This also helped in finalizing the composition of advanced casting powder for CMDB propellants. By 1985, CMDB propellants with good mechanical properties were processed in twenty kilograms grain size and static tested at static test facility, Thumba. Three more grains were static tested at Sriharikota for DRDO. These grains are to be used in Akash missile, being developed by DRDO. As an agreement could not be reached between ISRO and DRDO, the work was discontinued as the final booster grain weighs around 300 kg and this capacity was beyond the safety limits of quantity cleared for PED.

PED also developed a process for the manufacture of spheroidal or globular nitrocellulose for making Composite Modified Double Base Propellant (CMDB) by slurry cast technique. Spheroidal NC is made from fibrous NC with ethyl acetate as solvent in a mixer. The slurry cast technique with globular NC offers a variety of easily processed compositions with high performance potential as solid propellants. The key to the successful application of slurry cast CMDB propellant lies in the high quality source of fine particle, at least partially colloided spherical NC of average size ranging from 5 to 50 microns. This provides the continuum of uniformly plasticized binder in a propellant containing about 30 to 45% solid ingredients. Continuum binder is necessary to have acceptable mechanical properties and reproducible burning characteristics. Spheroidal NC was made in 500 g batches and propellant was demonstrated in 2 kg mixing. In essence, the slurry cast CMDB propellants were processed in a similar way as composite propellants with spheroidal NC.

The manufacture of advanced casting powder and NG were included in the proposal of SPROB in 1970 but could not be implemented. Hence, a new proposal for the manufacture of advanced casting powder was proposed at Nilgris in 1975 and submitted to the Chairman, Space Commission for approval. Since SLV-3 project was in full swing, the proposal was not considered and dropped. However, after 10 years, DRDO wanted a propellant with low smoke or smokeless propellant for a missile application. This calls for double base propellant but with increased energy comparable to the composite propellant. PED used its propellant expertise in developing a DB propellant containing HMX (Her Majesty Explosive) which is a nitramine and elder brother of well-known RDX. The casting route was used to make this propellant using DB powder containing fine HMX and NG. The propellant, meant for Nag missile under development at DRDO, was static tested both at VSSC and HEMRL, Pune. After successful static tests, the propellant was flown on 9th September 1997. The technology transfer document for the propellant and the advanced casting powder were handed over to General Manager, Cordite Factory, Aravankadu, by then VSSC Director, Dr. S. Srinivasan on 8th October 1996 in the presence of Dr. A. Sivathanu pillai, CCR & D. Thus, PED could use its expertise and the facility in helping DRDO to develop the missile under Integrated Guided Missile Programme, initiated by none other than Dr. APJ Kalam, the then Director of DRDL, Hyderabad.

The imported PBAN prepolymer was used in the first and second stages of SLV-3 flights and the first two flights of ASLV. The PSLV project as well as the Solid Motor Project wanted to continue the same propellant for the PSLV first stage as well as in the six strap-on motors. The first stage had about 140 tonnes of propellant. However, the imported stock of the resin with RPP was not enough to meet the requirements of PSLV developmental flights and there was an embargo to get the resin by imports. Hence, it was decided to go for indigenous resin being developed at polymer complex and scaled up in PFC. The PBAN resin is a ter-polymer of butadiene, acrylic acid and acrylonitrile and is made by solution or emulsion polymerization. The presence of acrylonitrile improved the spacing of the carboxyl groups which could be a factor in the more reproducible propellant cures and mechanical properties. Propellants based on PBAN also showed a lesser tendency to surface hardening, which is caused by oxidative attack at the double bonds. Since the development of the ter-polymer in 1957, more solid composite propellant has been produced from PBAN than from any other single prepolymer. Propellants based on this material have been used successfully from small tactical missile motors to the 260" diameter motor containing more than two million pounds of propellant. The space shuttle strap-on booster motors use PBAN propellants and contain around 500 tonnes of propellant in each motor.

Initially, a one to one substitute for imported PBAN based propellant, RCN-01, used in SLV-3 lower stages, was developed using indigenous ISRO PBAN by PED in 1985. This propellant, called Thejus-01, was developed for use in ASLV first and second stage motors in place of RCN-01. ASLV is a five stage vehicle with all solid propellant motors, with a gross lift-off mass of 41.8 tonnes and length of 23.85 meters. It is configured by the addition of two of the first stage motors as strap-on motors to the four stages of SLV-3. It is capable of putting a 140 kg satellite into a 400 km near circular orbit. Since ASLV insisted on an improved propellant for the stage motors, the solid loading was increased from 84% to 86% with 12% aluminum to 18% aluminum. The propellant, called Thejus-18(M), is a modified version of PEDPRO-1336 or ANPED-10. Thejus-18(M) used single epoxy curing agent with ferric acetylacetonate as catalyst to achieve the required cross-linking, while the PEDPRO-1336 used a combination of epoxy and aziridine curing agents. The propellant went through a quality assurance programme and a technology transfer document was made. The propellant was scaled up in RPP to 650 kg to evaluate the mechanical and ballistic properties. Based on the satisfactory properties achieved, the propellant was suggested for ASLV lower stages. As the project insisted on the use of imported PBAN resin and advanced formulation, the RCN-01 formulation was modified to increase the solid loading to 86% with 18% aluminum. This propellant, RCN-A18, was scaled up at RPP to get the propellant properties. With propellant properties meeting the requirements of the project, a full length S2 motor was cast. However, the motor showed some cracks and debonds. This called for some tailoring to reduce the modulus by reducing the curing agent ratio and the cure cycle. With reduced curing agent and cure cycle, in addition to loose flap laying, two S2 motors were cast and successfully static tested. The next S2 motor once again cracked.

Project felt that the propellant based on imported PBAN be shelved and go for indigenous PBAN or HTPB being developed for PSLV booster. The poor reproducibility, coupled with low mechanical properties of PBAN propellants also forced ASLV to look for alternate candidate propellant for the lower stages. This was further aggravated by the dwindling stocks of imported PBAN with RPP. Project ASLV was looking at PED to bail out of the situation and PED, as always, came to the rescue with its development on HTPB based propellant, which was being developed for PSLV booster motor, to replace the PBAN based solid propellants.

The AIAA solid rocket expert committee reported in September 1977 that from 1955 to 1975, the technology of solid propellants has come a long way in specific impulse, density and versatility of mechanical properties and burning rates. They also agreed that the next 20 years will not likely to produce dramatic improvements equaling those of the past 20 years. The Hydroxyl Terminated Polybutadiene (HTPB) propellants with AP oxidizer and high solids content appear to Thiokol to become the workhorse propellant. These formulations are under development for booster applications. Compared to today's PBAN propellants, HTPB formulations will improve the performance and mechanical properties and lower thermal coefficient of expansion at equivalent cost. Based on the reported findings, PED joined the race to develop HTPB based propellants with imported resin from M/S ARCO, USA and M/S Phillips Petroleum Company, USA. The HTPB resin from ARCO is based on free radical polymerization technique and the Phillip's is based on anionic technique. The propellant formulation work started in March 1979 in 4 kg level and scaled up to 40 kg in PED and 120 kg level at RPP. The formulation studies had 88% solid loading with 20% aluminum and cured with Toluene Di-isocyanate (TDI). The propellant can be cured at room temperature but to get desired mechanical properties, it takes more than 30 days. In order to get reproducible mechanical properties, the propellant was cured at elevated temperatures say 50 or 60°C for 5 to 7 days. The curative TDI is moisture sensitive and also causes breathing problems. Hence, its addition to propellant mix during mixing in horizontal mixtures had to be careful. In vertical change can mixers, it can be added remotely. Side by side, polymer scientists at Polymer complex started the polymerization of butadiene gas by solution polymerization technique in small steel bombs using hydrogen peroxide as the free radical initiator. The resin was characterized and found to meet the specifications of imported ARCO–R-45M. The resin was scaled up to one kilogram batch size at the pilot plant laboratory of Polymer complex using a 5 liter stainless steel pressure bomb fitted on a rocking device, to ensure that there were no scale up problems. More than 50 batches were made to ensure the reproducibility. All the resins were used in propellant formulation trials at PED.

The HTPB resin was scaled up at PFC in a 200 liter pressure reactor with a detailed parametric study to optimize the process conditions. Optimization of process parameters and product specifications were carried out in an interactive manner with propellant development team. At the end of the exercise, a HTPB resin of certain specifications, capable of producing propellants which meet the mechanical and ballistic property requirements of ASLV and PSLV projects, was available. When the resin requirements grew in tonnage, HTPB production was further scaled up to commercial level in PFC in a 1000 liter pressure reactor. To meet the requirements of projects, the production technology was transferred to M/S National Organic Chemical Industries Ltd. (NOCIL), Mumbai, on 29th October 1985. The plant with a production capacity of 60 tonnes per annum was commissioned in June, 1987. The NOCIL plant also used a 1000 liter reactor and made around 180 kg of HTPB resin per batch meeting the requirements of the Projects. When NOCIL stopped production, the technology was transferred to M/S Andhra Sugars, Tanuku, in Andhra Pradesh. The HTPB resin had 10 parameters to ensure the quality of the product.

The propellant developed with indigenous HTPB resin by PED was scaled up to 650 kg level in RPP to evaluate the mechanical and ballistic properties. The propellant was tailored to meet the specification of the project for the strap–on motors which are very similar to SLV first stage motor and first stage PS1 motor and PS3 motors of PSLV. All the propellants have a solid loading of 86%, though 88% loading was developed previously with both imported and indigenous HTPB. The compromise was to ensure that there are no undue viscosity problems during casting of big motors using large quantities of propellants (2.5 tonnes propellant per batch of mixing). PED finalized the propellant formulations for PS0, PS1 and PS3 requirements and passed on to RPP for further scale up and characterization. These were called Sthira-1211, Sthira-1211(M) and Sreyas-1311 respectively. The PS0 formulation used AP coarse and fine in the ratio of 4:1 and no catalyst, while PS3 and PS1 used coarse and fine in the ratio of 2:1 and used copper chromite catalyst as burn rate modifier depending on the burn rate requirements of the propellant. The first propellant scale up of HTPB propellant was carried out in December 1982. The HTPB propellant used Ambilink which is a mixture of butanediol and trimethyol propane as chain extender and cross-linker. Toluene di-isocyanate is used as the curing agent. Between 1982 and 1987, a large number of parameters in propellant formulations, processing technology, mechanical and ballistic characterization were studied in detail. Based on the results, necessary alterations in propellant composition and processing conditions were made, and their impact on final propellant properties was evaluated.

PED developed propellants based on ISRO Polyol, ISRO PBAN and Indigenous HTPB and offered as candidate propellants for PSLV. RPP and project preferred PBAN in view of SLV-3 experience and the use of these propellants in space shuttle strap-on boosters. The R&D group wanted to use HTPB based propellant in view of easy manufacturing and no imported chemicals for curing unlike the PBAN based propellants. Few new materials are in current systems because program offices don't demand them. Almost all development work on propellants in this country is dedicated to evolutionary improvements in existing system, because of chicken and egg problem. The developers don't want to use new materials because they are not readily available, not demonstrated and are considered high risk. So in the face of these problems, no risks are taken. The AIAA technical expert committee of Solid Rocket Motors also recommended HTPB for the future large boosters in 1977. To decide on the choice of the propellant, these propellants had to be compared on a common scale in respect of the desired propellant properties and process parameters. Based on motor design, its service environment and practical considerations, burning rate at 6.9 MPa, burning rate pressure index, propellant density, unloading propellant slurry viscosity, propellant mechanical properties and pot life were set as performance goals for finalizing the propellant choice. All the binders were used for propellant trials. The propellants were mixed in 4 kg, 120 kg and 650 kg levels and the process, mechanical and ballistic properties were evaluated by static testing Agni grains made from 650 kg mix. Further each of the propellant was cast in 1 meter diameter segment using both with WIMCO and APEP ammonium perchlorates. All the motors were static tested and ballistic properties evaluated. All the results were presented to the then Director, VSSC in end 1985 for a decision as project could not come to any conclusion as the propellant properties were almost the same. The meeting was conducted at SHAR guest house in Poes Garden, Madras. The meeting was attended by PSLV project team headed by Dr. Srinivasan, Propellant R&D and Production teams and the SPROB team. The Director, after hearing the pros and cons of the binders and propellants based on them, recommended HTPB as the choice for PSLV. This historic recommendation was a milestone in the annals of Indian launch vehicle project, as these projects have been moving with pride till date, with ASLV, PSLV and GSLV successfully launched with solid boosters employing HTPB propellant.

Once HTPB was chosen as the propellant for PSLV, further characterization of the resin and the propellant based on it were carried out before casting the first stage motor of PSLV. A systematic study of ageing was carried out on HTPB resin produced at PFC/VSSC and M/S NOCIL, Mumbai for a period of three years. Ageing of HTPB resin has no effect on the chemical properties of the resin except for a marginal increase in the viscosity of the resin. Also, ageing of the resin does not seem to have any systematic effect on the mechanical properties of the propellant. The storage effect of HTPB resins, on PS1 Propellant formulation, over a period of 3 years showed that processibility and pot life of the propellant slurry were not affected. Tensile strength and modulus increased by 25 and 33% respectively whereas elongation decreased by 10%, all for the initial periods of six months only and afterwards they remained more or less constant. Peel strength of propellant/ ROCASIN interface increased by about 30% over the resin ageing period. The study indicated that HTPB resins can be used safely over a storage period of 3 years. Large capacity mixes were carried out at RPP for both PS1 and PS3 formulations using various combinations of HTPB (NOCIL and PFC) and AP (APEP and WIMCO) from different sources. These mixings, done during mid 1987 and mid 1988, helped in finalizing the AP/HTPB source combinations for the three propellant formulations—PS1, PS3 and PS0. In April 1988, a subscale motor of S3 size (800 mm diameter) was made with PS1 formulation (900 kg mix). Along with this motor six Agni grains were also cast to evaluate burn rate and burn rate law. Mechanical properties were evaluated from different portions of the motor after dissecting propellant samples from various locations both in radial and axial directions to understand the scaling up effect due to grain size and the extent of cure on the mechanical properties.

With all the preliminary development, scale-up and qualification mix trials were completed in two years, the production of PS0 motors commenced at RPP, PS1 and PS3 motors at SPROB from second quarter of 1988. In May 1988, PS1 first segment was cast at SPROB. It took about sixty five hours at a stretch to complete the segment casting. The propellant formulation used NOCIL HTPB resin and AP from APEP. The first static test was conducted successfully at STEX on 21st October 1989 and the performance matched with the prediction. The second test was conducted in March 1991. One more static test was conducted when the propellant loading of the first stage was increased by about 9 tonnes which gave an additional payload of 90 kg to PSLV. This motor, called S139 motor, was static tested in April, 1997. The first flight of PSLV, PSLV D1, lifted off from Sriharikota on 20th September 1993. However, the launch ended up in the Bay of Bengal due to a small technical problem. The second flight of PSLV D2 was flawless and injected IRS P2 remote sensing satellite in to orbit on 15th October 1994. The successful productionization and flight testing of 129-tonne booster motor of PSLV, the third largest operational HTPB based solid rocket motor, established ISRO's capability in solid rocket technology. From then on the march of PSLV continued without a hitch and HTPB based composite propellants became the work horse propellant for the solid rocket motors of the Indian Launch Vehicles and sounding rockets.

While the PSLV was getting ready for the launch, ASLV project redesigned the lower stages and strap-ons to meet the recommendations of the failure analysis committees. HTPB based propellant developed for PS0 was used for ASLV strap-ons except in the head end segments which used a faster burning propellant (PS1) to achieve a reduced dynamic pressure during 40 to 50 seconds region where too many critical events like AS0 tail off, AS1 ignition and AS0 separation were occurring. The AS1 motor used the same PS0 propellants without canted nozzle. The AS2 motor also used the same propellant developed for PS0. The upper stages remained the same as SLV-3 upper stages and used HEF-20(1) propellant. With the modifications and to ensure minimum performance dispersion in AS0 motors, triple casting was adopted and the batch motor was tested prior to flight. With these changes, the third ASLV was launched with SCROSS-C1 satellite in to orbit on 20th May 1992, followed by the fourth flight in May 1994. Thus, HTPB propellants played a major role in the successful completion of ASLV project.

200 tonnes strap-on booster of GSLV Mk II with HTPB based propellant was static tested successfully in January 2010 and again on September 4th 2011. This is the biggest motor static tested so far in India. Today, all propellants used in all solid propulsion rockets are based on indigenous HTPB.

The pyrogen igniters for SLV and ASLV first and second stage motors are high burning rate, small sized, thin webbed and multi-star propellant grains. These propellant grains were based on the binder used in the main motor whose propellant is to be ignited. For example, for PBAN propellants, the igniter grains are based on PBAN with ground ammonium perchlorate and a burn rate catalyst like ferric oxide or copper chromite. These thin webbed grains were initially made by careful machining or pressure casting in batches. For Upper stages of SLV and ASLV motors, HEF-20 based propellant with ground AP and Ferric oxide was used. The SLV or ASLV first stage or second stage igniter grains weighed around 1.5 kg. Two such grains were used for first stage and one grain was used for the second stage. All the grains were charged into fiber glass motor cases. The spin motor used in S4 motor is the smallest motor made by machining HEF-20(1) propellant blocks. In the case of PSLV igniters, HTPB based propellant was used with ground AP and a combination of ferric oxide and copper chromite catalysts. The propellant developed by PED, called PEDPRO-2421, was used in PS1 main igniter, PS1 secondary igniter for main igniter and PS3 igniter had a burning rate of 17 mm/second. Apart from the binder, the propellant had 77% ground AP, 2% aluminum and 2% catalyst. The primary igniter grain for first stage motor (PS1) is 300 mm diameter and one meter long with a propellant web thickness of approximately 15 mm and containing about 30 kg of propellant. PS1 igniter was a free standing grain with paper tube insulation and charged into fiberglass motor case prior to test. First 10 grains were made by vacuum casting. Though the grains fired successfully, NDT revealed defects like voids, unfilled areas and porosity. The pressure casting technique gave defect free grains. The propellant slurry is deaerated after completion of mixing and the deaerated slurry is fed under pressure from the bottom of the assembled motor. The slurry rose through the annular space between paper tube and the mandrel, filling the space slowly and completely. The methodology was first used in August 1988 and found to give defect free gains. Later it was extended to make more grains simultaneously. A patent was filed on multiple pressure casting technique. The PS0 igniter propellant (Thejus-101) has a solid loading of 84% with 2% Aluminum and 0.15% copper chromite and had a burning rate of 8 mm/sec at 40 KSC pressure. The grain had a multi star port, was 1000 mm long with 5 mm web thickness and 4 kg weight. The grains were processed by multiple pressure casting technique, ten grains at a time.

Retro and ullage motors constitute the PSLV special purpose motors (SPMs). Retro motors are of two categories viz., Retro 1(RS1) and Retro 2 (RS2). Both are of same diameter but RS2 is shorter in length compared to RS1. Both are thin webbed grains. Ullage motors (US2) are similar to Agni grains with propellant web thickness of about 40 mm. Each PSLV flight required 8 numbers of RS1 and 4 numbers of RS2. The propellant formulation selected for Retro motors was PBAN based on RCN-860, based on mechanical and ballistic properties. The retro motors operate at a chamber pressure of 110–120 KSC. As the burn rate data at these pressures was not available, 12 Agni grain tests were conducted to evaluate the burn rate and burn rate law in January 1988. The first proof motor of RS1 and RS2 with actual star port grain configuration was cast during February 1989 by pressure casting technique. Based on the trials, the facility for casting 12 numbers of RS1 or RS2 motors from each mix was got ready. In March 1990, the first batch of RS motors was cast. Rocasin was used for motor insulation and the methodology of fixing the insulation to the case was different from conventional hand layup method for big motors. The motor size did not allow conventional hand laying as in large motors. 2 mm thick Rocasin sheet was laid over a non-stick mandrel to make a preform and the mandrel with preform kept inside the motor case which was sand blasted and lined with chemlok antirust agent and rubber solution. The preform sticks to the motor case and the mandrel was removed slowly. Then, inflating of the preform with nitrogen or air with the help of the bladder ensured the bonding of the preform to the casing. Liner was applied as usual before propellant casting. After the second PSLV flight, the propellant formulations for Retro and Ullage motors were changed to HTPB in view of the non-availability of PBAN resin and the curative MAPO. PS1 formulation based on HTPB was used. Conventional vacuum casting method was found suitable for ullage motors and insulation layup was similar to that followed for Retro motors.

While PS1 segments were being cast at SPROB, in some mixes there was increase in viscosity beyond the specified limits. This made the R&D team to think of an alternative to reduce the viscosity of propellant slurry during casting of segments. In one method, a slow reacting isocyanate, Isophorone Di-isocyanate (IPDI) was suggested in place of Toluene Di-isocyanate (TDI) as the reactivity IPDI with HTPB is comparatively slow compared to TDI. Propellant formulations were developed with long pot-life and suggested for S200 motor. However, the interface properties were not meeting the project requirements and the change was dropped. Also, IPDI is not available in the country unlike TDI. The second solution suggested was to add extra 3% rubber oil in addition to the DOA added to propellant formulation. That is, the propellant will have 6% plasticizer like HEF-20(1) propellant used in SLV and ASLV upper stages. These propellants showed poor aging characteristics and hence, had to be made just before use. This cannot be adopted for big motors like PS1. Also, the plasticizer migration may reduce the interface properties between propellant and insulation. A committee with Kurup as chairman with two outside experts was formed to recommend a solution for the problem. The committee recommended continuation of the existing practice as there was no unanimous solution. A third possible solution was to increase the present casting rate from 600 to 800 kg per hour to that used in Space Shuttle booster motor casting with multiple casting ports. This is being followed at present in S200 casting at SPROB.

4. POST PSLV DEVELOPMENTS

Hydroxyl Terminated Natural Rubber (HTNR) was prepared by depolymerisation of masticated natural rubber in presence of hydrogen peroxide in toluene for possible use as propellant binder. This binder was expected to be superior to polybutadiene binders in flow properties. In addition, the absence of pendant vinyl groups and the presence of a 1,4-cis configuration are capable of giving a lower viscosity at a comparable solid loading and an elastically more effective network. Also, this binder will be more energetic in view of higher H/C ratio in the polymer backbone chain. The resin was cured with TDI and contained all other ingredients as in HTPB propellant. The propellant formulation had a solid loading of 86% with 18% aluminum after a systematic study of formulation variables. Poor reproducible mechanical properties and poor ageing characteristic of the propellants coupled with poor humid ageing characteristics of the resin are the drawbacks of the binder.

Work on fuel rich propellants for air breathing propulsion started in PED in early 1986 using HTPB, Hydroxyl Terminated Natural Rubber (HTNR), PCPD (polycyclopentadiene) and naphthalene. These propellants carry less oxygen than normal propellants and heavily metalized. Air breathing engines can provide about 3 to 10 times higher effective I_{sp} values compared to the solid or liquid propellants at altitude below 30 km. A number of fuel rich propellant formulations based on HTPB/Magnesium have been developed and scaled up for production of dual thrust grain with two propellant formulations established at RH-200 level. More than fifteen grains were made and static tested. One grain was tested in connected mode. Magni 30 (8 mm/sec at 50 KSC) and Magni 40 (5 mm/sec at 40 KSC) were developed for booster and sustainer applications respectively for ABR-200 project. HTNR is better than other binders because of higher H/C ratio and lower decomposition temperature. A fuel rich propellant based on HTNR was developed during 1985–90 and contained 30 to 37.5% Magnesium or Magnesium-Aluminum alloy powder. The propellant was established in 40 kg blocks or Agni grains. The poor reproducibility of the mechanical properties and poor storage stability of the resin forced to abandon the HTNR binder usage in fuel rich propellants.

During AP based propellant combustion, large amounts of hydrogen chloride gas and other chlorine compounds are generated in the motor exhaust. These gases are highly corrosive and toxic in nature and form clouds in humid conditions. During the firing of space shuttle, more than 100 tonnes of chlorine compounds is produced and let into the atmosphere. This may lead to "heavy acid rain" polluting the environment and also cause depletion of the ozone layer. Efforts were made for development of alternative propellants which are environmental friendly and equally or more energetic to meet the future space requirements. An eco-friendly high performance solid propellant containing 20% HMX was realized in subscale mixing. Processability and mechanical properties are quite comparable to HTPB (blank). Burn rate and burn rate pressure index were found to be lower than those of the blank. The HMX based propellant is one such attempt to reduce atmospheric pollution. This propellant formulation is expected to give 30% less hydrogen chloride gas in the exhaust smoke. The propellant was tested in 2 kg motors to evaluate the performance.

As a continuation of eco-friendly propellants, propellants based on Ammonium Nitrate (AN) and on Ammonium Dinitramide (ADN) were initiated. AN has two disadvantages for use in solid propellants. AN has a crystal phase transition around 32°C which is accompanied by 3% volume change. This volume change of AN may lead to cracking of the propellant based on it. The phase transition temperature was shifted to higher temperatures by doping with potassium or other metallic salts. This was done in PED by melt doping where the potassium or metallic salts were added to melted AN and cooling the melt followed by grinding and sieving. This doping shifted the room temperature transition to around 50°C. The other draw back with AN is its hygroscopicity. The propellants made with this ammonium nitrate were difficult to ignite and also had low burning rates. Also, theoretical calculations showed the energetics to be much lower compared to AP based ones. These call for addition of energetic materials like HMX to boost the specific impulse and burn rate catalysts to whip the burn rate. In view of the difficulties, the attempt was stopped and work on alternative oxidizer like ADN was explored. ADN was synthesized and scaled up to ½ kg batch size in PFC. Beaker level hand mixings were done using ADN and HTPB resin. Cured propellant was tested for burn rate evaluation. An explosion during the synthesis of ADN made the activity to stop on safety considerations.

An igniter formulation, with 75% solid loading, for cryogenic upper stage (PICPED-75) was developed during 2000-04 which contained small quantities of aluminum (2%) and had low burning rate compared to solid propellant igniters. A number of cryogenic propellant static tests were done successfully for different durations using this igniter. PED also developed a very slow burning propellant for Reusable Launch Vehicle Technology Demonstrator (RLV-TD) during 2004–05. This HTPB propellant had 85% solid loading with 18% aluminum and a burn rate retarder oxamide (3%). The burn rate was found to be 3.11 mm/sec at 17 KSC or 3.85 mm/sec at 40 KSC. The propellant, designated as PED-RTB-85, was tested first in a subscale motor (RH-560) to ensure stability of combustion at low pressures on March 15, 2007. The test confirmed that the required burning rate, I_{sp} and other ballistic parameters are met. Subsequently, two more full size first stage strap-on motors were cast at RPP and static tested successfully on 19th November 2008 and 3rd August 2011. The mid and head end segments were cylindrical grains and nozzle end segment is ten lobe deep slotted stars to limit erosive burning. The nozzle end segment had composite insulation (alternate layers of carbon cloth and Rocasin) towards the dome side. For Chandrayaan mission, PED developed a state of the art, plasticizer-free propellant based on HTPB-IPDI resin system, which had a solid loading of 84% with 2% aluminum during 2005–07. This tubular propellant grain was used in a de-orbit motor so as to de-orbit the moon impact probe. The requirements of the propellant were quite challenging ranging from its remote ignitability, susceptibility to extreme conditions of space, the low degassing requirements and the need to ensure a particulate-free combustion exhaust. In addition, the propellant had to be non-aluminized to ensure that the motor exhaust on ignition is free from aluminum oxide so that the plume impingement effects are minimal. This propellant was also used in the spin rocket motor along with de-orbit motor. The propellant weighing 8 gm was inhibited with a non-degassing inhibition, 'chandini', based on viton rubber. Both the de-orbit motor and two spin motors worked as predicted to achieve the mission goal.

Glycidyl Azide Polymer (GAP) was synthesized by cationic polymerization of epichlorohydrin followed by condensation with sodium azide at high temperature. This polymer has hydroxyl groups in the terminals in addition to having energetic azide groups attached to the chain making it an energetic binder. The resin was scaled up to 5 kg batch level in PFC. The polymer had a molecular weight of 1800 to 2000. The low molecular weight product (~600) obtained, was used as a plasticizer in propellants. Initially, a non-aluminized propellant with 75% solid loading was made which had a burning rate of 20 mm/sec at 70 KSC with a burn rate pressure index of 0.22. GAP/AP propellants have poor mechanical properties especially elongation at low temperatures because of high glass transition temperature (T_g) of

GAP. Butyl ferrocene is added to propellant formulations to increase the burning rate without affecting the processibility because it is a liquid burn rate catalyst. However, the burn rate catalyst migrates to the interface and finally to insulation during storage, thereby affecting the burn rate and interface properties. To overcome this problem, the ferrocene was grafted to HTPB backbone by grafting. This binder was used to make fast burning propellants. Propellant studies in 4 kg level showed that ferrocene grafted HTPB based propellant gave matching burn rate requirements in comparison to HTPB propellant with ballistic modifier butylferrocene. Propellants with nano carbon tubes and nano ferric oxide were formulated to improve the mechanical properties and the burning rate during 2006–07.

In early seventies, PED started working on kerosene-Liquid Oxygen system under its cryogenic technique project. In the cryogenic laboratory, methods of handling, transferring and pumping of cryogenic propellants was attempted. The project was planned to be executed in 3 phases. After Sarabhai's death, the activity was shelved and the personnel working on the project were transferred to Liquid Propulsion Center. Almost after 25 years, PED tried to improve the performance of liquid propellants by the addition of metal powders. During 1993–97, UDMH-30% aluminum gel propellant and Kerosene-30% aluminum gel propellants were developed in an attempt to improve the performance of liquid propellants, very similar to aluminized composite solid propellants. Normally, metallic powders when dispersed in liquid propellants quickly settle down because of density difference and the dispersion becomes non-uniform. This was overcome by forming a gel after adding the metallic powder to the liquid propellant. The sprayability and ignitability of the gel propellants were established in collaboration with IIT, Madras.

5. SPIN OFFS OF PROPELLANT RESEARCH

The fallout and benefits from solid propellant research are numerous. Solid propellant development at PED gave two new polymeric binders viz. HEF-20 or lactone terminated polybutadiene (used in SLV and ASLV upper stage motors and in Apogee kick motor for APPLE satellite) and ISRO polyol(used in sounding rockets and a candidate propellant for PSLV booster) to the world propellant community. Many of the technologies transferred to industry were originally developed for use within space programme, but have found applications in the industrial mainstream for greater benefit of the society. ISRO polyol can replace the petroleum based PPG for making flexible and rigid foams as well as elastomers. One important application is the substitution of rigid foams for wood in the artificial foot/limb for amputed persons in Dr. Sethi's Jaipur foot. The experience gained in indigenisation of chemicals and polymers for use in solid propellants and processing helped in designing, fabricating, erecting and commissioning of a pilot plant for multi-base extruded double base propellants for Ordnance Factory Board at Ordinance Factory, Itarsi, during 1980–86. The process equipments indigenized included among others are extrusion press, a mixer, cutting machine and a coating pan, which were till then imported. The working of Itarsi pilot plant was demonstrated by extruding two different double base propellants and comparing with the performance of the same double base propellants extruded with imported machines. The second pilot plant was set up for HAL at Nasik for chemicals meant for MIG aircraft maintenance. While developing polymers and adhesives for bonding and inhibition purposes, the polymer group of PED developed expertise in formulating adhesives. At the instance of Sarabhai who had a request from Chairman, HAL, PED took up the development of the adhesives and sealants required for the MIG aircraft. These chemicals are required in small quantities and need low temperature storage. An aircraft had to be sent for bringing the chemicals from Russia and the procurement had to be in large quantities as that are the minimum packaging size available. Further to complicate the problem, these chemicals have limited shelf life for use. PED scientists at polymer complex made equivalents and these were sent to HAL for lab trials and finally trials in aircraft. The approved chemicals were made in Polymer complex and regularly supplied. When the number of import substitutes increased for supply, HAL made a request to set up a plant for this purpose. PED agreed to the request and set up a plant at HAL Nasik to make the adhesives and sealants and trained their personnel in making these materials. This not only saved foreign exchange and unnecessary trip of an aircraft to Russia to get the materials, in addition to honoring the Chairman ISRO's commitment to HAL.

The propellant community also indigenized the imported vertical mixer (300 gallon capacity) for mixing large quantities of composite solid propellants with the help of Central Machine Tools, Bangalore. During 1981–84, for non-destructive testing of rocket motors of one meter diameter, a 4 MeV Linear Accelerator (TIFROLINAC) was developed in collaboration with Tata Institute of Fundamental Research (TIFR). The telescopic azimuth elevation manipulation system was developed and fabricated by RPP. Except for a few devices and components, the system was wholly indigenous. A specialized machine was developed, fabricated and commissioned for end trimming propellant in motors by RPP during 1983–84 periods. This machine had distinct advantages like drastic reduction in machining time, absence of dust/chips generation resulting in greater safety and elimination of expensive dust collection system.

While studying the decomposition of Castor oil which is used in ISRO polyol making and in the end inhibitions of propellants, in presence of catalysts, it was found that the product had turned into a hydrocarbon. PED engineers extended this controlled thermolysis at relatively low temperatures and pressure in presence of suitable catalysts to other non edible oils like Sal, rubber, Mahua, etc. and found the oils converted into hydrocarbons. The liquid hydrocarbons on fractionation yielded petrol, kerosene, diesel and lube oils. The gaseous portion contained apart from carbon dioxide, C3, C4 and C5 hydrocarbons which are similar to cooking gas. This ISRO achievement of converting non edible oils in to hydrocarbons using a catalyst at low pressures is an efficient process. The liquid hydrocarbon portion, called SPACE CRUDE, was fully characterized at Madras Refineries Ltd., Chennai. Cars and scooters were run using the petrol fraction. This technology was patented both in US and Europe. The Center for Development Studies at Trivandrum, felt that the project, though exciting, is not economically viable because of the cost of seeds collection for getting the oils compared to the cost of imported crude. Here is an opportunity for converting Rocket Technology through space crude into national asset. Science makes one to dream and technology makes it possible to achieve the dream.

6. CONCLUSION

PED has stood by its slogan "To strive to seek to find and not to yield." Today solid propellant in ISRO is total, complete and self sufficient. In this journey from unsaturated polyester to hydroxyl terminated polybutadiene, the stress is on the role of pre-polymers as fuel binders. This became very significant when one considers that the binder system constitutes 12 to 15% of the solid propellant. This difficult task is achieved by strict quality control starting from resin manufacture. Solid propellants by its obvious advantages like simplicity, ruggedness, operational readiness and high thrust capability have come to stay. The development and proving of a solid propellant take 4 to 5 years. The future of propellant technology will concentrate on the reduction of cost of the propellant so that the payload placed in orbit costs less. No major achievement in improving energetics will be achieved in the next 20 years. HTPB and AP will continue to be the work horse of solid propellants. Rocket propulsion is not a mature area contrary to popular opinion. This means, there is a need for more intense R&D in solid propellant area. Energetic materials are enabling technologist and ISRO can achieve significant performance advantages in rocket propulsion, explosives and pyrotechniques by using new energetic materials and developing an understanding of their behavior and properties. The neglect of developing new propellant systems contributes to the extremely high costs of getting payload to orbit. The new materials will give significant performance improvements along with improved safety. When a new system wants to buy its propulsion unit 'off the shelf', it will not be available and will be buying a very old technology. It will not take advantage of the results of currently available in research laboratories. The commercial sector will not be the leader in developing this technology.

The concept of tiny rockets to space vehicles as a self-propelling phenomenon and ISRO's planning on propellant technology to maintain a lead over needs started together. In each of the tiny motors during the early days, therefore, one saw a prelude of tomorrow's giant vehicles and the successful firing of each of this insignificant motor were burning away gradually decades of nation's backwardness in this vital field.

Growth of ISRO Rocket Motors

If propellant weight is considered as representative of the size of rocket which it is, the weight increased from a stagnant figure of 10 kg in 1967 to 220 kg in 1970, 700 kg in 1972 to 10 t in 1979, to 130 t in 1989 and to 200 t in 2010 as a result of ISRO's efforts. Considering the diameter of solid motors, the growth is from 75 mm to 100 mm, to 125mm, to 200 mm, to 300mm, to 500 mm as in sounding rockets and then to 1 meter dia (10–12 tonnes as in boosters of SLV-3, ASLV and strap-ons of PSLV), 2 meter dia (7.6 tonnes as in third stage of PSLV), 2.8 meter dia (129–139 tonnes as in booster of PSLV) and 3.2 meter dia (207 tonnes as in GSLV MK-III).

ISRO Launch Vehicle Flights

The first successful flight of RH-75 using indigenous composite propellant (Mrinal) on 21st Feb 1969 marked the beginning of an era in the development of solid propellants in ISRO. The first all solid propellant satellite launch vehicle SLV-3 was successfully flight tested on July 18, 1980 by putting a 40 kg Rohini satellite into a near earth orbit, followed by two more successive flights of SLV-3 in 19th June, 1981 and 7th April 1983. The next two successive flights of ASLV (SLV-3 upgraded with two strap-on boosters) in 20th May, 1992 and 4th May 1994 carried SROSS-1 (stretched Rohini series of satellites) satellites into the near earth orbit. This was followed by the first successful PSLV flight in 15th Oct 1994 using HTPB booster motor of 129 to 139 tonnes, HTPB upper stage motor 7.6 tonnes and with 6 strap-on motors,each of 10–12 tonnes to put Indian Remote Sensing Satellite (IRS) into the sun synchronous orbit. During the period 1994 to 2015 (21 years), PSLV has performed 27 numbers of various successful missions which includes the Chandrayaan-1 mission launched on 22nd Oct 2008, Mangalyaan mission to Mars orbit, launched on 5th November, 2013 and also for launching Indian Regional Navigation Satellite System (IRNSS-1D) launched on 28th March, 2015 (IRNSS-1A, 1B and 1C are already on the orbit). The Chandrayaan mission using the Moon Impact Probe (MIP) had a key role in the finding of water on moon. India became the 6th nation to be part of moon mission, other nations being America, Russia, China, Japan and European space agency. It is estimated that moon has two crore tone He-3 deposits which is an energy source for future. In the Mangalyaan mission, Mangalyaan orbiter travelled a distance of 66.6 crore km to reach the Mar's orbit and taken pictures to study the Mar's atmosphere. India is the first country to achieve this in the first attempt, that too with the lowest cost. India is the fourth country in the chronological order to achieve Mars mission, other countries being USA, USSR and Europian space agency. Using IRNSS-1A, 1B, 1C and 1D, navigational survey around India can be done. In one of the PSLV missions (PSLV C9), PSLV has successfully launched 10 satellites (two numbers of India and 8 numbers of nano satellites of foreign countries) into varies desired orbits, thus India became the second nation to launch multi-satellites for remote sensing. PSLV was also used to launch GSLV-D-5 with indigenous 12 tonne cryo on 6th Jan 2014 to put 1982 kg GSAT into the geo-stationary orbit. Another step ahead in launch vehicle technology is the successful launch of GSLV-MK-III with two numbers of 200 tonne boosters for the CARE—Crew module atmospheric re-entry experiment, conducted on 18th Dec., 2014.

The state of development of all major components—propellants, insulation, ignition systems and the processing and characterizing of these systems—has reached the point, a new programme can be undertaken with very high confidence regarding their outcome both from schedule and performance viewpoint. Solid propellant propulsion systems will continue to offer the designer of launch and deployment vehicles options, worthy of the most serious considerations.

Glossary

ADN	–	Ammonium dinitramide
AET	–	Acoustic emission testing
BAMO	–	Bis-azidomethyl oxetane
Ballistic modifiers	–	Materials like copper chromite, ferric oxide, etc. which enhance the burn rate of solid propellants
Bayonet casting	–	The de-aerated slurry in the casting can is gradually pumped into the motor case from top under nitrogen pressure using bayonets
BDNPA	–	Bis (2,2-dinitropropyl) acetal
BDNPF	–	Bis (2,2-dinitropropyl) formal
Bonding agents	–	They can improve the oxidizer-binder bond
Binder	–	Acts as a fuel as well as the matrix to hold the oxidizer and other additives to give structural properties to the propellant
Bulk modulus (k)	–	It is a measure of incompressibility of a material when subjected to hydrostatic pressure. It is the ratio of stress to volumetric strain
BTTN	–	Butane triol trinitrate
Calorific value of the propellant	–	It is the quantity of heat in calories evolved when 1 gm of the propellant is completely burned (cal/g)
Cardinal principles	–	Principles used in establishing hazardous operations involved in solid propellant handling which has proven to be hazardous both from the point of fire and explosion
CAT	–	Computer Aided Tomography
Centralite-1	–	Symmetric dimethyl diphenyl urea
Centralite-11	–	Symmetric diethyl diphenyl urea
15 CDV 6	–	It is a high strength low alloy steel containing Cr, Mo, V and carbon
Characteristic velocity (C^*)	–	Chamber pressure × nozzle throat area/propellant mass flow rate
Chuffing	–	Sudden fall in chamber pressure and delayed regain of varied pressure
CMDB	–	Composite modified double base
Coefficient of viscosity	–	It is the force in dynes/cm^2 required per unit area to maintain a unit difference of viscosity between two parallel layers of liquids one cm apart
Colloidal propellant	–	Nitrocellulose colloided with non-volatile solvents called gelatinizers/plasticizers

Combustion instability – Oscillatory combustion

Composite propellants – Both oxidizer and fuel are in distinct molecules

Coefficient of thermal expansion (CTE or $\propto$) – It is the ratio of increase in length to the original length for unit rise in Temp. °C.

Creep Compliance – Linear visco-elastic materials subjected to a constant stress will exhibit increase in strain with time

Cryogenic propellants – They need very low temperature working condition for their storage in liquid form

CTPB – Carboxyl-terminated polybutadiene

Curing – It is the process of converting a liquid slurry into a solid mass

Darkening agents – Radiant heat absorbers like carbon black

De-coring – It is the process of removal of mandrel from the motor safely after curing

De-wetting Strain (ε_d) – The volume change of propellant specimen under uni-axial tension

Dilatant – Liquids whose viscosity increases with increase in shear rate

Discharge Coefficient (C_D) – Mass discharge rate/throat area × chamber pressure

DMA – Dynamic Mechanical Analysis

Double base propellant – Contains 2 major explosive components, NC and NG as major ingredients

DTA – Differential thermal analysis

DSC – Differential scanning calorimetry

Dynamite – Mixture of nitroglycerine and kieselguhr

EGDN – Ethylene glycol dinitrate

Elongation – It is the percentage extension per unit length of the sample at break point. It is represented by $\in b$ as %

Equilibrium pressure (P_{eq}) – The pressure at which the rate of gas production exactly balances the rate of gas discharge

Erosive burning – The increase in burn rate caused by a high velocity gas stream across the burning surface

Explosives – They decompose by a detonation process with high detonation velocity

Exhaust velocity (V_e) – Specific impulse multiplied by acceleration due to gravity

Fire Arrow – A rocket used by Chinese in 1232 AD

Flash suppressors – Flash and flame temperature reducers like dinitro toluene, potassium compounds

Fuel rich propellants – They contain maximum amount of fuel (metallic or polymer) with least amount of oxidizer

GAP	–	Glycidyl azide polymer
GAPA	–	Azide terminated glycidyl azide plasticizer
Gel propellants	–	A new class of liquid propellants containing metallic powders, in gel form
Glass transition temperature	–	It is the temperature below which an elastomeric material becomes hard, brittle and glassy
Globular/Spheroidal Nitrocellulose	–	Fibrous nitrocellulose converted to spheroidal form and stabilized using wetting agents and gelatin
Heat of formation (ΔH_f)	–	It is the quantity of heat evolved or absorbed when 1 gram mole of a substance is formed from its elements
Hess's Law	–	The amount of heat evolved or absorbed in a chemical process is the sum of the heats of reaction of several subsequent partial reactions starting with the same reactants and leading to same products
Heat of combustion (ΔH)	–	It is the quantity of heat evolved when 1 gram mole of a substance is completely combusted
HMX	–	Cyclo tramethelene tetra nitramine
HNF	–	Hydrazinium nitro formate
HNIW or CL-20	–	Hexanitro hexaazaiso wurtzitane
HP	–	Hydrazine perchlorate
HP_2	–	Hydrazine diperchlorate
Homogeneous propellant	–	Ingredients contain oxidizer and fuel components in the same molecule
HTNR	–	Hydroxyl terminated natural rubber
HTPB	–	Hydroxyl terminated polybutadiene
Hybrid propellants	–	Generally contains a solid fuel and a liquid oxidizer
Inhibition	–	It is given to restrict the combustion of certain specific areas of the propellant grain by forming a non-inflammable layer
Inspection	–	It means checking of a product or component or material for the acceptance or rejection of it
Insulation	–	It is for protecting the motor case from high temperature
Interstitial casting	–	Casting solvent containing NG and additives are fed from bottom through the interstices of filled casting powders for solvation
EPDM	–	It is a Ter-polymer of ethylene-propylene-diene (cyclopentadiene) random copolymer
First law of thermodynamics	–	States that energy may neither be created nor destroyed in a reaction.
Idealized P-T curve	–	Pressure remains constant throughout burning

ISRO polyol	– Self condensation product of 12-hydroxyl stearic acid, followed by condensation with Trimethylol Propane (TMP)
Liquid bi-propellants	– Both the oxidizer and fuel are liquids.
Loose Flaps	– They are insulator flaps laid in the highly stressed regions-dome and cylindrical regions of a case bonded motor to minimise the damages such as debond/crack due to thermal stresses during processing of large grains
Liner	– It is a thin layer of elastomeric polymer material applied uniformly on the insulator surface for proper bonding of propellant to the insulator
LTPB or HEF-20	– Lactone terminated polybutadiene or High energy fuel-20
Mesa burning	– Burn rate decreasing with increase in pressure
Maraging steel	– It is a special steel which contains Ni, Co, Mo, Al and Ti
MS	– Margin of safety = SF-1, where SF is the factor of safety which is capability/induced load
Mass discharge rate (M_D)	– Discharge coeff × Throat area × Chamber pressure
Mass flow rate or mass burning rate (M_b)	– propellant density × propellant burn rate × propellant burning area
Mass ratio of a vehicle (MR)	– Ratio of the final mass m_f (after the consumption of all usable propellant in the vehicle or stage) to m_0 (before rocket operation)
Micro structure of HTPB	– Cis, trans and vinyl content of HTPB
Modulus	– It is the ratio of applied stress and the resulting strain. Its unit is Kg/cm^2 or MPa and is represented by E.
Monopropellant	– Both fuel and oxidizer groups and elements in the same molecule
Multiple pressure casting or bottom casting	– Evacuated slurry is pumped into a ray of motors from bottom side using nitrogen pressure
NC	– Nitrocellulose
NDT	– Non-destructive testing (Radiography)
Newton's Third Law of Motion	– To every action, there is an equal and opposite reaction
Newtonian liquids	– Liquids whose viscosity is independent of shear rate and remains constant
NG	– Nitroglycerine
Nitramine propellants	– Propellants containing HMX and/or RDX
RDX	– Cyclo trimethylene tri nitramine

Non-Newtonian liquids	–	Liquids whose viscosity is dependent on shear rate
NR	–	Neutron Radiography
OC	–	Octa azacubane
ONC	–	Octanitrocubane
Optical holography	–	Holographic NDT which is basically an optical interferometric technique which is very sensitive to inner cracks, de-bonds, de-lamination, etc.
Oxidizer	–	Supplies oxygen for the fuel to burn in a propellant.
PBAA	–	Poly butadiene-acrylic acid copolymer
PBAN	–	Poly butadiene-acrylic acid-acrylonitrile terpolymer
Penetrometric pot-life	–	It is the time up to which the propellant slurry already cast remains in a semicured or compressible state
Plateaunising agents	–	Compounds like lead salts which reduces the pressure index of DB propellants to zero, i.e. burn rate remains constant with increase in pressure.
Pot-life	–	It is the time from the End of Mix (EOM) viscosity to the time up to which the propellant slurry is amenable for easy casting
Poisson's ratio (υ)	–	It is the ratio of lateral contraction to longitudinal elongation when the sample is subjected to tension
Pre-polymers	–	They are low molecular weight polymers which can be further polymerized or cross-linked to network structure
Poly dispersity of the binder	–	It is the ratio of weight average molecular weight to Number average molecular weight
Poly NIMMO or PLN	–	Poly-nitrato methyl methyl oxetane
PGN	–	poly glycidyl nitrate
Process aids	–	They improve the processability of solid propellants
Propellant	–	Energy source for propelling a rocket
Propellant mass fraction '&'	–	Fraction of propellant mass m_p in the initial mass m_0, i.e., m_p/m_0
Pseudo plastic liquids	–	Liquids whose viscosity decreases with increase in shear rate.
Pyrogen igniter	–	It is basically a small rocket motor having a fast burning propellant grain
Pyrotechnic igniter	–	It contains pelletised pyrotechnic charge as main charge in vented/perforated containers
Quality	–	"Fitness for use" or "conformance to requirements"
Quality control	–	It is for improving the quality of the product
Reactive insulator	–	It contains reactive groups for direct bonding with propellant without a liner

Relaxation modulus — Modulus calculated from stress relaxation curve

Reliability — It is the probability that a system will perform successfully under given conditions for a specific period of time. Reliability = Quality now + quality later

Resonance burning — Irregular secondary peaks in the P-T curve of only perforated grains

Rheology — It is the study of deformation and flow of materials

Rochetta — Rocket

Rohini — The first Indian satellite

RTR — X-ray Radiography—Real Time Radiography

Shear modulus (G) — It is the ratio of shear stress to shear strain

Single base propellant — Contains Nitrocellulose (NC) as the major ingredient

SLV-3 — The first solid propellant satellite launch vehicle

Specific heat (C_p) — It is the quantity of heat required to raise the temp of unit weight of the substance by 1°C.

Specific Impulse (I_{sp}) — Thrust produced per unit weight of propellant burned.

Stabilizers or anti-oxidants — They protect the polymeric fuel from thermal and photo degradation

Stress relaxation — Stress decay with time at constant strain, represented by $\in_0 f(t)$

Tensile strength (TS) — It is the minimum load required to break a specimen of the material of unit cross section area. It is represented by σb in kg/cm^2 or MPa

Thermal conductivity $(\lambda$ or $k)$ — It is the quantity of heat conducted in unit time per unit area per unit temperature gradient

TGA — Thermo gravimetric analysis

Thixotropic — Certain fluids like uncured propellant slurry when subjected to a constant rate of shear shows a decrease in viscosity as a function of time of shearing.

Thrust coefficient (C_F) — Total thrust divided by (average chamber pressure × throat area)

Thermoplastic binders — They are softened or melted with heat but on cooling, they become rigid and retain the shape of the mould

Thermosetting binders — They are initially liquids or meltable solids but after reaction with curing agents form a cross-linked network

TMETN — Trimethylol ethane trinitrate

TEGDN — Triethylene glycol dinitrate

TNAZ — 1,3,3-Trinitro-azetidine

Total Impulse (I) — Specific Impulse multiplied by weight of the propellant

Triple base propellant – Contains 3 major explosive components NC, NG and nitro-guanidine

UT   – Ultrasonic Testing

Vacuum casting   – Propellant slurry is passed through a slit plate and falls as thin strips into the rocket chamber either enclosed in a vacuum bell or serving as its own bell.

Visco-elastic   – Materials which exhibit the deformations of both elastic solids and viscous liquids

Viscosity   – It is the resistance to flow of liquids

Zero Flow Time (ZFT) – It is the time from End of Mix (EOM) to the time up to which the slurry shows amenability to flow

www.ingramcontent.com/pod-product-compliance
Lightning Source LLC
Chambersburg PA
CBHW082140210326
41599CB00031B/6045